AF752295

Earth Sciences in the 21st Century

nova
science publishers
www.novapublishers.com

Earth Sciences in the 21st Century

The Basics of Groundwater
Melissa Clutter, PhD, Chloé Fandel, PhD and Ty Ferré, PhD (Authors)
2022. ISBN: 978-1-68507-740-2 (Hardcover)
2022. ISBN: 978-1-68507-874-4 (eBook)

Geo-Information Technology in Earth Resources Monitoring and Management
Dr. Varun Narayan Mishra, Dr. Praveen Kumar Rai
and Dr. Prafull Singh (Editors)
2021. ISBN: 978-1-53619-669-6 (Hardcover)
2021. ISBN: 978-1-53619-796-9 (eBook)

Recent Developments in Geomorphology Research
Ronald M. Reale (Editor)
2021. ISBN: 978-1-53619-445-6 (Softcover)
2021. ISBN: 978-1-53619-506-4 (eBook)

Walking Science's Narrow Path: Zavisa Janjić
Dragutin T. Mihailović, Slobodan Nickovic and Tijana Janjic (Editors)
2021. ISBN: 978-1-53618-899-8 (Softcover)
2021. ISBN: 978-1-53619-054-0 (eBook)

The Rare Earths: Their Occurrence, Chemistry, and Technology
S. I. Levy (Editor)
2020. ISBN: 978-1-53618-682-6 (Hardcover)
2020. ISBN: 978-1-53618-725-0 (eBook)

More information about this series can be found at
https://novapublishers.com/product-category/series/earth-sciences-in-the-21st-century/

Bruce B. Hicks

Observing Micrometeorology

A Personal Tour through an Evolving Science

nova science publishers

www.novapublishers.com

Copyright © 2022 by Nova Science Publishers, Inc.
https://doi.org/10.52305/IBEJ6475

All rights reserved. No part of this book may be reproduced, stored in a retrieval system or transmitted in any form or by any means: electronic, electrostatic, magnetic, tape, mechanical photocopying, recording or otherwise without the written permission of the Publisher.

We have partnered with Copyright Clearance Center to make it easy for you to obtain permissions to reuse content from this publication. Simply navigate to this publication's page on Nova's website and locate the "Get Permission" button below the title description. This button is linked directly to the title's permission page on copyright.com. Alternatively, you can visit copyright.com and search by title, ISBN, or ISSN.

For further questions about using the service on copyright.com, please contact:
Copyright Clearance Center
Phone: +1-(978) 750-8400 Fax: +1-(978) 750-4470 E-mail: info@copyright.com.

NOTICE TO THE READER
The Publisher has taken reasonable care in the preparation of this book, but makes no expressed or implied warranty of any kind and assumes no responsibility for any errors or omissions. No liability is assumed for incidental or consequential damages in connection with or arising out of information contained in this book. The Publisher shall not be liable for any special, consequential, or exemplary damages resulting, in whole or in part, from the readers' use of, or reliance upon, this material. Any parts of this book based on government reports are so indicated and copyright is claimed for those parts to the extent applicable to compilations of such works.

Independent verification should be sought for any data, advice or recommendations contained in this book. In addition, no responsibility is assumed by the Publisher for any injury and/or damage to persons or property arising from any methods, products, instructions, ideas or otherwise contained in this publication.

This publication is designed to provide accurate and authoritative information with regard to the subject matter covered herein. It is sold with the clear understanding that the Publisher is not engaged in rendering legal or any other professional services. If legal or any other expert assistance is required, the services of a competent person should be sought. FROM A DECLARATION OF PARTICIPANTS JOINTLY ADOPTED BY A COMMITTEE OF THE AMERICAN BAR ASSOCIATION AND A COMMITTEE OF PUBLISHERS.

Additional color graphics may be available in the e-book version of this book.

Library of Congress Cataloging-in-Publication Data

ISBN: 979-8-88697-369-3

Published by Nova Science Publishers, Inc. † New York

To Laurel —
a tower of forebearance.

"Look at the innermost causes of things, stripped of their husks"
– Marcus Aurelius, 'Meditations.'

The skills of our community are many. The following is the product of the fertile brain of David Beardsmore, a major contributor to the early science of the Division of Atmospheric Physics of CSIRO at Aspendale and reproduced (with permission and with David's editing) from 'Winds of Change' (Garratt et al. 1998).

Among a group of men sincere
Were famous people whose career
Was studying the atmosphere.
They talked of weather foul and clear
Until a layman standing near
Said "Can I ask a question here?"
"What is meteorology?"

A gent who knew synoptic maps
Explained inversion rates and lapse
And jet stream winds and thunder claps.
He also said "One day, perhaps,
We'll have predictions of cold snaps
From photos of the polar caps.
That's meteorology."

As if he'd been shot from a gun
A second stood and said "Well done:
But is it known by everyone
The weather patterns are begun
By radiation from the sun,
The measurement of which is fun.
It's meteorology."

An Eastern gentleman did sigh,
"But what of ozone in the sky,
Without which man would quickly fry?
To check its thickness we must try
By letting off balloons to fly
Up through the atmosphere on high.
That's meteorology."

Another chap of great repute
Said "Really, don't you think it's cute
To use a program to compute,
With data tailor-made to suit,
The water use of leaf and root
Or estimate the growth of fruit
That's meteorology."

Cried yet another, "Seems to me
That on this I disagree,
For 'tis a lovely thing to see
The stomates working constantly
And varying but in degree
Among potato, grass and tree.
That's meteorology."

Another still then quickly rose
To say "Now let us just suppose
That you're the only one that knows
Evaporation rates from snows
And where the run-off water goes.
What exultation this bestows!
That's meteorology."

Through uproar loud that did ensue
Another rose and gave his view,
"If nothing's done 'bout CO_2
The world will ultimately stew.
Thus, you'll agree it is quite true
Our search for every tiny clue
Is meteorology."

The layman viewed these with disdain
And said, "For all this show of brain,
What satisfaction does each gain
By praising but his own domain?
From what I hear, it seems quite plain
You need the whole lot to obtain
Meteorology."

Contents

Preface

This is NOT a conventional textbook. It is a summary of observations reached over the last many decades, serving to strengthen early doubts about the ways convention describes the relationships between fluxes and gradient in air near the surface. The material presented here represents the conclusions of the author alone. Many of these conclusions vary from what is taught in today's meteorological classrooms and differ greatly from 'contemporary understanding.' Some of them will withstand the test of time; others will not. One of the conclusions is that the atmosphere lacks the ability to read textbooks.

As a young Experimental Officer at the famed CSIRO Division of Meteorological Physics in 1962, the author was introduced to the basics of air-surface exchange by some of the masters — Priestley, Swinbank, Deacon, Webb, Dyer and Taylor. However, with a background in statistics and physics there was an intellectual conflict that developed, involving the way in which scarce data were analyzed and claimed to support theories and intellectual constructs then being popularized, in particular the similarity theory put forward by Monin and Obukhov (1954). In this, the author was not alone. Many illuminating discussions took place about the Monin-Obukhov Similarity Theory (MOST) and related matters, usually over morning tea with at least three of the six above. Many decades later, the author discovered that the concerns so central in the Australian discussions were shared by many in North America. Frank Gifford, in particular, warned loudly about the dangers of reliance on dimensional analysis when clear evidence of the roles of assumed central variables was not available. It seems likely that these concerns were held by many researchers who preferred to keep them quiet because (a) the mainstream bandwagon was rolling and (b) there was nothing better to offer.

At the time the present author was entering the micrometeorological community at Aspendale, there was a growing and very active atmospheric sciences program centered in the American Midwest. Starting during WWII, the Institute of Meteorology at the University of Chicago offered courses by such eminences as Rossby, Byers and Starr. From the micrometeorological perspective, the modern sonic anemometer owes its development to initial work by Vern Suomi and colleagues at the Institute of Meteorology in the late 1940s and subsequently at the University of Wisconsin. Suomi's contributions to the science have been reviewed in most informative detail by Lewis et al. (2010). The importance of the Midwest community was well perceived by the Australian group. It was while visiting the University of Chicago in 1956 that Priestley wrote his monograph (Priestley, 1959). The present author first entered the United States on the payroll of the University of Chicago in 1965 and was introduced into the US meteorological community by luminaries like David Fultz, Harry Moses and Paul Frenzen. Before arriving in Chicago, the author spent a month or so with the Air Resources Laboratory (now of the National Oceanic and Atmospheric Administration) in Washington, DC, to which he returned in 1990 as its Director.

In eastern Europe and the Soviet Union there was a parallel universe of micrometeorology, well described by Foken (2006). Rarely did the two universes meet, but one special occasion was the International Turbulence Comparison Experiment of 1970, at which Thomas Foken and the present author were two of the youngest participants. It was during this experiment (at Tsimlyansk, in southern Russia, near Rostov) that personal interaction with Obukhov was possible.

There was always a strong awareness that what we know is limited by what we can observe and that good observational data are essential. In Australia, the scatter in observations at Deniliquin, Kerang, Hay and Gurley was disturbing and thought due to measurement deficiencies. However, as time progressed and instrumentation improved the scatter was not greatly reduced. The early opinion voiced by Geiger (1965) that micrometeorology is essentially statistical and any description of it must also be statistical was largely ignored. Fully deterministic relationships were sought, to connect such quantities as fluxes and gradients, but it quickly became obvious that the variability of the atmosphere prohibited prediction for any specific circumstance. Attention was directed instead to addressing ensembles; it was thought that the way to handle variability was largely through averaging. Turbulence intermittency was noted but was rarely examined because it lay outside the accepted framework. Simulations were viewed as little more than representations of the imagination of modelers and the lack of field experience among many communities of model practitioners was a reason for dismay. Nevertheless, it was acknowledged that useful models were a societal necessity and that an end goal of field research was their development.

There was a common recognition that the measurement methods then available were rudimentary and likely to generate poor and scattered data. A great advance in measurement technology was the development of sonic anemometers, introduced into the meteorological community through work by Suomi, Businger, Kaimal and coworkers. Sonic anemometry quickly attained a maturity that spawned the advances of 70s and 80s. Concern about data quality and analytical constructs remained and came to the fore with the reports of the ground-breaking Kansas study (Haugen et al. 1971) which, among other results, yielded an unexpected new value for the von Karman constant. Review of the Kansas analysis revealed that the cause for this seemingly anomalous result probably lay in the way discrepancies in the data were rationalized, a point that was transmitted personally by this author during a visit to the Kansas team, in Boston, in 1971.

The present author's early lack of enthusiasm about MOST has been moderated over the subsequent years by periodic reminders that in spite of its difficulties MOST remains a useful and convenient construct for modeling. However, the concern continued as it became increasingly clear that many workers ignored the fact that the MOST framework was undeniably susceptible to the syndrome now commonly referred to as 'shared variable' (see Hicks, 1978). Early researchers were confronted with a shortage of data and in the lack of good data the MOST construct gained popularity among experimentalists because the appearance of good results was then often assured. Within the modeling community, the cautions expressed by critics appear to have been overwhelmed by the theory's intellectual beauty and simplicity. With the development of advanced sensing systems (ultrasonic anemometers, infrared hygrometers, etc.,) the capability now exists to make measurements on a routine, ongoing basis that exceed, by far, the quality and quantity of the data on which the original MOST expressions were based. However, many models continue to uphold the creed of legacy in spite of available evidence.

The present document is a personal review of what is actually known, guided by experimentation rather than expectations. It is not the intention to undermine any of the models presently serving society well, since the contemporary and legacy expressions are easy to handle and if they work well — then so be it. Rather it is the intent to draw attention to how thoroughly the science has often misled itself and to the errors of the common fallacy that understanding of ensemble averages is adequate to explain specific circumstances. Likewise, it is not the purpose to reiterate the detailed descriptions of the science that have been presented elsewhere. There are many excellent textbooks that present the basics of the science in a way that will not be replicated or challenged here. Among the many superb treatises that examine the science and its origins, the books identified in the text-box seem especially relevant and readable.

The present purpose is to review some of the early conclusions, to examine how they compare with recent data and to provide a view of the evolving surface boundary layer in a way that might generate further investigation. If it is deemed that my analysis is sloppy, my response is "Sorry." My intention has been to simplify rather than be diverted by details when such details are not necessary. I anticipate the criticism that much of what follows constitutes heurism, or hand-waving. To this I respond simply — "Yes." In answer to the anticipated charge that I have cherry-picked the data, I have the same response. The intent has been to illustrate the points that are made and rather than use other peoples' data to illustrate interpretations with which they would likely disagree, I have made use of a wide variety of data available to me, generated largely by my colleagues. I have selected from these sources the cases that best illustrate the arguments presented, but in all cases I have searched to find evidence that the points being made could be contradicted by other data. The process is complicated by the fact that no single dataset provides information on all of the variables of possible relevance, compounded by the realization, now becoming increasingly apparent, that the atmosphere cannot always be viewed as a fluid medium isolated from the biosphere.

It is appropriate to express appreciation for the fellowship that I have enjoyed with the community of fellow-thinking researchers, and others. Very rarely have there been occasions of acrimony, although many frustrations have cropped up and opportunities have certainly arisen. In the text that follows, I have tried to exemplify the humanity and humor of the major figures who have (often unintentionally) served as mentors. If any of my comments are found objectionable, then so be it. If any of my conclusions are wrong, then also so be it.

The Early Cast

C. H. B. (Bill) Priestley accepted the position of Director of the new Atmospheric Physics Chapter of the Australian Commonwealth Scientific and Industrial Research Organization in 1946 and constructed a powerful cadre of researchers starting with the colleagues he brought with him from England and the local talent he attracted. The Section became well known through its subsequent location at Aspendale, a beach suburb of Melbourne, Victoria. His task was simple — provide new information that would help refine the way in which Australian agriculture made use of its limited supply of water. Although a theoretician at heart, he immediately embraced experimentation by all available means: evaporation pans, lysimeters, gradient methods and eventually eddy covariance.

W. C. (Bill) Swinbank followed Priestley to Aspendale in the first years of existence of the Meteorological Physics Section. He was an unabashed experimentalist, who developed a theoretical wind profile that fell to the wayside when eddy covariance measurements of the momentum flux became possible. He is also well known for his work on incoming infrared radiation, modifying the founding relationship put forward by Sir David Brunt. In later years he became a senior contributor to research programs at the National Center for Atmospheric Research, in Boulder, Colorado.

E. L. (Len) Deacon was a mainstay of atmospheric physics research at Porton, on the Salisbury Plain in England, before moving to Australia to join the new Meteorological Physics Section. Deacon's enhancements of the power-law wind profile and his association with its 'Deacon number' elevated him to a position well known in the international community. A major interest in air-sea exchange dominated much of his research in later years.

A. J. (Arch) Dyer had just obtained his doctorate in electrical engineering when he joined the CSIRO team at Aspendale, with the goal of developing field instrumentation for measuring eddy covariances. The success of his instrumental skills led to a long interest in turbulence of all kinds. His 'Evapotron' and his later 'Fluxatron' became essential components of all of the Australian micrometeorological programs of the 1960s and 1970s and led the way to many of the methods used today.

R. G. (Reg) Taylor was involved in some of the earliest studies of atmospheric dispersion in forests, conducted under the leadership of Frank Pasquill in Queensland in the wartime 1940s. He developed the heated wake fine wire directional anemometer that became a feature of the eddy flux instrumentation used in field programs until about 1966.

E. K. (Eric) Webb focused on two major aspects of air-surface exchange — the evaporation from lakes and the log-linear wind profile (Webb, 1970). An accomplished pilot, he had a hands-on familiarity with turbulence in the lower atmosphere and parleyed that understanding into a long examination of the nighttime atmosphere. He is also well known for his examination of the role of water vapor fluctuations in determining the fluxes of trace gases through the atmosphere, particularly carbon dioxide, and in specifying the slow atmospheric concentration trends of CO_2 with time.

The Meteorological Physics Section had a total staff of less than thirty in 1961, when the present author was hired as a young Experimental Officer by Dyer and Swinbank. The subsequent growth of the organization has been summarized by Garratt et al. (1998). A review of the life and times of Priestley has been presented by Garratt et al. (2011). The contributions of Dyer have been summarized by Hicks (1988) and those of Swinbank by Frenzen (1974).

To this list must be added the cast of colleagues at Argonne National Laboratory in the 1970s, in particular Paul Frenzen, Marv Wesely, Dale Hess, Jimmy Sheih and Ted Yamada. And finally the list must include those dedicated experimentalists who have provided most of the measurements considered here, in particular Will Pendergrass, Dennis Finn, Temple Lee and Alan Weber.

The text to follow has been prepared with the intention to focus on what has been observed, with limited attention to theories not yet verified. Most authors addressing similar topics start by explaining the current understanding and then present results to substantiate it. Here, the starting point will be the observations and it will then be seen how current understanding agrees with them, or otherwise. In this, I claim to be unusually qualified. I have received limited academic training in meteorology. Instead, I rely on observations, experience, logic, and the after-effects of extensive collaboration with colleagues not so academically challenged. It must be expected that there will be mistakes in the text presented here, although I have tried to eliminate them. So be it.

What I conclude and present here is based on observations, with acceptance of those descriptions presented elsewhere that agree with my experience. I have come to appreciate the view of one of the early masters of observational science:

> "Experience has been the mistress of whoever has written well, and so as mistress I will cite her in all cases." (Leonardo da Vinci)

As this writing progressed (over a period of at least fifteen years) the viewpoints expressed by Richard Scorer in his 'Natural Aerodynamics' have been faithfully followed (except when I have departed). It is my hope that, as Scorer said, a reader "will glimpse a little of the beauty of the world of conjecture and become less fierce in his dogmas." In the present text, there will be very little attention to numerical modeling, some examples of which are viewed by me as somewhat like taking "my way or the highway" from the psychological arena and transferring its basics into computer sciences.

The documentation to follow starts with some views of how the micrometeorological specialty developed and convention evolved (Part I), followed by examinations of field experiments and how their results might cause some rethinking of legacy assumptions (Part II). Part III takes a first step towards consideration of surface complexity. Part IV extends considerations to a land surface with a vegetative canopy. Part V addresses two specific aspects of contemporary interest – deposition and dispersion. Part VI looks forward and Part VII says goodbye. Finally, appendices provide basic information regarding symbols and sources of observations.

Check out –

Garratt, J. R. 1992. *The Atmospheric Boundary Layer*, Cambridge University Press, 316 pp.
Lorenz, E. N. 1993. *Chaotic Behavior in Systems*, Washington University Press, 345 pp.
Monteith, J. L. and M. H. Unsworth, 2013. *Principles of Environmental Physics*, Elsevier, Amsterdam, 401 pp.
Munn, R. E. 1966. *Descriptive Micrometeorology*, Academic Press, New York, 245 pp.
Scorer, R. S. 1958. *Natural Aerodynamics*, Pergamon, London, 313 pp.
Sutton, O. G. 1953. *Micrometeorology*, McGraw-Hill, New York, 333 pp.

Part I: The Origins of Convention

Chapter 1

Airplanes, Airships and Gas Warfare

"If you want science to develop quickly, start a war" — anon

Wind tunnel studies provided the basis for many of the developments in understanding of turbulence in the early 1900s. Earlier studies of fluid dynamics (e.g., of flow in pipes) provided useful background and the role of the Reynolds number as an indicator of the transition from streamline flow to turbulence was well known. The rapid development of aircraft required better understanding of flow and turbulence in the atmosphere. Transferring accepted understandings to the case of airfoils required new experiments, which were undertaken with enthusiasm and alacrity before and especially during World War I.

All of this was taken as a starting point by the earliest micrometeorologists, who recognized that the aeronautical and pipe-flow formulations related to velocity regimes differing greatly from those of the lower atmosphere. Moreover, the effects of density differences were not readily amenable to laboratory study. In the lower velocity regimes of micrometeorological interest, effects of density gradients in the air were (correctly) anticipated, but the effort was first to consider those situations in which such density effects could be ignored. The fundamental reliance on neutral stability as a benchmark arose accordingly. In general, field studies were designed to quantify and explain observed departures from neutral.

Micrometeorology was elevated in importance in 1915, when gas warfare entered the conflict in Europe. In WW-I, meteorologists provided guidance about where and when to release clouds of gas to cause the greatest damage to adversaries. Related research took place in many countries, not limited to the well-known combatants. A review of such literature as is now accessible reveals that fears of gas attacks on battlefields were widespread and the discipline now known as micrometeorology blossomed. Terrain complexity was a dominant concern, since dispersion was influenced not only by the wind fields predicted synoptically but by the influences of whatever variations the landscape imposed. It was as a result of the unpredictable consequences of terrain and local winds imposed on a prevailing westerly flow that gas attacks by Germany in the fields of Flanders were curtailed. In 1916, the German gas warfare capabilities were transferred to the Russian battlefield, where the battlefields were flatter and wind fields were more conducive. In the meantime, the allied forces continued to exploit gas warfare in Europe, as they attacked from the west with the prevailing wind behind them (q.v. McCamley, 2006). In the following decades, related research continued despite popular revulsion of gas warfare. Extensive studies were made of the most effective agents to be used and of the optimal methods to deploy them and the specialty of micrometeorology has profited. Much of the related literature is hidden behind a cloak of secrecy imposed to avoid alarming the public about the nefarious thinking of their 'protectors.'

The expertise required to optimize the use of gas releases was developed on the basis of experiments made elsewhere and through the experiences of those affected. In Germany, Fritz Haber (the 'father of gas warfare' and a subsequent Nobel Prize winner) insisted on personal involvement on the western front and worked with meteorologists to design the most effective release methodology — mostly centered around two questions: when and where? In England, specialized research groups developed, such as the group at Porton Down that conducted micrometeorological and dispersion studies over the Salisbury Plain. There is little available documentation of these early activities, but a vivid picture results from the reminiscences of participants in such English field work (e.g., Len Deacon) and the attributions by luminaries in the developing discipline, particularly Sutton at Porton Down through his textbook (Sutton, 1952a) and in his contributions to the International Symposium on Atmospheric Turbulence in the Boundary Layer (Hewson, 1952). The papers collected by Friedrich et al. (2017) provide a general view of the gas warfare business, but with little attention to the roles of atmospheric physics and turbulence.

Interest in the dispersion of gases in warfare continued, largely unabated, during the period between WW I and WW II. Italy employed gas as a weapon in Ethiopia in the 1930s and Japan used chlorine in China. Shortly after the Japanese incursion into Southeast Asia in 1941 the allies combined to address the issues of gas warfare with increased intensity. Australia and India were threatened. Japan was not a signatory to the Geneva Convention and had already deployed gas agents in the conflict in China. It was recognized that understanding of the dispersion of gas inside a forest canopy (characteristic of Southeast Asia) was exceedingly limited and as a result a collaborative experimental investigation was conducted in Queensland, Australia. Details of this study are difficult to obtain but based on materials in the Australian War Archives it is apparent that the leader of these studies was Frank Pasquill (of later dispersion fame) with assisting personnel that included Reg Taylor (subsequently a colleague of Priestley, Swinbank and Deacon in the 1950s and onwards).

Until the emergence of gas warfare as an issue, most military interest in the atmospheric boundary layer was associated with the effects of the wind on the trajectories of projectiles. One can imagine the interest of Roman Legions in the flight of their arrows. The concerns were typically related to the wind field at greater heights than affected the release of gases. Tall towers were a common feature of artillery test ranges and tall towers have always served as a magnet (perhaps more appropriately, a 'strange attractor') for meteorological researchers.

It remains unclear as to how much the airship programs in Europe contributed to the understanding of the lower atmosphere, but it is expected that considerable related research was conducted in Germany in support of their airship programs of the 1920s and 1930s.

Parallel research commenced in England in 1924, with the setting up of the Royal Airship Works at Cardington. Lempfert (1931) reviewed the extensive research on wind fields conducted in the first five years of this activity and Frost (1948) reported support for the power law wind profile based on tall tower studies conducted at this location after strong interest in airships declined (due to numerous costly accidents):

$$\bar{u}_{z2}/\bar{u}_{z1} = (z_2/z_1)^n. \quad (2.1)$$

Here, $\bar{u}_z$ is the wind speed at height z and n is a constant found (by Frost) to be about 1/7 in neutral conditions. Without confirming evidence being available, it seems reasonable to assume that the tall-tower wind profile studies reported by Hellman (1915) had relevance to the German military programs of that time, regarding both dirigibles and poison gas.

The decline of interest in airships following several costly disasters caused Cardington research to be limited while that at Porton continued to grow through the years of WW II. It was the Salisbury Plain that spawned the modified wind profile, proposed by Deacon (1949), in which atmospheric stability and surface roughness were introduced as modifiers of the basic power law relationship of Equation 2.1. The Deacon profile was tested in many subsequent field studies, notably including the Great Plains project in the USA (Davidson and Barad, 1956).

At much the same time as military and aeronautical considerations focused on the wind profile above the region of immediate association with the surface, theoretical expectations favoring a logarithmic profile in the surface boundary layer (SBL, initially defined as that layer that influences life and subsequently as the region of the atmosphere where changes in turbulence regimes are dominated by changes in the surface). Deacon's (1949) analysis addressed the lowest 10 m of the atmosphere and it was on the basis of these observations that the exponent n in Equation 2.1 was postulated to be a function of the surface and stability. On the other hand, the logarithmic dependence on the height of measurement (above some zero plain that remained obscure, d) was substantiated by observations reported by Best (1935) and subsequently endorsed by a cacophony of researchers (q.v. Sutton, 1953). Johnson (1959) conducted and reported the obvious comparison and found that ". . . the logarithmic law may be valid at low levels while the power law may give better agreement at levels above 8 m . . ."

Although modern micrometeorology is dominated by the logarithmic profile and its consequences, the power law description of the wind profiles remains a useful method to extrapolate from the SBL into the layer affected by convection above it (the planetary boundary layer: PBL). In this regard it is noteworthy that considerations of dispersion and especially of the imposition of controls on emissions have continued to make use of the power law, but with reservations. For example, Touma (1977) emphasizes that reliance on the power law approximation imposes the need to consider the consequences of surface changes and the departure from neutral stability – precisely what energized Deacon in his 1948 presentation.

The fine detail of how the wind profile varies in the immediate vicinity of the earth's surface remained the subject of research that blossomed only when more advanced wind sensors became available. The foundation was invariably the logarithmic wind profile in the surface boundary layer:

$$d\bar{u}/dz = u_*/kz \tag{2.2}$$

where k is the von Karman constant derived as about 0.4 by aerodynamicists. From this, the wind speed at height z is derived as

$$\bar{u}_z = (u_*/k) \cdot \ln([z - d]/z_0), \tag{2.3}$$

which introduces the displacement height, d, of the logarithmic profile and the familiar roughness length z_0, essentially a constant of integration.

It was quickly obvious that the SBL relationships (Equations 2.3 and 2.4) require that n' cannot be independent of height, since n' must approach unity as the surface is approached. An open question remained as to the height at which the SBL relationships transformed into the PBL-scale power law that extended to greater heights. The matter remains contentious.

Chapter 2

Pipe-Flows and Flate-Plates

"More power laws are due to the availability of log-log graph paper than to any process of nature" — W. C. Swinbank

The interaction between the air and surfaces dominated much of the research in the early stages of aircraft development. Research tended to focus attention on particular aspects of the general problem, starting with studies of laminar flow over flate-plates, evolving into studies of flows over curved surfaces and eventually considering turbulence and its origin as a central issue. Many textbooks summarize the developments that arose from laboratory studies of this kind (see Sutton, 1953; Garratt, 1992). From the SBL perspective, studies of the way in which a modified laminar boundary layer develops downstream of the edge of a smooth plate has led to analogies with leading-edge situations affecting the natural surface, with changes from one crop to another marking the boundary of relevance. While the micrometeorologist learns from the results of a laboratory investigation of laminar-layer development, the extension to the outside world entails consideration of differences that continue to be somewhat elusive.

The transition from laminar to turbulent flow was initially studied in flows through pipes. These investigations led to the identification of the Reynolds number ($Re \equiv Vd_p/\upsilon$) as a means to identify the demarcation between laminar and turbulent flow. Here, V is the flow velocity, d_p the diameter of the pipe and υ kinematic viscosity. Once again, the relevant textbooks are numerous and uniform in their depictions of the behaviors involved: laminar flow when Re is small, transitioning to turbulent flow when Re is large, with an indeterminant regime between. Pipe-flow investigations resulted in a criterion that remains a part of the disciplinary basis: that turbulent flow tends to exist when Re > about 2000. In subsequent developments, chemical engineers have expanded the pipe-flow universe to the extent that its fundamentals are accepted and practiced in modern constructions. Integrating the fluid dynamics and chemical engineering results into micrometeorology entails a number of steps that should be taken with care. This extrapolation is hindered by an exasperating language difficulty. Whereas the physics background of meteorology (and especially of micrometeorology) imposes the requirement that the relationships that result must be expressed in terms of contributing physical properties. In contrast, other disciplines (like those contributing the pipe studies) manipulate quantities comprising two or more of these fundamental properties. The relevance of the physical properties can be quickly lost. Full understanding of the results of the relevant studies requires attention to the choir of contributing researchers contributing to the discipline. Their names are forever associated with specific dimensionless combinations of physical quantities — Froude, Nusselt, Schmidt, Lewis, Grashof and Sherwood, to name a few. In this choir, Reynolds, Richardson and Prandtl stand out as micrometeorological favorites.

Relating the results of pipe-flow studies with the open-air natural situation is not straightforward. The situation is compounded by the methodology employed in analysis of laboratory fluid dynamics experimental observations and in the reporting of results. A common approach appears to be regression analysis, not among contributing physical parameters but instead among the dimensionless quantities made up from them. Correlations are commonly in search of power-law dependences, so that observations are first subjected to a logarithmic transformation. The frequent reporting of intuitively unlikely exponents is a consequence that challenges extrapolation of the results and invites re-analysis using more advanced statistical approaches.

A major consequence of pipe-flow studies was the empirical result now referred to as the Chilton-Colburn analogy. This invokes a dependence of exchange between trace substances carried by a fluid flowing through a pipe with the surface, with a power law relationship involving the ⅔ power of the molecular diffusion characteristics of the substance in question — usually referenced to either the fluid viscosity through the Schmidt number or its thermal diffusivity through the Lewis number. The resulting Chilton-Colburn analogy (Chilton and Colburn, 1934) has been verified repeatedly in pipe-flow and conduit studies. It has found ready acceptance in micrometeorological considerations of air-surface exchange of trace chemicals, as will be discussed below (in Part V) under the heading of "Dry Deposition." However, questions remain about the extrapolation of results obtained from studies of controlled flow through confined channels to outdoor circumstances of open-air exposure.

Conclusion

Laboratory studies provide valuable guidance, but not necessarily answers.

Chapter 3

The Agricultural Incentive — Water Resources

"Water is the driving force of all nature" — Leonardo da Vinci.

While military requirements and aeronautics dominated the thinking of a particular community of boundary-layer meteorologists in the period embracing World Wars I and II, civilian focus on agriculture energized a separate community and led to the development of agricultural meteorology as a recognized specialty. From the start of the 20th millennium to present days expanding populations and increasing utilization of dry regions for agricultural purposes, frequently requiring irrigation, have raised water availability to the level of national concerns. Much of the early relevant micrometeorology was devoted to quantifying the rate of water loss through evaporation. It was hoped that some treatment would be found to limit water loss, from lakes and reservoirs as well as from field crops. In recognition of this need, a standardized method for assessing the daily evaporation from a water surface was promoted and became widely employed in networks at a national and even continental scale (e.g., across the USA, Europe and Australia). The Australian situation (being better known to the present author) presents a useful example of how the science evolved.

In the 1950s, the evaporation regime of the Australian countryside was monitored using a network of Class-A evaporation pans, installed according to international standards. Figure 3.1 shows the guidelines for deployment of a Class-A pan, based on the specifications of the United Nations Food and Agriculture Organization and the World Meteorological Organization (q. v. Brower and Heibloem, 1986). The daily loss of water (after allowing for precipitation) was a direct indication of the potential evaporation, i.e., the evaporation rate if the surface happened to be water. Modifying this assessment to the case of the actual evaporation from the surrounding area when the surface was vegetated caused a number of analytical methods to be developed.

A major conceptual problem arising with the use of Class-A pans (as were standardized at the international level) is that they were installed on stands slightly above the terrain and hence provided an obstacle to the wind. Moreover, solar radiation striking metal sides caused the water to be heated, so elevating evaporation rates (especially at low latitudes). In response to these perceived problems an improved evaporation pan was developed — a square pan buried so that the edges were flush with the surface. This improved 'Colorado pan' was also widely deployed. Considerable research was conducted to relate the findings of one pan design to another and to the evaporation rate measured for a nearby lake. On average, the Class A system yielded values about 20% higher than those given by the Colorado pan and about 25% higher than from a nearby lake (Rosenberg, 1973; Allen et al. 1996; 1998; Linacre, 1994). To protect from the ravages of local fauna, pans were normally protected either by fencing or wire screens. Although it was found that these often failed to protect against thirsty and persistent wildlife, the data produced by networks of evaporation pans of various configurations have proved to be useful in a variety of applications (e.g., Robinson, 1998).

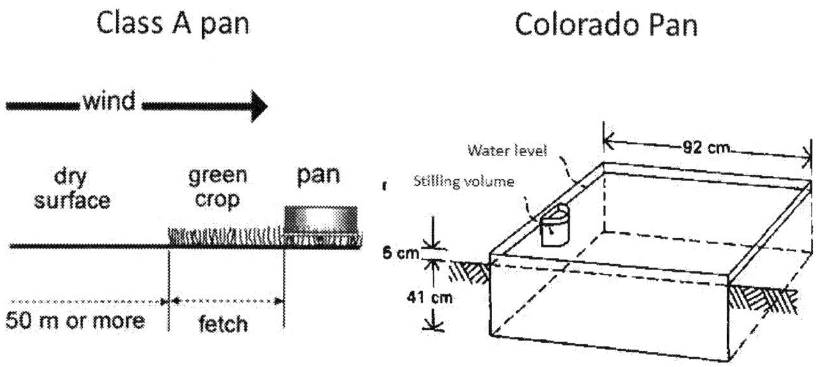

Figure 3.1. Two of the ubiquitous pans for measuring potential evaporation, i.e., the evaporation from an exposed water surface located where such data are needed. To the left is a depiction of the installation of a Class-A pan, as prescribed by the World Meteorological Organization. The drawing to the right shows an alternative configuration, referred to as the "Colorado pan." The two differ in their deployment. The Class-A pan stands on a wooden frame whereas the Colorado pan is buried so that its water surface is about level with the surrounding surface.

The pan methodology was intended to yield a measure of the potential evaporation in a selected area. If the intent was to quantify the rate of evaporation from surrounding vegetation, then a calibration point was required against which the pan results could be compared. The favored approach required the use of a weighed lysimeter, typically more than 1 m in diameter and of depth such that the entire root zone of the crop under consideration could be captured (e.g., Ritchie and Burnett, 1968). The container was filled with soils arranged to replicate the surrounding soil matrix, with a crop sown to merge the test surface with its surroundings. The utility of weighed lysimeters was limited by two major factors. First, they could not be moved; their utility was primarily in site-specific research programs and as reference points for evaluating other methodologies. Second, they required the accurate measurement of small changes in weight, necessitating continued dedicated attention by highly skilled technicians. The experimental demands were sufficient that use of lysimeters was infrequent. However, the many intensive studies of alternative methods for deriving evaporation rates from field measurements spawned a wide range of relationships, some of impressive complexity facilitated by the new availability of digital computers.

Conclusion

The simplest methods continue to hold water.

Chapter 4

Interpreting the Surface Heat Energy Balance

"The earth has received the embrace of the sun and we shall see the results of that love"
— Sitting Bull.

While the use of evaporation pans satisfied the demands of many meteorologists and climatologists, agricultural research imposed more stringent demands on the determination of evaporation rates, especially in the case of test areas with different specialized crops. In the early twentieth century, micrometeorology became popular as a practical tool for assessment of the evaporation rate, E. The many methods developed shared a common feature — they all relied on a determination of the total heat available for transfer from the surface (vegetated in this case) from which the part of the total heat transfer contributed by the latent heat of evaporation is then derived. The focus was then not solely on the evaporation rate itself, E, but on the latent heat of evaporation (L_wE).

The sun is the source of most of the heat energy driving motions in the atmosphere. Other sources, such as geothermal, are commonly assumed to be small in comparison, an assumption likely to be the cause of concern only to residents of some specific locations like Rotorua or Reykjavik. Radiation arriving at the top of the atmosphere is reflected, scattered and absorbed by the air and the aerosols it carries, eventually reaching the surface as a combination of direct and diffuse components. In total, these make up the incoming solar radiation, $R_\downarrow$, conventionally expressed as the energy flux per unit area of a horizontal surface. Some of this incoming energy is reflected and some of it is emitted back as infrared radiation following the Stefan-Boltzman expectation but with an emissivity that depends on the characteristics of the surface. The consequence is that in daytime the surface benefits from an acquisition of radiative energy, referred to as the net radiation, R_n. In the terrestrial case, some of this net radiation is used to heat the ground: the ground heat flux, G. Some of it is used by plants in photosynthesis, P. The plants also make use of the incoming energy to promote transpiration, resulting in the evaporation of liquid water at a rate E (in terms of mass of water per unit time and per unit surface area). The energy associated with the loss of water through evaporation is then L_wE. The remaining part of the surface heat energy balance is dominated by the heating of the surface in direct response to the incoming radiation and the generation of buoyant packets of air. The resulting convective heat transfer, due to the fact that hot air rises, is the 'sensible' heat flux, so-called because it is a property with real physical relevance – heating of the air.

In practice, there is often a layer of vegetation occupying the ground surface. The heat energy balance that results is then conventionally written as:

$$R_n - G = H + L_wE + P + S \qquad (4.1)$$

where S is a heat storage term contributed by whatever canopy (vegetative or otherwise) occupies the terrain. There have been many assessments of the contribution of P in Equation 4.1 – the heat energy associated with photosynthesis. Photosynthesis makes preferred use of short wavelength radiation, itself a small part of the radiation spectrum. The result is that plants make use of less than 2% of the incoming radiation, corresponding to 3% to 4% of R_n. This estimation can be interpreted as an indication of the kind of error that might be anticipated if attempts are made to balance the energy budget while omitting P. The storage term S is also often omitted. The matter of S will be of interest later, in discussion of Bowen ratios and their experimental applications. A frequently-detected imbalance in the surface energy budget is sometimes attributed to the omission of S and/or P, or alternatively to advection or inadequacies of the instrumentation (q.v. Leuning et al. 2012).

In early research, the terms S and P were intentionally ignored, because the focus was then on the central daytime situation in which other terms dominate. The simplified energy balance equation

$$R_n - G = H + L_wE \tag{4.2}$$

then provided a convenient starting point for a number of methodologies to derive estimates of H and L_wE by apportioning the available heat energy (R_n – G) according to some other measurements, such as the temperature and humidity gradients above some area of specific interest. The Bowen ratio ($\beta \equiv H/L_wE$) can be quantified as $(c_p/L_w)\delta T_a/\delta q$, where δT_a is the difference in air temperature between two heights above the surface of interest and δq is the change in specific humidity between the same two heights. L_w is the latent heat of vaporization of water and c_p the specific heat of air at constant pressure. The ratio c_p/L_w is often written as the psychrometric constant ($\gamma \equiv c_p/L_w$). If measurements are made of the net radiation and the ground heat transfer, then Equation 4.2 yields quantifications of H and L_wE provided the fluxes H and L_wE share the same eddy diffusivity (see Bowen, 1926). If the Bowen ratio $\beta \equiv H/L_wE$ can be specified, as above, and if R_n and G are measured, then the assumption of a simple surface energy balance (Equation 4.2) leads to

$$L_wE = (R_n - G)/(1 + \beta) \tag{4.3}$$

This Bowen ratio energy balance (BREB) method has been mainstream within the agricultural meteorology community since its inception and remains common today. Later, the matter will be addressed in more detail.

Conclusion

Watch out for the details.

Chapter 5

Introducing Plant Physiology

"We need the tonic of wildness and nature" — H. D. Thoreau

There have been more than three centuries of studies of stomata (Meidner, 1987). They serve as the route by which CO_2 is extracted from the air to facilitate growth by photosynthesis, with aperture controlled by external conditions such as the intensity of photosynthetically active radiation (PAR) and water availability (q.v. any relevant compendium, e.g., Zeiger et al. 1987, Hopkins, 1999). Water is integrally involved in the photosynthetic process. These same stomata provide a pathway for the water contained in the leaf biomass to evaporate into the air. The rate of transport of water vapor (outwards) and CO_2 (inwards) is classically described by a conductance, k_{st}, which modern instrumentation can measure. To describe the exchange between a leaf and the air an additional term was needed, k_a, to quantify the exchange within the air itself.

As in the case of the analogy with an electronic circuit, conductances are no more than the inverses of resistances, so that $k_{st} = 1/r_{st}$. Conductances are favored for the present discussion, because they underpinned the development by Penman (1948) of a method to estimate L_wE while avoiding the need to measure in-air gradients δT_a and δe. An easy assumption was to equate the vapor pressure of water inside stomata to the saturated value at the temperature of plant tissue (clearly an approximation, because of Raoult's law), and then to relate the loss of water to the vapor pressure difference with the air in which a leaf resided. The considerations that follow are well presented in many textbooks (e.g., Stull, 1988; Monteith and Unsworth, 1990; Garratt, 1992) and there is no need to repeat them here. The end result is that

$$L_wE = [s(R_n - G) + \rho c_p k_a(\text{де})]/[s + \gamma(1 + k_a/k_{st})] \qquad (5.3)$$

where s is the slope of the saturated humidity curve at temperature halfway between the surface and the air (these are usually close enough that the latter is assumed), c_p is the specific heal of air at constant pressure, γ is the psychrometric constant ($\gamma = c_p/L_w$) and де is the familiar vapor pressure depression ($= e_s(T_a) - e$).

Equation 5.3 is like the familiar Penman-Montieth equation. It provides a method for assessing evaporation rates from routine meteorological data, at the level of a leaf for which the stomatal conductance can be specified. Relating the understanding of the behavior of stomata of a particular plant species to a field of the same vegetation presented (and continues to present) major difficulties. The leaf area index *LAI* thus provided the basis for a useable first-order approximation when the requirement was for a measure of the rate of water loss from a field.

The 'big leaf' model arose (q.v. Jarvis and McNaughton, 1986) to couple physiological and aerodynamic considerations in a simple visualization. The main components of this model were an aerodynamic resistance describing the coupling of the air above a canopy with the air in which the foliage is bathed, R_a, and a stomatal resistance serving the same role as for a single leaf but representing the entire plant stand, R_{st}. (Here, as elsewhere in this document, lower-case symbols relating to air-surface exchange characteristics refer to a unit area of the components of the surface, whereas upper-case relate to the integrated consequence of these as viewed from the perspective of the atmosphere above the canopy.) If the resistance associated with stomata r_{st} is expressed as per unit of leaf area and the leaf area index is *LAI* (the area of leaves per unit horizontal area of the plant canopy), then a first-order estimate of the effective stomatal resistance for the vegetated surface as seen by the atmosphere can be estimated as $R_{st} = r_{st}/LAI$ (but this will be reconsidered later). There have been many refinements of this simple big leaf model, such as arise from the need to consider evaporation from the ground separately from that from the foliage (Shuttleworth and Wallace, 1985). A two-big-leaf system has also been promoted (e.g., Dai et al. 2004), one 'big leaf' being shaded and the other unshaded.

The appropriate index of canopy leaf area is not always the single-sided leaf area, because in some vegetation stomata populate both sides of leaves. For other circumstances (e.g., the exchange of sensible heat), the one-sided *LAI* is never as appropriate as its two-sided equivalent. Later work (as presented in discussion of deposition processes, to follow) has shown that the scaling of resistances by *LAI* is not as simple as was earlier considered.

In its various forms deriving from Equation 5.3, the Penman-Monteith relationship constitutes a well-recognized method for assessing evaporation from extensive crops. In the 'big leaf' context it requires a quantification of the two contributing resistances, R_a and R_s. Both determinations impose the need for approximations that may or may not be appropriate in any particular circumstance. The aerodynamic resistance R_a cannot be measured directly but micrometeorological considerations indicate that it can be approximated as $\bar{u}/u_*^2$, a matter that will be discussed later. The wind speed can be measured, but determination of u_* is demanding.

To provide an even simpler methodology, Priestley and Taylor (1972) examined the likely ranges of the Bowen ratio over extensive saturated surfaces and proposed a relationship for the latent heat flux:

$$L_wE = \alpha(R_n - G)s/(s + \gamma). \tag{5.4}$$

Using many datasets then available, the constant α was found to be about 1.26. While the applicability of Equation 5.4 is obviously limited by its assumptions, it has found widespread use and remains a favored simplification in climate modeling. Many papers have examined the constant α in detail, using more recent (and presumably more detailed) datasets, with some providing strong supporting evidence (e.g., Stewart and Rose, 1977) and others finding a need to incorporate other variables such as soil conditions (Burba and Verma, 2005).

Prior to the 1960s, quantification of the evaporative water loss from landscapes relied on interpretation of the surface energy balance, using a variety of methodologies designed to satisfy specific requirements. Bowen (1926) proposed that the available heat energy ($R_n - G$) can be apportioned according to differences in temperature and humidity between two heights above the surface of interest. This Bowen Ratio Energy Balance (BREB) methodology has been favored for agrometeorological studies for nearly a century. The Penman-Monteith formulation was developed to simplify the BREB approach and make it more widely applicable using more conventional meteorological observations. However, these methods all yielded estimates of *E* or of L_wE vulnerable to the relevance of the assumptions to the area and time of concern. A method for directly measuring the turbulent exchange of water vapor in the air was needed. To this end, eddy covariance techniques were explored extensively throughout the long early histories of micro- and agro-meteorology.

Conclusion

Vegetation is a central contributor to the health of the atmosphere.

Chapter 6

The Birth of MOST

"Never take anything for granted" — Benjamin Disraeli.

The basis for all physical descriptions of the SBL wind profile was the understanding of neutral stratification and Fickian diffusion — that fluxes are proportional to gradients. As micrometeorology evolved, the two paths considered above were followed largely independently. One path centered on the description of the wind profile, the other on the apportionment of available heat energy. While the search for reliable datasets continued, the door was open for new theories.

Following the productive application of dimensional analysis in fluid dynamics, the Buckingham Π theorem was adopted by many micrometeorologists as an entry point for consideration of the formal interactions among the central micrometeorological variables — the momentum flux (formulated as $\rho\overline{w'u'}$), the heat flux ($\rho c_p\overline{w'T'}$), height (z), the thermal/buoyancy parameter (g/θ) and the corresponding gradients $\partial\bar{u}/\partial z$ and $\partial\bar{\theta}/\partial z$). In particular, Monin and Obukhov (1954) presented a framework, largely based on dimensional analysis, for addressing the matter of flux/gradient relationships in the air. This similarity theory was intended to apply to the simplest case imaginable — that of a spatially homogeneous surface when conditions are not changing with time or space. When field studies made use of sites that satisfied these conditions, tests of this theoretical framework usually met with convincing success. In the following many decades, Monin-Obukhov Similarity Theory (MOST) has dominated research studies and the evolving field of numerical simulation of the atmospheric boundary layer (q.v. Foken, 2006).

The matter of exchange between the atmosphere and its lower boundary, the 'surface', becomes a major consideration when one is interested in such matters as the local wind speed, the loss of water by evaporation, the fate of atmospheric contaminants, or the dispersion of hazardous materials. In such applications, it is common to take the view from aloft – the surface is seen as the lower boundary affecting the atmosphere (the micrometeorological 'surface') represented using a few decisive and deterministic properties such as the roughness length (z_0) and displacement height (d). From a different perspective, the same exchange pertains to damage to surface structures by the drag force and harm to ecosystems (both terrestrial and aquatic) due to the deposition of pollutants carried by the atmosphere from sources that may be far upwind. These two considerations are inevitably linked. Damage to structures due to exposure to air pollutants is rife, especially in urban areas but commonly in the case of bridges and other complex structures that are susceptible to the failure of some specific exposed component (e.g., the accelerated rusting of metallic components). It is well accepted that exposure to air pollutants weakens plants, animals and structures, making all of them more vulnerable to any extreme insult that might then be blamed for their eventual failure. In the case of exchange with some particular component of the surface (e.g., a specific obstacle), it is local considerations of the area under examination that are crucial. The linkage between this view from below and the atmospheric view from above has been the subject of decades of research, most recently relying on numerical simulations. In the present discussion, there will be no simulations presented. Instead, the focus will be on the way in which randomness and determinism interplay, with the conclusion that an all-embracing description with the precision desired by many may be an unanswerable prayer.

The conclusion by Geiger (1965) that formulations describing the lower atmosphere are essentially statistical is a central consideration, amplified by the observation of Lorenz (1963) that chaos is a feature of the atmosphere. As a pragmatic response to the need to extract meaning from sets of observations often dominated by unresolvable variability (aleatoric uncertainty) averaging of data within conveniently defined ensembles became an accepted practice. In consequence, the relationships now seen in almost all numerical simulations are associations among averages. In the simulations of today, the resulting equations are then employed as if they apply to all circumstances, an assumption that is clearly at odds with the constraints originally imposed (e.g., by Monin and Obukhov, 1954). It is not clear how the resulting fully deterministic depiction of the surface layer of the atmosphere mirrors the real world. Furthermore, it is not clear whether any differences are important. The matter has been of concern for decades, with many authors (e.g., McNaughton, 2009) presenting evidence to underline the insecurities hidden within the MOST construct. Principal among these are

1. The requirement for a good site: in the terrestrial "real world," horizontal uniformity is unlikely except for situations that are of little practical interest.
2. The assumption of time stationarity: in practice, the atmosphere is always evolving with time. The central concern is whether this rate of change affects the interpretation of experimental data and the application of the resulting formulations.
3. The focus on "neutral" as a foundation: original work grew out of such considerations as aerodynamics, with velocities large enough that atmospheric stability was of little concern. In air near the surface, the atmosphere tends to have two dominant stability regimes, stable and unstable, and neutral tends to be transitional and therefore potentially in violation of the time stationarity constraint.
4. The dimensional analysis approach: in concept, the underlying principle for non-dimensionalizing a variable is to sequentially reduce the unexplained variance. In micrometeorological expressions, the most common normalizing factor is the ubiquitous loss of momentum (wind speed decreases as the surface is approached), usually introduced by using the friction velocity, u_*. Although u_* is a familiar and important quantity, it is highly variable and is subject to all of the factors that influence the atmosphere near the ground. Use of u_* as a normalizing variable generally does not succeed in reducing the unexplained variance. Instead, in MOST the inclusion of u_* can result in relationships that are a product of the details of the formulations and not of any contributing physical behavior (q.v. Hicks 1978).

Early researchers were confronted with a shortage of even poor data and the powerful analytical construct of MOST inevitably gained popularity. With the development of advanced sensing systems (ultrasonic anemometers, infrared hygrometers, etc.) the capability now exists to make measurements on a routine, ongoing basis that exceed, by far, the quality and quantity of the data on which the original MOST thinking was based.

Even in the early days of micrometeorology, mathematicians reminded readers of the problems associated with dimensional analysis. For example, Saint-Guilhem (1960; translation by F. A. Gifford, 1967) concludes that "dimensional analysis is an art full of dangers" due to the often-overlooked requirement that one must first be convinced that the selected variables do indeed contribute. A further complexity was discussed by Hicks (1978), who showed the consequences of relating two dimensionless quantities that have a shared variable. This was not a new discovery. The shared variable syndrome was a topic of continuing interest to the early developers of the flux-gradient relationships now considered to be 'legacy'. Many workers have since alerted the micrometeorological community to the inherent dangers of common micrometeorological analytical methods, for example relating turbulence quantities like σ_w / u_* to stability when the latter is formulated as $\zeta = z/L = -kgHz/(\rho c_p u_*^3 \theta)$ (q.v. Garratt, 1992). (Detailed discussion of relationships of this kind will follow.) Many data presentations have plotted the first of these variables against the second, so demonstrating that a plot of u_*^{-1} against u_*^{-3} yields a behavior corresponding to a cubic, regardless of what the other variables might be.

Following the guidance of MOST, the conventional method for examining the stability dependence of gradients has been to relate the dimensionless gradients to stability (e.g., relate $\phi_m = [kz/u_*][\partial\bar{u}/\partial z$ ζ as the appropriate stability index, in which case u_* and z are worrisome shared variables. In more recent times, plotting against the Richardson number ($\mathrm{Ri} \equiv \rho c_p[\partial\bar{\theta}/\partial z]/[\partial\bar{u}/\partial z]^2$) has become popular, but once again the shared variable syndrome is a concern since $\partial\bar{u}/\partial z$ is then an offending quantity. The corresponding analyses for heat and moisture are also affected, with $\partial\bar{\theta}/\partial z$ (and $\partial\bar{q}/\partial z$) serving as additional shared variables.

The assumption of exchange by a continuum of turbulence was conceptually attractive, yet field experience generated considerable early concern. At first, experimental methods involved analog recording and instrumentation working at the limits of reliability. Understanding of short-term processes and interruptions in the turbulent transfer phenomenon was limited to what was observed by actual on-site watchers, in real time.

Even in the earliest days, there was active discussion among the proponents of the simplifications represented by Equations 2.1 and 2.2 and the micrometeorological expectations associated with MOST. Both sides agreed that atmospheric stability would play an important role. Two of the main players at that time were Deacon and Swinbank. Len Deacon proposed that allowance for the effects of atmospheric stability could be captured adequately by adjusting the basic power law relationship (in Equation 2.2) to allow the exponent to vary with stability. Deacon's (1949) extension of the power law wind profile led to a description of the predicted wind profile near the surface as

$$\bar{u} = (u_*/k)\cdot(1-\alpha)\cdot[(z/z_0)^{1-\alpha} - 1] \tag{6.1}$$

which provides a transition between the power law expression for the wind profile aloft and the logarithmic law in the SBL (see Sutton, 1953). The constant α isn't; it varies with stability.

Deacon's name was commonly associated with this approach, to his chagrin. Swinbank, had a different approach, based on MOST. He assumed that there was a length scale that captured the effects of atmospheric stability, in unstable conditions. The result was an exponential wind profile (Swinbank, 1964), supported (erroneously) by results obtained using the then-standard method of applying a friction coefficient computed using neutral wind profiles to a low-level wind speed measurement. (In passing, the values of C_f were derived by the present author, back in 1963, but were then applied to a wind speed measured at a different height, an error that was not recognized until long after publication.)

In the absence of buoyancy, turbulent energy is generated as the result of the momentum flux working against the local gradient of the wind. The production rate of turbulent energy is then

$$\varepsilon(M) = -\rho\overline{w'u'}\cdot(\partial\bar{u}/\partial z) \qquad (6.2)$$

Convective heat also generates turbulence, at a rate

$$\varepsilon(H) = \rho(g/\theta)\cdot\overline{w'T'} \qquad (6.3)$$

The flux of water vapor also contributes, but to a smaller extent. In practice, if we have a flux H expressed in W m^{-2} and a flux of water vapor E causing a latent heat of evaporation L_wE also expressed in W m^{-2}, then the flux H in the simplest expressions that neglects the evaporative contribution should be replaced by the term $(H + L_wE/14)$. (The factor 14 derives directly from consideration of the fundamental relationships. Go, look.)

A measure of atmospheric stability can be defined in readily understood physical terms — the ratio of the production rate of turbulent energy by convection to that by the wind shear

$$\zeta' = \varepsilon(H)/\varepsilon(M) = -[gH/(\rho c_p\theta)]/[u_*^2(\mathrm{d}\bar{u}/\partial z)] \qquad (6.4)$$

from which the physical meaning of the z/L stability index of MOST becomes evident.

The dominant goal of early experimental micrometeorology was to utilize such indices of stability to describe departures of gradients from neutral expectations, in which case the relevant stability measure should be made up entirely of either flux quantities or gradient information. If the form is mixed, containing both gradient and covariance terms as in Equation 6.4, then it contains within it some of the variation in fluxes or gradients that it is intended to explore. Two different but equally applicable derivations of Equation 6.4 arise, one using flux information only and the other using gradients. The former replaces the gradient term in Equation 6.4 by the neutral expectation of the logarithmic wind profile, leading to

$$\zeta = -kgHZ/(\rho c_p\theta u_*^3) = Z/L, \text{ where } L = -\rho c_p\theta u^3/(kgH); \qquad (6.5)$$

the latter introduces the near-neutral relationship describing the transfer of heat, leading to

$$Ri = (g/\theta)\cdot(\partial\bar{\theta}/\partial z)/(\partial\bar{u}/\partial z)^2 \qquad (6.6)$$

which defines the Richardson number, Ri. There is nothing new in any of this. In reality, the discussion above is a simplification. Garratt (1991) gives a much better treatment, in which the role of water vapor buoyancy is inherent, through an emphasis on the virtual temperature which incorporates immediate consideration of the buoyancy of water vapor.

The Richardson number was developed in the early days of micrometeorology and remains a favored measure of atmospheric stability. The definition of L in Equation (6.5) is also familiar. L was initially proposed by Obukhov (1946) along with others, such as Lettau (1951) as a length scale indicative of atmospheric stability and became a key component of MOST. Hence, the ordinate in the plots commonly presented for unstable conditions was either $-\zeta \equiv -Z/L$ (as defined above) or its logarithm, whereas in earlier analyses the stability parameter of choice was the Richardson number, *Ri*, as in Equation 6.6. A number of variations on the formulation of stability were common; *Ri* as given here is the gradient Richardson number. By shuffling the various quantities involved, the flux Richardson number is readily derived.

In the 1970s, micrometeorologists, forest meteorologists, agricultural meteorologists and soil scientists finally got together and their interests and specialties started to converge. Eddy covariance methods were used to extend micrometeorological understanding beyond the perfect-site focus of the 1960s, with rapidly improved ability to formulate flux-gradient relationships over surfaces of gradually increasing complexity — e.g., a vineyard (yes, the roughness length is indeed a function of wind direction relative to the rows), a swamp and forests. Throughout this period, there was an increasing interest in the oceanic case, about which more will be said later.

The 1980s brought the need to address the air-surface exchange of air pollutants and so-called dry deposition studies dominated much of the literature. Complex terrain started to be addressed using aircraft eddy covariance systems. In this period, there was rapid expansion of related studies in Europe and these studies have continued to current times. Today, the results of experiments in Asia frequently appear in mainstream journals.

Studies of climate change and its anthropogenic causes led to a concentration on the air-surface exchange of carbon dioxide, starting in the 1990s. Forest ecosystems were of special interest, including many in demanding terrain. The driving consideration was the view that growing trees represented a sink for CO_2 that was known to be accumulating in the atmosphere. The reality that any CO_2 sequestered in forest biomass would be released into the air once trees died was sometimes overlooked in the face of the more immediate need for funding. There is an overriding concern that is rarely acknowledged. While it is true that planting of seedlings in areas where old forests have been harvested is often portrayed as mitigating the adverse climate impacts of deforestation, too often the harvested old growth hardwoods are being replaced by other non-native species more appropriate when rapid growth helps assure profits.

Following several terrorist attacks early in the 21st century, concern about the consequences of an intentional release of hazardous gases caused studies of urban areas to become more frequent and more extensive. Contemporaneously, the new millennium brought with it the development of meteorological models that required accurate depiction of air-surface exchange. However, these meteorological models were based on the formulations of the 1960s and their application to urban areas is obviously limited. The urban situation represents an extreme departure from the outdoor laboratory studies of the 1960s and 1970s.

Conclusion

Always examine the assumptions.

Chapter 7

Measurements of Fluxes and Profiles — The Early Days

"Mad dogs and Englishmen go out in the midday sun" — Noel Coward.

Application of Reynolds decomposition to the problems of heat exchange resulted in a search for sensors responding quickly enough to enable direct measurement of atmospheric fluxes by eddy covariance. In this, the instantaneous value of each variable of relevance (*w, U, x*, where *w* is the vertical wind speed, *u* the horizontal along-wind velocity and *x* some other variable of interest, e.g., *T*) is decomposed into mean and fluctuating parts, e.g.,

$$w = \overline{w} + w'. \tag{7.1}$$

so that the instantaneous vertical rate of transfer of the property x can be expressed as

$$w \cdot x = (\overline{w} + w')\cdot(\bar{x} + x'). \tag{7.2}$$

The time average of this fluctuating exchange rate yields the flux of micrometeorological interest –

$$F_x = \overline{w} \cdot \bar{x} + \overline{w'x'} \tag{7.3}$$

Since $\overline{w}$ is necessarily zero (iff variations in density are not critical, otherwise look at Webb et al. 1980), the product of the averages disappears and the familiar eddy correlation (or covariance) form results. In the most familiar applications, we can consider the fluxes of momentum, sensible heat, latent heat of evaporation and carbon dioxide as

$$\tau = -\rho\overline{w'u'} \tag{7.4}$$

$$H = \rho c_p \overline{w'T'} \tag{7.5}$$

$$L_w E = \rho L_w \overline{w'q'} \text{ and} \tag{7.6}$$

$$F_{CO2} = \rho\overline{w'C'} \tag{7.7}$$

where C is the mixing ratio of CO_2. There are many niceties that arise, involving the need to consider fluctuations in air density (as contributed by, for example, fluctuations in water vapor content associated with the evaporation rate), the requirement that the fluxes refer to the exchange across a plane defined by the criterion that there is no gain or loss of dry air across it, the time scales by which instantaneous values are measured and the length of the period over which time averages are computed. Whole textbook chapters have been devoted to these topics, often to the dissatisfaction of the present author who remains doubtful about the relevance of the discussions presented when applied to terrain that is not horizontal. The presentation by Garratt (1992) starts with understanding of the thermodynamic factors controlling the relationships, with emphasis on the role of buoyancy (and of the water vapor contribution to it) as the dominant factor when departures from neutrality and the associations between fluxes and gradients are considered. There is no need to revisit these deliberations.

It was not until the late 1920s that Scrase (1930) demonstrated that eddy correlation methods worked. It then took more than two decades to develop the instrumentation required for routine field exploitation of the technique. In the interim, attention was mainly devoted to studies of dispersion (e.g., Sutton (1947), following studies conducted during World War I. English studies (Calder, 1952; Pasquill, 1956) were accompanied by experiments elsewhere that were not well reported. In the 1950s, new technology permitted the science to be explored in more detail and widespread interest was generated by the results presented by Record and Cramer (1958) from the famous Prairie Grass experiment conducted in Nebraska.

Measurement by eddy correlation was a focus of the 1950s. The main obstacle was the lack of suitable sensors for u' and w'. The need to measure evaporation (as discussed in Part I) triggered the development of eddy correlation techniques in Australia. Directional hot wire and heated wake anemometers (Taylor, 1958) served as early devices for measuring velocity rather than speed. Ball-and-disc mechanical systems were used to integrate the product of two input signals derived from visual tracking of traces made by a chart recorder (Taylor and Webb, 1955). Work in other countries led to the eventual development of sonic (Kaimal and Businger, 1963) and pressure sphere (Martinot-Legarde et al. 1959; Thurtell et al. 1970) anemometers but in the interim the availability of helicoid propeller anemometers provided a welcome step forward (Gill, 1973). Bunker (1955) demonstrated that aircraft can provide a useful platform for measurement of covariances.

Much earlier, aeronautical engineers had quantified the value of the von Karman constant k as in Equation 2.2 as about 0.4 (see various works, e.g., by von Karman) and a first challenge for micrometeorologists using the new technologies was to test the applicability of this constant to the atmospheric case of much lower velocity regimes. Field studies reported by Sutton (1947; 1952) and by Swinbank (1951; 1952) led to the evaluation: $k = 0.41$. In subsequent work the value of $k = 0.4$ was widely accepted but with a number of variations proposed in later years as new data became available, presumably obtained using ever-improving methodologies. (The present author's review of the experimental approaches and especially of the instrumentation used in many of these evaluations of k have resulted in a preference to disregard many of them, particularly those derived using sonic anemometers with non-orthogonal paths. This matter will be examined later. A value of $k = 0.41$ has been assumed in all analyses presented here).

As discussed by Hicks and Baldocchi (2020), development of eddy covariance methods evolved rapidly in the 1950s and 1960s, largely in response to the continuing need to measure evaporation rates. In the Australian studies, a more practical methodology was sought that permitted immediate analysis of the resulting covariances with real-time read-out. To this end, heated wake hot wire anemometers were arranged to derive the vertical and horizontal wind components. These were mounted on a wind vane as illustrated in Figure 7.1. When coupled with signals derived from a fine-wire thermometer, values of $\overline{w'T'}$ were then computed using a dedicated analogue computing system — the Evapotron (Dyer, 1961).

Figure 7.1. The sensor head of the Evapotron, as used in the Australian Kerang and Hay experimental programs of the early 1960s. Note that the vertical wind component was not measured directly; it was inferred from measurements of two tilted heated-wake anemometers. The sensor head also shows the wet- and dry-bulb fine-wire resistance thermometers.

As its name implies, the Evapotron was designed specifically to measure the flux of water from an area of particular interest, to which end the measurement of sensible heat was augmented by a parallel measurement using a fine wire wet bulb sensor. Two independent analog computing systems were then used, one with a fine wire temperature system yielding $\overline{w'Ta'}$ and the other with a wetted fine wire thermometer yielding $\overline{w'Tw'}$. (T_a is air temperature, T_w is the wet bulb temperature.) From these, $\overline{w'q'}$ values were derived as $\overline{w'q'}$ $s + \gamma)(\overline{w'Tw'}) - \gamma(\overline{w'Ta'})$, with quantification of L_wE resulting (see Dyer 1961). (Here, as above, $\gamma = c_p/L_w$. Note that agrometeorologists tend to favor the use of vapor pressure, e, as the humidity measure, whereas micrometeorologists generally prefer the specific humidity q. Conversions are easy.) The central feature of the Evapotron was its use of modified commercial kilowatt-hour meters to perform the necessary computation and integration of products.

In most Australian micrometeorological field studies of the 1960s, three Evapotrons were deployed — one to measure H, a second to measure $(H + L_wE)$ and a third attempting to measure the momentum flux. In practice, the first two were reliably successful, yielding results that were as good as could be expected in real-time comparisons against measurements of $(R_n - G)$. However, the Evapotron was incapable of yielding reliable estimates of the friction velocity, because of the need to derive both u' and w' measurements from the same sensors and the resulting propagation of errors that arose in the operation of either. As shown in Figure 7.1, two heated-wake anemometers were employed, facing into the wind and inclined so that one measured $(u + w)$ and the other $(u - w)$. The two signals were then combined electronically to provide signals proportional to u and to w. While independence of these two resulting signals was achievable in theory, in practice the demands on instrumentation and analog circuitry could not be overcome. This source of error did not affect the heat flux measurements, because any causative anemometer errors were not shared in the calculation of either the $\overline{w'T'}$ $\overline{w'q'}$ covariances (i.e., an error in *w'* was not correlated with an error in *T'* or *q'*).

The Evapotron enabled tests to be made of a number of fundamental properties of the daytime surface boundary layer. (Nighttime operation was far more difficult, requiring careful watch of the instrumentation and the resulting payment of overtime.) Before using the instrumentation in studies of the relationships between fluxes and gradients, extensive field trials were conducted to demonstrate that the requirements imposed by the surface energy balance were satisfied. Moreover, the statistical nature of the covariances being measured and the existence of unresolvable (aleatoric) uncertainty were studied in considerable detail, with a verification of the expectation that the standard error associated with the determination of an average covariance would vary inversely with the square root of the sampling duration. Figure 7.2 is a plot of relevant field results reported by Dyer and Maher (1965).

The findings presented in Figure 7.2 were the basis for the strategies underpinning the many Australian field studies of the 1960s. The half-hour sampling interval was selected as an optimum, providing some protection against the need to consider changes occurring within the interval (i.e., satisfying the basic requirement for time stationarity) while also satisfying the need to sort data into ensembles containing enough samples to yield averages with less than 10% statistical uncertainty. Data were examined to ensure that the consequences of changes with time were negligible, e.g., that air temperatures and humidities at the end of a measurement period did not differ from those at the beginning to an extent that the resulting in-air heat storage term became a concern. Issues of such kind have surfaced in many guises since the tests that yielded Figure 7.2. Contemporary concerns related to fetch, footprints and the ability to evaluate model predictions all involve the ability to measure a statistically varying quantity so that the average has adequate certainty.

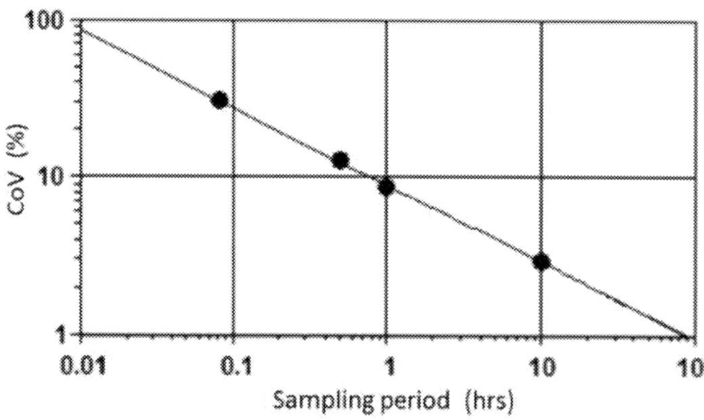

Figure 7.2. Results reported by Dyer and Maher (1965), showing that the coefficient of variation (CoV $\equiv \sigma_H/H$) associated with measurements of *H* by eddy correlation over a period (t) varies, as expected, with $t^{-0.5}$. The line drawn is the result of a power-law regression, yielding a slope of −0.49 (c.f. the expected value: – 0.50).

In work conducted in the 1980s and 1990s, the interpretation of eddy covariance observations made at a single location was examined often. Several studies are reviewed by Finkelstein and Sims (2001), who conclude that the cases of $\overline{u'w'}$
different susceptibility to scatter.

In the 1970s several fast response water vapor sensors were developed — the quartz crystal hygrometer (Hicks and Goodman, 1971), the infrared sensor (Hyson and Hicks, 1975; Raupach, 1978) and the Lyman-alpha hygrometer (Buck, 1976). Much earlier, Russian workers had developed similar instruments (Elagina, 1962), unknown to the Australian community. Today, the experimental field is dominated by fast-response infrared hygrometers with carbon dioxide measurement as an added feature (Auble and Meyers, 1992). In parallel, anemometry has evolved from fine wire heated wake anemometers, to propellers, pressure spheres and now to sonic devices. Fast response temperature measurement remains best if fine wire resistance thermometry is used, but often the ability of sonic anemometers to provide a fast response temperature signal as a by-product provides an exceedingly attractive alternative. (A cautionary note is often overlooked: sonic anemometers measure the virtual temperature, T_v, not the absolute temperature T_a).

In the 1960s, the new covariance methods were employed in outdoor-laboratory conditions (as shown in Figure 7.3), to explore the relationships between fluxes and profiles in ideal conditions. The focus on ideal conditions was so that the central problems confronting micrometeorology at that time could be examined in the most simple and addressable situations. That is, experimental conditions were sought in which variations in time and in the horizontal plane were negligible, so that the diffusion relationships could be considered one-dimensional, in the vertical. The fact that the results are appropriate for such limited situations is often overlooked in modern applications. Also overlooked is the fundamental circularity involved: the goal is to describe fluxes and profiles in terms of stability, which in turn is quantified in terms of fluxes and profiles.

Figure 7.3. The field site at Hay, 1965, showing the 16 m wind profile tower on the left and a group of participants watching the approach of the CSIRO DC3 about to take a low pass over the site. The grouping of people at the center of the scene shows C. H. B. Priestley to the left, R. G. Taylor at the center and W. C. Swinbank (arms characteristically akimbo) to the right.

In the many Australian field programs of the early 1960s great pains were taken to ensure that all of the measurements were of the highest possible quality. Temperature gradients, for example, were obtained using horizontally-aspirated platinum resistance thermometers connected in bridges and protected from radiation errors by chromed tubes at the end of arms projecting horizontally upwind from the supporting tower. Wind speed was measured with conventional three-cup lightweight and fast-response anemometers, with jewel bearings assembled and maintained by a professional watch maker. Calibrations were conducted before and after a field study, using a rotating arm system to extend calibrations into a low-speed regime not accessible in wind tunnels. (A lecture theater was forfeited to permit rotating arm calibrations in an undisturbed atmosphere).

Field studies involved teams of researchers, giving attention to all measurement systems and of necessity imposing a need for monitoring sensor performance and the prevailing atmospheric conditions. The resulting datasets provided both good data and an encyclopedic collection of visual observations that proved a valuable adjunct to the data themselves. These observations contributed to the resulting understanding in several ways. For example —

- Watching the performance of mechanical wind sensors (wind vanes or propeller anemometers) led to awareness that sensible heat exchange occurs in bursts, especially in light winds. It did not matter too much whether the observations were in daytime or at night; infrequent bursts of turbulence occurred in both cases.
- In the most extremely unstable circumstances, energetic but coherent dust devils ('willy willies' in local parlance) moved slowly across the landscape, following paths of least resistance at the surface. If the paths of two of these cells crossed, two could combine into a single somewhat larger cell. Such crossing resulted because the twisters tended to follow sheep tracks in which surface roughness was minimal.

Outside the near-neutral range, the exchange phenomenon was visualized as being dominated by intermittent strong events, not correlating well with the alternative view of a coherent spectrum of eddy sizes. It was this thinking that led to the development of various running-mean removal methods to simplify the eddy correlation analysis. Tests made using side-by-side systems with different high-frequency band pass filters demonstrated that in most daytime conditions the choice of a mean removal time constant was not overly critical provided the time constant exceeded 30 seconds, for the common measurement height then employed: 4 m. Patience and continuing observation led to the conclusion that these strong convective cells over-rode a continuum of less effective turbulent exchange. As time progressed, the association with the concepts of forced and free convection became solidified.

In the lack of a field-deployable eddy correlation capability to measure the $\overline{u'w'}$ in the earliest Australian studies, an alternative methodology was employed to derive values of the friction velocity. Assuming that a value of 0.4 was appropriate for the von Karman constant, estimates of the friction velocity were derived from wind measurements made close to the surface after calibrating the site using neutral wind profiles. The approach adopted was to await a period of high cloud cover and strong winds, when the net radiation (easily measured) was near zero. Measurements of $\partial\overline{T}/\partial z$ and $\partial\overline{u}/\partial z$ were then made through such a period of interest and 30–min 'runs' of data for which $\partial\overline{T}/\partial z$ (and hence $\overline{w'T'}$) was close to zero were selected for further study. From the resulting assemblage of neutral runs, plots of $\overline{u}$ against ln(z) values of u_* and z_0 were derived. On some occasions, no suitably unchanging neutral conditions were encountered, so that the site could not be so calibrated. Data from these particular studies were discarded. To avoid uncertainties associated with the need to account for a finite displacement height, all studies in this series were conducted over either grazed land with minimal vegetation, recently harvested cropland or bare soil. (The landscape of the series of studies at Hay, in New South Wales, was arid grazing land, maintaining Merino sheep at a population density below one per hectare, as illustrated in Figure 7.3) After checking that the displacement height was appropriately zero (i.e., that the plots of $\bar{u}$ against ln(z) yielded a straight line in the selected neutral circumstances), two representations of the site roughness were then evaluated — the roughness length z_0 and the friction coefficient $C_{fs} = u_*/\bar{u}_s$ where $\bar{u}_s$ was the wind speed recorded by a specially installed anemometer at a height of (typically) 0.5 m. For all analysis that followed, u_* was evaluated as the product $C_{fs}\bar{u}_s$. Hence, the flux-gradient relationships derived from the Australian field studies of the early 1960s represent, in actuality, the effects of stability departures from neutral as heights increased from 0.5 m up to, usually, 16 m.

In the earliest field experiments attention was solely on the wind and temperature gradients and their associated fluxes, all within the SBL where fluxes are assumed to be constant with height. There was little attention given to the fact that in (debatable) theory the definition of the various covariances should be made with reference to a plane arranged such that there is no transport of dry air across it. Effects associated with the latent heat flux were not overlooked, but were largely disregarded since the dominant factors influencing the acceptance of neutral runs were the constancy with time and the acceptance of a logarithmic wind profile (produced on the spot, in real time).

The dominant trace gas in the atmosphere is certainly water vapor. Water vapor is lighter than air (its molecular weight is 18, versus 29 for air) and hence modifications are necessary in the description of all processes in which buoyancy or density are factors. Atmospheric stability is accordingly and properly quantified in terms of virtual potential temperature (θ_v), rather than air temperature itself (e.g., Brutsaert, 1982; Garratt, 1991). The resulting development of the conventional micrometeorological relationships, following thermodynamic lines, is also well presented by Garratt (1991). There is no need to reproduce these derivations here. However, it is relevant to note that the Richardson number is better expressed as

$$\grave{R\imath} = (g/\theta)\cdot(\partial\overline{\theta v}/\partial z)/(\partial\bar{u}/\partial z)^2 \tag{7.8}$$

which differs from Equation 5.6 solely by the change from potential temperature to virtual potential temperature. The consequences are usually too small to be of concern. Distinction between the alternative measures of temperature is not always clear cut. An important example is the derivation of temperature fluctuations from measurements of the speed of sound: sonic anemometers report virtual temperatures. Indeed the first measurement devices relying on sound propagation were specifically designed to measure temperature with rapid response (Suomi, 1957). The extension of the sonic methodology to wind vector measurement came later (Kaimal and Businger, 1963).

Note that the treatment so far, while conventional, is decidedly lackadaisical. More detail is provided in early examinations of the issues arising, such as by Priestley and Swinbank (1947) and Montgomery (1948). Two considerations are of immediate relevance. First, these papers give a clear explanation of why the equations assume the relevance of the specific heat of air at constant pressure rather than that at constant volume. In essence, the reason relates to the intent of the exercise — to assess the transfer across a layer such that there is no exchange of dry air across it. Second, the buoyancy of water vapor must be taken into account, at the outset. The relevant consideration is more appropriately the transfer of heat energy, occasioned by the combination of temperature changes and humidity, both of which affect convection in daytime and stratification at night. Montgomery (1948) couches his analysis in terms of specific enthalpy, but notes that others refer to this as 'total heat.'

Taking the higher-level view that it is the specific enthalpy exchange that is relevant, then it follows immediately that specific heat and water vapor serve as no more than two contributors to the convective exchange process. Expressing the flux-gradient relationships in terms of specific enthalpy and the 'equivalent temperature', T_η (defined as the temperature of dry air that would have the same specific enthalpy, η) would lead to a quantification of the relevant diffusivity that would be appropriate for all of the total heat and its two contributing components, H and L_wE. The present author is still puzzled about why, in early work, there was great debate about whether sensible heat and latent heat were transferred by the same eddies and so with the same eddy diffusivity; i.e., does $K_w = K_H$? The matter was resolved, to the satisfaction of most doubters, when field studies demonstrated the equality (Swinbank and Dyer, 1967).

Conclusion

Encourage sitting and watching — seeing is believing.

Chapter 8

The Legacy Flux/Gradient Relationships

"We suffer from a profileration of flux/gradient relationships" — W. C. Swinbank

The micrometeorology described in modern literature arose from early field studies using poor instrumentation and from interpretations of the resulting data in the light of theories then favored. The intent of the early experiments was to quantify the relationship between fluxes and gradients, focusing first on (a) the temperature gradient and sensible heat flux and subsequently on (b) the wind speed and momentum case and (c) the humidity gradient and the latent heat flux. It was acknowledged that terrain complexity would impose uncertainties and obscure the processes that it was desired to study. Hence, great pains were taken to ensure minimal changes in upwind fetch. Much of the earliest field work took place in Australia, under the leadership of C. H. B. Priestley and the field direction of W. C. Swinbank. The Australian experience will be emphasized here because (a) it paralleled and sometimes led similar enterprises elsewhere and (b) it is best known to the present author (having participated in most of the Australian field studies of the 1960s).

The early datasets consisted of observations made within the surface boundary layer (SBL), in the region of the PBL identified as the constant flux layer. Meteorological towers were rarely more than 16 m in height. In the Australian experiments, great care was taken to ensure that the towers were as sturdy as possible while also imposing minimal obstruction to the wind. Single lengths of steel tubing (about 6 cm diameter) were used, guyed at the 8 and 16 m levels. Sensors were normally placed on arms, extending upwind and well away from the supporting mast and at intervals determined logarithmically (typically at 1, 2 4, 8 and 16 m above the surface). The separation of cup anemometers from the supporting vertical mast was always a minimum of six times the diameter of the cup arrays, this having been determined by detailed experimentation.

In the case of the SBL wind profile, it was assumed that the principal role of atmospheric stability is to modify the neutral profile. Some workers advocated that heat was transferred in the same way as momentum, with the same eddy diffusivity. Others anticipated a role for buoyancy, so that heat would be exchanged with greater diffusivity than is the case for momentum whenever the air is convective. Discussions therefore revolved around the equality or otherwise of K_H, K_m and K_w. The last of these three was of special interest, since it pertained to the rate of loss of water from crops and water was a scarce commodity in Australia. These turbulent diffusivities were specified as in their one-dimensional Fickian definitions:

$$K_H = \overline{w'T'}/(\partial\bar{\theta}/\mathrm{d}z) \tag{8.1}$$

$$K_m = -\overline{w'u'}/(\partial\bar{u}/\partial z) \tag{8.2}$$

$$K_w = \overline{w'q'}/(\partial\bar{q}/\partial z) \tag{8.3}$$

In retrospect, it is surprising that there were questions raised about the equality or otherwise of the three diffusivities. Two dominating considerations were acknowledged, but were viewed as matters to be experimentally verified rather than absolutes on which future advances could be based. First, the difference in the direction of transport and the inherent difference in the spectra of eddies involved in the cases of daytime sensible heat and momentum led to the expectation that K_H and K_m would then be different. Second, the thermodynamic principles describing buoyancy required that H and L_wE must be viewed in concert, as two contributions to the buoyant transfer of total heat (as already mentioned). Hence, K_H and K_w must be seen, almost by definition, as the same as the result obtained when the total heat exchange is considered. The experimental program served to prove all of these expectations.

The early methods for quantifying the terms in the flux-gradient relationships were discussed already. In the absence of a method for measuring u_* directly, the issues were avoided by relying on a wind speed measurement made very close to the surface. To start with, the neutral association between the surface drag and the wind speed was accepted:

$$\partial\bar{u}/\partial z = u_*/kZ. \tag{8.4}$$

Here, Z is the height above some zero plane (often the ground itself, but close to the top of any vegetative cover that is present; $Z = z - d$ where d is the zero-plane displacement height and z is the height above the ground); as before, k is the von Karman constant: $k = 0.41$. Rudimentary measurements verified the expectation that Equation 8.4 oversimplified matters except for the periods when there was no change of potential temperature ($\bar{\theta}$) with height and hence when the effects of convection and/or buoyancy could be ignored. Given Equation 8.4 for the neutral wind profile, the stability-modified dimensionless wind profile was then quantified as

$$\phi_m = (kZ/u_*)\cdot(\partial\bar{u}/\partial z). \tag{8.5}$$

The experimental challenge was then to determine how ϕ_m varies with stability, usually $\zeta = Z/L$.

A limitation on the experimental program was at that time not fully appreciated. Previous experience had indicated that the period following dawn was highly intermittent and hence the early morning condition was not routinely studied. There was no search for neutral conditions in the morning; it was assumed that any 'neutral' occurring then would be transitional and therefore irrelevant. The arguments that caused a lack of attention to the early morning situation were, however, completely overshadowed by the requirement that the research staff enjoyed a convenient and cultured breakfast. The hunt for "neutral" was limited to the evening case, with strong winds, complete cloud cover and a satisfied appetite.

Integration of Equation 8.4 yields the familiar log-law:

$$\bar{u} = (u_*/k)\cdot\ln(Z/z_0), \tag{8.6}$$

so introducing the well-known roughness length z_0 as a constant of integration analytically derived assuming a neutral wind profile indicated by a straight line when plotting the wind speed versus ln(Z).

A key and recurring property in all relationships addressing the association between fluxes and gradients is the *friction velocity*. This is true not only for the case of the wind profile as considered above, but also for the relationships associated with heat and water vapor exchange. Details of this (and other aspects that are glossed over here) will be found in familiar sources such as Businger (1988), Stull (1988) and Garratt (1992). In order to examine the way in which changes in stability influence the wind profile, the generally accepted approach is to find an appropriate function $F_m(\zeta)$ or $f_m(Ri)$ such that

$$\phi_m \equiv (kZ/u_*)\cdot(\partial\bar{u}/\partial z = F_m(\zeta) \text{ or } f_m(Ri) \tag{8.7}$$

Using a similar philosophy, the case of the dimensionless temperature gradient was also addressed, leading to an experimental goal to explore

$$\phi_H \equiv (kZu_*/\overline{w'T'})\cdot(\partial\bar{\theta}/\partial z) = F_H(\zeta) \text{ or } f_H(Ri). \tag{8.8}$$

In passing, observe that the overall philosophy of exploring the way in which stability influences the various profiles as multiplicative departures applied to the relevant quantity at neutral raises some intriguing complexities in the case of the temperature gradient, since by definition $\partial\bar{\theta}/\partial z$ would be precisely zero at neutral if there was no heat exchange occurring at the same time, and hence their ratio is indeterminate. Also, since water vapor is buoyant, its role in the determination of stability needs to be taken into account. In the mathematical expressions above the absolute temperature gradient $\partial\bar{T}/\partial z$ has been replaced (correctly) by the gradient in potential temperature $\partial\bar{\theta}/\partial z$, but the extension to account for the additional modification accounting for water vapor buoyancy (leading to consideration of the virtual potential temperature, θ_v) has not been indicated. From the perspective of specifying stability using Ri, the more appropriate gradient to consider is $\partial\overline{\theta v}/\partial z$. However, details of this kind are generally considered either unwelcome details or nuisances of negligible practical importance.

In the early Australian work, exploration of ϕ_m involved normalizing field measurements of wind gradients using friction velocities derived as a proportion of the wind speed measured close to the surface. In reality, there was a dichotomy in the analysis, since in the mid 1960s a theoretical description of the surface wind profile had been developed and was a focus of part of the Australian wind profile investigations – the exponential wind profile (Swinbank, 1964). Readers will note that the exponential profile lapsed into disciplinary history after first being promoted as in agreement with observations. (The underlying problems were discussed earlier.) The matter came to a head when the first measurements of u_* by eddy correlation became available.

At the time of the early micrometeorological field studies, the familiar KEYPS formulation (associated with Kazansky and Monin, 1956; Ellison, 1957; Yamamoto, 1959; Panofsky, 1961; and Sellers, 1962 – hence the acronym) gained considerable acceptance. There are several derivations of the KEYPS equation, most originating with considerations of kinetic energy but all resulting in

$$[\phi_m(\zeta)]^4 - \ddot{g}\zeta\,[\phi_m(\zeta)]^3 = 1 \tag{8.9}$$

where ğ is a constant, proposed to be 14.8 by Panofsky (1961). Equation 8.9 has the desired attributes for ϕ_m. It approaches unity as $\zeta \rightarrow 0$ and trends to a dependence on $(-\zeta)^{-1/3}$ as $-\zeta$ increases.

The availability of the Evapotron to measure the sensible heat flux enabled the relationship for ϕ_H to be examined. The studies were confined to daylight hours because (a) the eventual intent was to address the issue of water loss and (b) the instrumentation was not good enough to do otherwise. The results were few and scattered, however when ϕ_H was plotted against $-\zeta$ on log-log graph paper the order among them was clear. A line through the points was drawn using a French curve that happened to be available. Later it was shown that the line drawn could be approximated by a simple expression of the form

$$\phi_H = (1 - a.\zeta)^{-b} \tag{8.10}$$

The advance from the Evapotron (which was unable to measure the momentum flux covariance to the Fluxatron (which was, see Dyer and Hicks, 1965) enabled a new examination of the situation for ϕ_m. When eventually analyzed in the same way as for ϕ_H, it was concluded that a similar curve could be prescribed for ϕ_m. The scatter in the data permitted a wide variation in how the constants β and γ could be specified in an equation for ϕ_m like that for ϕ_H. The exponent b was approximated as 0.25 for momentum, with the same constant ($a \approx 15$) within the bracketed term. The convenience of these choices was that the relationships resulted in numerical equality between ζ and Ri in unstable stratification.

The now-familiar integral forms relate to the need to express the wind speed itself in terms of roughness, height, u_* and stability. At the time of first publication of the initial flux/gradient experimental results, the integral forms were not promoted because they were viewed as the results of mathematical exercises involving expressions that were initially poorly founded — the ϕ-functions. The ψ-functions arise from a generalization of Equation 8.6, to account for the effects of stability in a manner readily compatible with the descriptions focused on gradients:

$$\bar{u} = (u_*/k)\cdot(\ln(Z/z_0) - \psi_m(\zeta)) \tag{8.11}$$

where

$$\psi_m = \int(1 - \phi_m(\zeta))d(\ln\zeta). \tag{8.12}$$

The corresponding approach for the sensible heat case introduces ψ_H:

$$(\bar{\theta}_0 - \bar{\theta}_z) = (\overline{w'T'}/ku_*)\cdot(\ln(Z/z_{0T}) - \psi_H(\zeta)) \tag{8.13}$$

which relates the temperature difference between the effective surface and the air at height Z to the heat flux, with the recognition that the roughness length appropriate for the exchange of heat might well differ from that for momentum. This will be explored later, but for the present note that a new function is introduced:

$$\psi_H = \int(1 - \phi_H(\zeta))d(\ln\zeta) \tag{8.14}$$

The integrals expressed by (8.12) and (8.14) are intended to compare the wind speed and temperature at two different levels within the range of applicability of the ϕ-functions, conventionally taken to be from $Z_1 \approx 10 \cdot z_0$ to $Z_2 \approx |L|$. Later, the requirements of the modeling community overrode this initial conservatism and formal integral forms are now widely employed for the unstable case, following presentation of them by Paulson (1970).

$$\psi_m(\zeta) = 2 \cdot \ln[(1+x)/2] + \ln[(1+x^2)/2] - 2 \cdot \tan^{-1}x + \pi/2 \tag{8.15}$$

$$\psi_H(\zeta) = 2 \cdot \ln[(1+y)/2] \tag{8.16}$$

where

$$x = (1 - a\zeta)^{0.25} \tag{8.17}$$

$$y = (1 - a\zeta)^{0.5} \tag{8.18}$$

These are integrals from $Z = 0$ to the height Z, provided $0 > Z/L > -1$, but computer models frequently extend the formulations to greater heights. It might be noted that these expressions are integrals from $Z = 0$ to the height Z of relationships that are known to become inapplicable as the zero plane is approached. Therefore, they are of most practical use when considering differences between two heights well above the surface, e.g., for $Z > 10 \cdot z_0$ or alternatively $Z > h$ where h is the height of the canopy. Such criteria differ among researchers, depending on individual preferences (and sometimes on experience).

Instead of formally integrating the empirical functions for ϕ_m and ϕ_H (summarized by Equation 8.10) the original data were also analyzed (using planimetry) to derive the integral forms ψ_m and ψ_H directly. When plotted on log-log paper, it was found that a simple quadratic form fitted quite well. The best fits (derived by regression) were initially given as

$$\ln(\psi_m) = 0.032 + 0.448\ln(-Z/L) - 0.132(\ln(-Z/L))^2 \tag{8.19}$$

$$\ln(\psi_H) = 0.598 + 0.390\ln(-Z/L) - 0.090(\ln(-Z/L))^2. \tag{8.20}$$

A literature search reveals several different forms of these equations, which can probably be used interchangeably. In modern applications, the integral forms presented by Paulson (1970) are universally favored. The log-quadratic forms (as above, see Kanemasu et al. 1987) lapsed into obscurity, but remain useful expressions for simplified modeling. As in the case of the empirical relationships so far considered, these relationships were not intended for application outside the range of applicability of the ϕ-functions.

Studies of the stable case proved to be far more demanding. First, it was immediately clear that scatter was larger in the stable case, almost regardless of what variable was being considered. However, it was accepted that a Taylor expansion of the fundamental relationships yielded a useful approximation to observations of the near-neutral regime, extending further into stable than into unstable. The first term approximation was widely accepted as the most justifiable description of the data — the familiar log-linear expression:

$$\phi_m = 1 + \acute{\alpha}\zeta \tag{8.21}$$

with the value of *ά* varying with variables such as the site, the observer, and the way in which the analysis was conducted. Most frequently, a value in the range five to six is assumed and such seems compatible with averaged field data obtained for conditions such that $0 < \zeta < 1$. Webb (1970) points out that the log-linear expression appears to extend into the unstable regime, as far as $\zeta \approx -0.03$.

It is conventionally assumed that the same expression holds for all properties being transferred in conditions of stable stratification, there being (in the absence of buoyancy as a driving force) no difference in the eddies involved in the transport. This argument is outwardly satisfying, but upon contemplation it appears somewhat deficient. One basis for concern is the assumption that the zero planes are the same for all of the variables being considered (momentum, heat, moisture, carbon dioxide, etc.). This assumption is certainly valid in many cases, but if the surface is forested (for example), then all bets are off (but perhaps there is more logic involved than is outwardly apparent, as will be discussed later). It is also true that displacement of a parcel of air in the vertical imposes a small thermally-generated restoring force, such that the role of buoyancy cannot be completely ignored. This will become apparent in considerations, later, of the flux-gradient relationships as revealed by modern data.

In neutral conditions, the summary just given explains why u_* enters the formulation of the wind profile. It is the Fickian expectation, verified by experimentation in high winds, that the wind gradient and the momentum flux are directly interconnected. The light wind case requires a bit more thought, as will become apparent below. Contemporary data tend to support the early results. As an example, Figure 8.1 shows results obtained in the Ocotillo (Texas) field study of 2011 (White et al. 2013; Hicks et al. 2014). Data used are for 15–minute runs. After sequencing according to stability, every twentieth point is plotted. Hence, the number density of data points is a fair indication of the spread of stabilities encountered. The presentation is for the range $-1 < \zeta < 1$, this being the range of applicability of the flux-gradient relationships as originally documented. For the case of ϕ_H there is considerable lack of conformity near neutral, doubtlessly a direct consequence of both the temperature gradient and the sensible heat flux fluctuating around zero but not necessarily in synchronization on a 15–min basis.

In neutral conditions, the summary just given explains why u_* enters the formulation of the wind profile. It is the Fickian expectation, verified by experimentation in high winds, that the wind gradient and the momentum flux are directly interconnected. The light wind case requires a bit more thought, as will become apparent below. Contemporary data tend to support the early results. As an example, Figure 8.1 shows results obtained in the Ocotillo (Texas) field study of 2011 (White et al. 2013; Hicks et al. 2014). Data used are for 15–minute runs. After sequencing according to stability, every twentieth point is plotted. Hence, the number density of data points is a fair indication of the spread of stabilities encountered. The presentation is for the range $-1 < \zeta < 1$, this being the range of applicability of the flux-gradient relationships as originally documented. For the case of ϕ_H there is considerable lack of conformity near neutral, doubtlessly a direct consequence of both the temperature gradient and the sensible heat flux fluctuating around zero but not necessarily in synchronization on a 15–min basis.

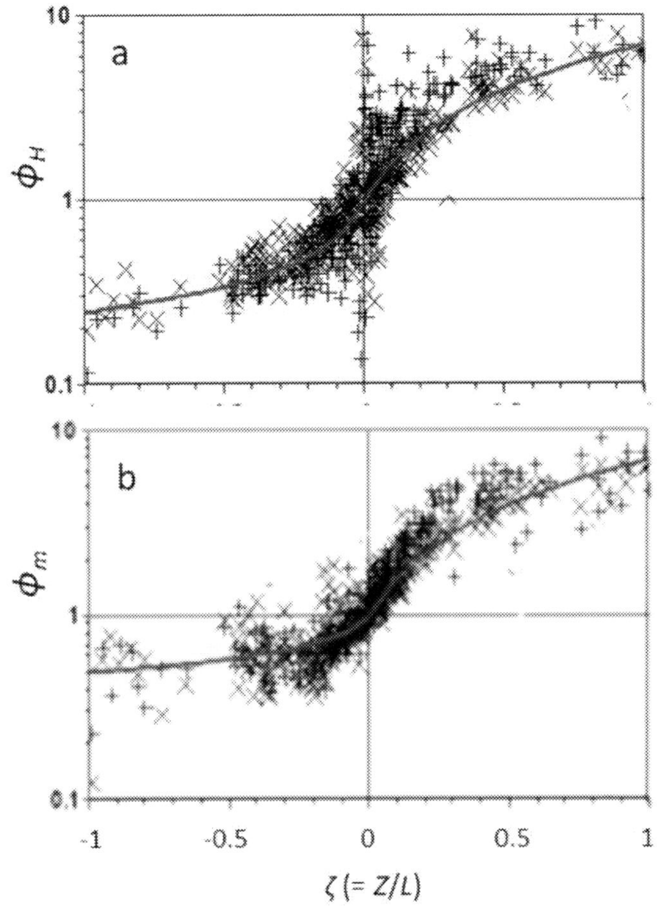

Figure 8.1. An example of the agreement between field data and the legacy expectations (the line drawn) relating the stability dependence of the dimensionless temperature and wind gradients ϕ_H and ϕ_m. Data are from the Ocotillo field site in Texas (Hicks et al. 2014). 8.5 to 14.75 m differences have been used to compute data represented by + ; 14.75 to 21 m by x.

Conclusion

There are many reasons to dispute the assumptions underpinning the ϕ–functions and even more reasons to dispute ψ. Use with care.

Chapter 9

Modeling and Tribalism

"The main beneficiary of numerical modeling may be the computer industry."
— Noel E. Ridere

It has been said that meteorologists are artists and like all artists meteorologists tend to fall in love with their models. The enthusiasm with which early researchers adopted the Monin-Obukhov Similarity Theory (MOST) is an example of this syndrome. Whether or not MOST is a good basis for describing the lower atmosphere, its simplicity and adaptability to many applications have resulted in its acceptance as a basis for almost all contemporary simulations involving the SBL. This is despite early awareness that dimensionally based analyses (of which MOST is a good example) risk generating apparent behaviors that are reflections of the statistics involved rather than of the contributing physics. Upon scrutiny, it becomes apparent that many of the MOST-based early conclusions about turbulence and turbulent exchange in the lower atmosphere continue to be used in today's models, even though subsequent experimentation with vastly improved instrumentation has shown that refinements are warranted.

The challenge of SBL and PBL modeling is to simulate processes and other characteristics that are often influenced by randomness, using expressions that are inevitably deterministic, using limited data and incomplete understanding. The fundamental requirement is for a method to describe the region of the atmosphere below the level of strong synoptic control, since it is in this region that people live and society is most influenced. The central considerations are then that (a) the wind aloft is adequately predicted by existing numerical models, (b) the wind in contact with the surface is zero (the no-slip condition) and (c) the rate of exchange of momentum at the surface depends on the characteristics of the surface (conveniently described in term of z_0 and d, or C_f as discussed earlier). The need is to interpolate between these extremes and to extrapolate so that the region above the level of current understanding is adequately described. Similar considerations apply in the cases of temperature, humidity and any other atmospheric quantity of special interest.

There are countless models of the behavior of the lower atmosphere. Many are integrally involved in weather forecasting, where the familiar flux-gradient relationships are often used to interpolate between the large-scale flow aloft and the lower boundary of the atmosphere. In such applications, the widely accepted flux gradient depictions are often assumed to apply far beyond their actual range of applicability ($-1 < \zeta < 1$). There are two widely divergent options available: either extrapolate the MOST description of the SBL beyond its level of applicability or adopt the earlier PBL description of the change of wind speed with height (such as the power law profile or one of its modifications) and merge it into the SBL below. Modeling preference for the former is common but is wrong.

There is no need for the recent discussion to be diverted by considering 'models' based on simple multivariate correlation analyses – these are no more than best fits to existing site-specific data. Expecting them to operate in a predictive mode is an intellectual stretch that will be let pass, here. However, applying a regression analysis to the results of a deterministic simulation offers a solution with considerable attraction. This approach has been adopted in some situations (e.g., Marzban et al. 2014), but it remains susceptible to the criticism that it assumes the completeness of the initial selection of important factors and the veracities of their formulations. To refine weather predictions when the relative roles of several contributing factors are not well known, a procedure of adaptive tuning has evolved (q.v. Wabba et al. 1995; Gong et al. 1998), but once again an assumption is inherent — that the processes involved are known and adequately formulated. Predicting the contributions of chaos remains a challenge.

A simulation is essentially an assemblage of relationships describing relevant processes and their interactions. If the results obtained are found to disagree with observations, then something is wrong. It is well known that some of the relevant deterministic expressions can result in consequences that are then chaotic (q.v. Lorenz, 1963); if such issues arise, it is often not obvious how to interpret results and/or avoid the problem. A common response is to look at the details of the contributing formulations and to adjust these formulations or their interactions so that agreement with reference observations is then improved. A standard example of this process is the refinement of data to satisfy model requirements – e.g., data initialization as used in weather forecasting models. The intent is then to avoid the development of inconsistencies and unwelcome randomness in the products that result. However, in other situations the goal is often to force predictions to agree with observations of some kind. This is 'tuning'. After tuning, agreement with the reference observations is certainly improved but the cause of the initial problem persists. The disagreements have merely been obscured and are likely to arise again when comparisons are made with observations outside the set of conditions of the reference dataset. The literature is well populated with instances in which models are promoted because their predictions agree with a specific set of observations, sometimes using 'verification' data from the same circumstances as were previously used to tune the simulation (although characteristically representing a different time frame). A common strategy is to sequester part of the data obtained in a particular experimental program and to use this dataset later to demonstrate how well the resulting model works. Such strategies might well prove satisfying to some, but they do not lead to a product shown to work well elsewhere.

The conclusion that model predictions agree with observations that helped in its development is not a result of great practical utility. At the opposite extreme, the finding that model predictions fail to represent observations constitutes an opportunity for additional research, because certainly something is being missed. The obvious question that arises is often overlooked – "Is my selection of important factors and processes complete and sufficient?" Unfortunately, this step imposes a need for a researcher to look at the completeness of his own understanding. There is an element of reluctance sometimes involved, but if the intent is to improve understanding of the processes involved then it is clear that most attention must be given to situations in which models do not work well.

The strategies involved in testing model performance are worthy of further consideration. Consider a model with results relating to a complex surface, such as a forest. Testing the adequacy of the model requires a comparison with observations of some kind. In the case of wind speed, the simulation yields an average across a grid cell of considerable size and any measurement using a single tower might not be an appropriate benchmark. Suppose the grid cell is populated with a network of towers, each carrying a reporting anemometer. The model will yield an average value across the cell and the appropriate experimental measurement that would serve best would be the spatial average derived using measurements from a large number of such towers and anemometers. But by definition the spatial average provided by the network is made up of readings that in the best of all possible worlds would be distributed such that half of them would yield speeds greater than the spatial average and half lower. There is an obvious corollary — that the time sequence of a spatial average is necessarily smoother than that of any of the contributing measurements. The consequence is a simple reality of sampling: it is not possible to locate a single instrument such that its measurements will be precisely representative of a grid cell. The obstacle this presents to model verification is regretted but is unavoidable. The practical challenge is to minimize the uncertainties involved.

Given that the major processes involved in the lower atmosphere are the results of turbulence and that the correlations associated with the eddy fluxes of energy, momentum, and trace gases are inevitably small, the answer to the dominant question has always been to step past the issue of aleatoric variability and to address, instead, the averages in both space and time. This leads to an obvious conclusion. We can never forecast the precise behavior of the atmosphere on an event basis, i.e., over any short averaging time. As the averaging time gets smaller, the contribution of uncertainty increases. Put another way – as the grid size decreases the ability for predictions to agree with observations will also decrease. So why the continuing effort to reduce model grid sizes? The dominant reasons appear to be threefold – job security, industry lobbyists and a religious belief that more detail results in better performance. Management support for these ever-expanding computer efforts is guaranteed by the belief that bigger is necessarily better and the energy of researchers and lobbyists.

It must be expected that model predictions and observations will disagree. The operational goal is to minimize the magnitude of the disagreement. The interpretation of such disagreement is not only a matter of examining the assumptions of a model, but also of where and how observations are made. The role of research is not only to improve the descriptions of those processes shown to be in need of improvement, but also to identify, examine and formulate any processes that are missing. If it is believed that all of the processes are adequately included in a particular simulation, then the questions of importance must be related to the way in which these processes are formulated, including consideration of any constraints upon them and how these affect extrapolations.

There are many purposes for which meteorological models are constructed. If the intent is weather forecasting, then the role of micrometeorology is to describe the boundary condition at the surface. Synoptic-scale considerations enable the atmosphere above the level of reach of surface influences to be described, provided the drag affecting the flow field and the heat flux affecting its stratification and mixing are well enough formulated. The no-slip condition ordains that the wind approaches zero as the height of measurement decreases. (How does this translate into the circumstance of an alpine forest?) The role of micrometeorology is then to describe the gradients of wind, temperature and other atmospheric constituents so that the coupling between the air aloft and the surface can be formulated. The size of the areas of concern is governed by the model, in many cases having a grid size of 4 km or larger.

In weather forecasting models, the reference height of choice is usually in the range 10 to 20 m. The challenge is then to relate the usual state variables (u, v, w, T) at this reference height to surface fluxes and to the atmosphere aloft. By extension, other atmospheric quantities can be considered similarly — e.g., water vapor, carbon dioxide and pollutants of many kinds. A moment's reflection and perusal of the data presented here will reveal that a height as low as 10 m might still be too great for classical relationships to be accepted. In the case of air quality studies, the main height of interest may well be very close to the surface, often at or below a height of 2 m above the ground (this being where people are mostly exposed). Furthermore, examination of the assumptions of many common models shows that in some cases the height scale is referred to the ground and not to any specific zero plane. In these models, forests do not exist and their heat storage capacity is ignored. The presence of a forest might then be simulated by no more than an increase in the assumed roughness length, with consequent (and obvious) unwelcome repercussions on predictions near dusk and dawn.

The matter of achievable verification is a fundamental consideration. Experience tells us that uncertainty pervades all attempts at prediction. A model prediction is usually of one value representing an average across an area of specific size at a particular time. The model grid size could be anywhere between meters and many thousands of meters, often about 4 km. As has been emphasized above, finding value-by-value agreement between model predictions and the measurement made at the same time at any place within a grid cell would be fortuitous (or manipulated). There is no alternative but to consider the matter of model evaluation in terms of statistics and ensembles. Likewise, there should be no expectation that any model can yield a prediction accurately mirroring observations made at any specific time and place. This conclusion certainly applies to the terrestrial daytime surface boundary layer. Its extension to other situations remains an issue about which there must be more thought and additional experimentation.

> The use of the word 'verification' is intentional. The simulation community often prefers to discuss "validation," which is a more forgiving consideration. "Verification" means that the answers are right. "Validation" indicates that the results follow logically from the assumptions. An alternative meaning of "validation" is that some authority has accepted the simulation as being appropriate for some particular purpose. The Latin roots are interesting and the preference of the simulation community is revealing.
>
> There are THREE fundamental steps in the simulation testing process.
>
> 1. Does the coding do what is intended? – validation.
> 2. Do its products agree with external data? – verification
> 3. Is there no simpler form that will do as well? – parsimony.

The addition of parsimony is an obvious step towards keeping things as simple as is possible.

As a dedicated experimentalist, the author cannot avoid the opportunity to point out a major disconnect between the approaches of the conservative and progressive sections of our community. The former comprises those who accept what is already in place and promote activities to improve it. The latter anticipates the need for changes in the legacy understanding and promotes anticipatory studies to accelerate its evolution. Early offenses include the avowed philosophy of funders in the late 20th century to support field studies only when modelers said they were necessary. The alternative, to conduct experiments to address issues not yet considered by modelers was not favored. A regrettable consequence was that repeated entreaties for small observational arrays of advanced SBL instrumentation was not heeded. These advanced sites were proposed, for example, to provide real-time measurements of the fluxes presently estimated (poorly) in numerical simulations. Instead, it was deemed best to rely on instrumentation and locations selected only to make the existing data resource more reliable and not to permit an expansion of the dataset as was then anticipated as a need for models yet to come.

Conclusion

A numerical simulation is a deterministic representation of a researcher's understanding of relevant processes, which is/are sometimes chaotic.

Chapter 10

Consider the Oceans

"To hell with the torpedoes" — Admiral David Farragut.

The earth surface is dominated by oceans and yet the emphasis of the present material has been overwhelmingly terrestrial. The reason is obvious – we are not aquatic animals. But, in the light of the considerations summarized here there is an obvious need to inject a modicum of balance. Moreover, it is the oceanic case that most clearly supports the need to consider the atmosphere as more than an entity in its own right. The atmosphere is merely the gaseous part of a physical, chemical and biological system that extends to well beneath the surface. This is obviously the case for land surfaces and is equally obvious for the oceans. It seems likely that the structure of wind-driven waves has been of interest to sailors since boats were invented. Certainly, by the time of Leonardo da Vinci the physics of wave breaking was of relevance to sailers; his codices present detailed illustrations of how waves break along their downwind crests. While the generation of waves complicates the relevant physics of the air-surface coupling, the situation is not as complicated as when the surface carries a canopy of vegetation, as in the common terrestrial case. However, plant life cannot be ignored. The typical ocean surface is biologically rich, unaffected by lack of water and with ready access to incoming radiation.

From the perspective of the SBL, the central questions revolve around the specification of the roughness length and its equivalents for the cases of momentum, heat and mass transfer. Despite decades of investigations, there are questions remaining. The classical description of the roughness of the ocean formulates the drag coefficient as a function of wind speed, with appropriate allowance for the effects of atmospheric stability. The common conclusion is that the drag coefficient increases linearly with wind speed beyond a threshold value that varies according to the circumstances and the preferences of the investigator. However, Charnock (1955) postulated that the surface roughness of the sea is determined by the stress upon it and so proposed that the roughness length could be approximated as

$$z_0 = \dot{a} \cdot u_*^2/g \tag{10.1}$$

where $\dot{a}$ is a constant. There is the appearance of a circularity in the Charnock proposal, in that the drag on the surface is certainly determined by the roughness of the surface, i.e., by z_0. However the Charnock proposal says that z_0 is determined by the drag on the surface. The resolution to this apparent circularity is that the surface is a component of the exchange between the air and the ocean below, not a rigid boundary; the transfer of momentum to the surface is the same as the transfer of momentum from the surface to the water below. That is, the surface is not accelerating or decelerating.

Conventional formalism leads to an expression for the drag coefficient at neutral ($C_{dn} \equiv (u_*/\bar{u}_{10})^2$), conventionally representing the interpretation of data obtained at 10 m height above the average surface,

$$C_{dn} = (k/\ln(z_{10}/z_0))^2. \tag{10.2}$$

Garratt (1977) provided substantial support for the Charnock proposal, by reviewing a large number of experimental datasets. The conclusion reached from his Table 4.1 is that the best estimate of $\dot{\alpha}$ is 0.019 x/÷ 1.44 (computed geometrically). However, there are indications that the assumption of constant $\dot{\alpha}$ might not be correct (Guan and Xie, 2004). Field data are alternatively well represented by a linear fit relating the drag coefficient to the wind speed: e.g.,

$$C_{dn} = 0.001(1 + 0.05\bar{u}_{10}). \tag{10.3}$$

Figure 10.1 shows results from some field studies, selected to illustrate aspects that will become apparent later. The values 0.001 and 0.05 enter as the constants in Equation 10.3 because they fit the data of Figure 10.1. In general, Equations 10.1 and 10.2 appear to represent the open-ocean data for wind speeds above a lower limit for $\bar{u}_{10}$, varying according to the experimental circumstance from about 4 to more than 8 m s^{-1}. At lower wind speeds, there is little evidence to indicate that the ocean surface is any rougher than would be expected of aerodynamic smoothness. Consequently the original Charnock relationship has been augmented (for the open-ocean case) by adding a term to indicate that Equation 10.1 works as an addition to smoothness. Smith (1988) proposes that

$$z_0 = \dot{\alpha} u_*^2/g + 0.11\nu/u_*. \tag{10.4}$$

The results shown in Figure 10.1 are from four different studies: (a) over Bass Strait (Hicks, 1970, 1972; with more than 100 km upwind fetch), (b) over Lake Michigan (Hicks, 1972; also with fetch > 100 km), (c) using the same instrumentation as was used in (a) and (b) but at a site in an area potentially affected by a coral reef in Papua-New Guinea (PNG), and (d) at the same PNG location but using an experimental sonic anemometer (Hicks et al. 1974). Each point plotted represents an average of results sorted according to wind speed, grouped into convenient ranges and corrected for the effects of atmospheric stability. There are many studies that report more data than are shown here. The present selection is designed to reveal some considerations that otherwise would be obscured if the instrumentation or researcher differed. The datasets are presented here on the assumption that they truly indicate some of the uncertainties arising in historic experiments, in spite of the authorship.

The PNG experiment arose as a test of a sonic anemometer developed at the University of Papua New Guinea, in Port Moresby. As a field demonstration of the new instrument measurements were made of eddy fluxes using a tower of convenience, marking a live coral outcropping at the entrance to Port Moresby harbor. The water was shallow in the immediate vicinity of the tower, but the upwind fetch was considered to be representative of much deeper water, offering no obstruction to shipping. Two sets of results from the Port Moresby study are depicted in Figure 10.1. The first made use of the same mechanical instrumentation as had been used in other studies, such as those also depicted in the figure. The second shows results from use of the sonic anemometer.

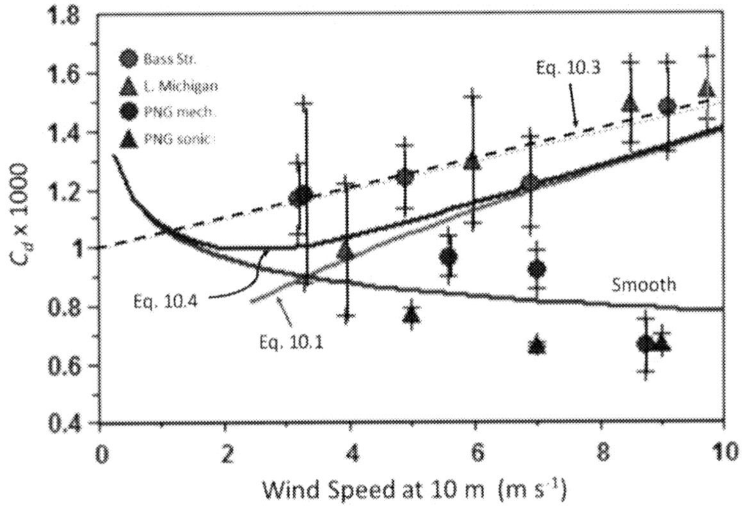

Figure 10.1. The variation with wind speed of the drag coefficient measured in several field experiments conducted over water. Stability corrections have been applied so that these results represent neutral conditions with wind speed measurements made at 10 m height.

It is acknowledged that the error bars shown in Figure 10.1 are larger than appear in most other reports of similar results, however the present selection of data is shown for a distinct and relatively unique purpose. The results obtained with the new sonic anemometer were consistently like those obtained using the mechanical wind sensors used previously. However both sets of instrumentation yielded results differing substantially from expectations; they varied much as would have been expected if the surface were aerodynamically smooth out to a wind speed of about 10 m s^{-1} (the greatest wind speed experienced in the PNG study).

The open water results of Figure 10.1 are in line with observations made by others in oceanic conditions. The PNG results are not; they appear to be more in synch with the aerodynamically smooth expectations. Deacon (1979) offered an explanation — that live coral exudes oils that serve to smooth the water, thus somewhat protecting the coral itself from extreme wind speed ravages. Actual measurements were made of the surface tension of the seawater upwind of the PNG site, during a follow-up visit intended to explore Deacon's hypothesis. The results were indicative of a departure of surface tension from conditions further upwind but were inconclusive because the method of measurement was not well suited to open-ocean exposure. A survey of related biological literature indicated that indeed exudations by coral were the causative factor (Deacon, 1979).

There is considerable scatter evident in the results plotted in Figure 10.1. It is now recognized that much of this scatter resulted from the corrections applied to reduce the observations to 'neutral' equivalents (q.v. Hicks et al. 2018). Given contemporary understanding and with awareness that reliance on ψ-functions imposes assumptions about the relevance of z_0, a re-analysis would be appropriate. However, the original data have been lost.

Note that the line representing Equation 10.1 in Figure 10.1 assumes that $\dot{a}$ = 0.016 (as is the result derived from examinations of the many field studies conducted by the present author). Field data (e.g., as summarized by Garratt 1977) indicate that different experiments lead to different conclusions regarding both the rate of increase of C_{dn} with wind speed and the wind speed at which there is a transition from the aerodynamically smooth situation to the Charnock expectation (or similar). That is, Equation 10.4 is a simplification that might be refined by weighting the two extreme expectations differently. The transition from aerodynamically smooth flow is anticipated to occur in the range 3 to 5 m s^{-1}. In Figure 10.1 the transition is illustrated as being in this range.

Any consideration of the effective surface roughness in light winds deserves attention to a long-standing observation. The situation of interest is one of flow within a two-phased system, the upper component gaseous and the lower liquid, but both responding to all imposed forces. One such imposed force is the momentum flux. In the case of an undisturbed water surface, the assumption that the momentum flux is constant with height across the boundary imposes the result that the velocity of the surface fluid should be about equal to u_*. Indeed this is the case, as has been verified by countless observers who have dropped the requisite peel of an orange overboard and measured its drift velocity with a stopwatch. The repercussions of this are uncertain, but clearly the coupled-media aspect of the overall problem must be a major consideration.

Formulations of the oceanic drag coefficient have been revisited and refined on many occasions. Fairall et al. (2003) provide an extensive review of progress, refined by others to take such aspects as wave age and ocean currents into account (Kara et al. 2007). However, as time has progressed there has been increasing awareness that C_d does not continue to increase linearly with wind speed beyond the range of $\bar{u}_{10}$ observed in ocean tower micrometeorological studies. A study of the tracks of tropical cyclones (Powell et al. 2003) led to the observation that C_d levelled off at some point "above hurricane force." Moon et al. (2004) report the results of tests based on examination of cyclone tracks, showing that C_d appears to maximize at about 0.002 ± 0.0005 beyond about $\bar{u}_{10} = 40$ m s^{-1}. In this regard, the earlier discussion of aerodynamic smoothness is worthy of revisiting. In extremely violent seas, the sea surface is not an easily recognizable feature. As in the discussion of the smooth case, it seems likely that a practical simulation of the behavior would be to consider the 'boundary' between the air and the ocean to be a layer through which momentum from the wind is being exchanged and shared with any spray that results. A concept like the conventional roughness length is difficult to equate with observations, because the spray zone extends far into what would conventionally be seen as the SBL. In fact, using the 10 m height as a reference point could be inappropriate. Considering the combined air/water mix to be an expanded constant flux layer (as in the work by Deacon, 1977) seems an appropriate first step. Identifying the location of the relevant zero plane could be a major problem.

The exchanges of heat, water vapor and trace gases across the air-sea interface have been the subject of great attention, with a variety of models being proposed and with the detailed characteristics of the surface itself, such as its chemical composition, often serving as an unresolved complexity. As in the case of the coral exudate film affecting the exchange of momentum, as postulated by Deacon, the presence of a surface microlayer has been widely accepted by the community addressing the exchange of other properties.

The presence of a thermal film is readily demonstrated for the heat exchange case using infrared surface temperature radiometry while disturbing the surface (e.g., by sparging). The thermal characteristics of this surface layer are also widely appreciated and continue to be a cause for concern about reliance on ocean temperatures measured with immersed thermometers when details of the coupling with the atmosphere are being investigated. Early assessments of the thermal coupling between the atmosphere and the global oceans made use, for example, of engine cooling water intake temperatures reported by ships, which necessarily imposed a positive error on the derived sensible heat fluxes. The thermal film is such that the molecular-scale surface of the water is cooler than the subsurface, by an amount typically in the range 0.2 to 0.3 K and across a layer of the order of 1 mm thick. Relevant data from open ocean conditions are few; the requisite measurements are difficult. However, Khundzhua and Andreyev (1974) reported a range from 1.6 to 2.4 mm for the Black Sea. Wesely et al. (1975) examined the matter using data obtained in studies of a strongly heated and expansive inland lake (the cooling pond of a nuclear power plant) and derived an average value of 0.38 mm. The direct relevance of the latter results to the open ocean is admittedly questionable. However the overall message is clear — that even in exceeding high thermal regimes with heat fluxes many orders of magnitude above those of the open ocean, the thickness of the subsurface thermal boundary layer is of the order of a millimeter. The magnitude of the temperature drop across this layer is a function of both the rate of heat transfer and the momentum flux.

> Published papers sometimes contain diagrams guaranteed to excite the interest of readers like yours truly. Some recent papers in leading journals show plots of the oceanic drag coefficient C_d against quantities that reduce to $C_d^{1/2}$. To be more specific would be to denigrate authors whose work is, on the whole, a valuable contribution. However, there must be continuing awareness of the risk of producing impressive diagrams that are essentially meaningless. If any criticism is warranted, then it must be shared not only among authors but also with reviewers.

Assessments of the global heat energy balance sometimes rely on ocean surface temperature data uncorrected for this surface film effect. While the consequences on calculations of the sensible heat flux can be readily estimated, the consequences on the calculation of the latent heat exchange are not so easily addressed. In some marine assessments air in contact with the surface is assumed to be saturated, which is wrong even if the temperature is correctly reported (the NaCl molarity of sea water is the key factor).

Different scientific communities look at surface sublayer issues with different emphases. Those concerned with heat and enthalpy consider the thermal film. Those focusing on gas transfer deal with microlayers of chemical and biological activity (as presented by Liss and Duce, 1997). The observations made at Port Moresby, as interpreted by Deacon (1979), relate to the smoothing of the surface and reduction of drag on it, attributed to organic exudates from coral that modify surface tension. In all cases, it is the presence of a surface film that is a critical factor. Earlier work by Garrett and Bultman (1963) and Harvey (1966) demonstrated that the surface film of seawater contained organic chemical and biological populations of (e.g.) plankton that served to damp surface ripples, presumably through their influence on surface tension. Broecker and Peng (1974) reported on radon exchange across the air-sea boundary, reinforcing their stagnant film model of air-sea gas exchange and supporting earlier predictions that the gas exchange rates between the air and the sea vary with the square of the wind speed (see Kanwisher, 1963). Watson et al. (1997) and Frey (1997) provide further evidence. The complexity of the situation has been emphasized more recently by Cunliffe and Murrell (2009), who suggest that the microlayer could be gelatinous. On the whole, the surface tension proposal of Deacon (1977) starts to ring true. Pervading all discussion of this kind is a focus on relatively calm seas. Intuition says that rough seas are more likely to be important factors than calm seas, whatever the context.

Immediately above the surface there is conventionally postulated to be a thin layer of air through which transfer is dominated by molecular diffusivity or viscosity. Coupling the above surface and below surface layers in a numerical simulation presents difficulties that are certainly not insurmountable; it is perhaps more difficult to accept that these layers actually exist in open-ocean conditions. In concept, any such layers can only persist if the forces striving to disrupt them do not exceed those defining their existence. The disrupting force is the surface stress due to the wind. It should be expected, therefore, that any high-resistance microlayer will be susceptible to the wind stress, which relates as the square of the wind speed. This is in accord with the early observations of Kanwisher (1963) and the observers who followed him.

The availability of satellite imagery subsequently elevated ocean whitecapping to a level of considerable importance, since the obvious dependence of whitecap coverage (usually reported as a percentage of the observed area — *W*) on wind speed provides a pathway to derive oceanic wind speed estimates from satellite data. Figure 10.2 presents an interpretation of data published by Kondo et al. (1973). The intent is to show that increasing whitecapping depends on the square of the wind speed beyond a transitional velocity. In the present case, the transitional velocity is assumed to be 6 m s^{-1} and the resulting correlation coefficient and power law exponent derived by regression are 0.88 and 2.08 respectively. If it is assumed, instead, that the transitional velocity was 5 m s^{-1}, then the resulting values are 0.90 and 1.96 respectively. The conclusion to be drawn from the data considered here does not appear to be greatly influenced by the transitional wind speed assumption made. Other workers have looked at the whitecapping issue purely pragmatically, by plotting the ln(W) against $\ln(\overline{U})$ and using regression to obtain a best fit of the form $W = \acute{a} \cdot \overline{U}^{b}$. The value of $b \approx 3.5$ has become a widely accepted conclusion (Monahan and Muircheartaigh, 1981).

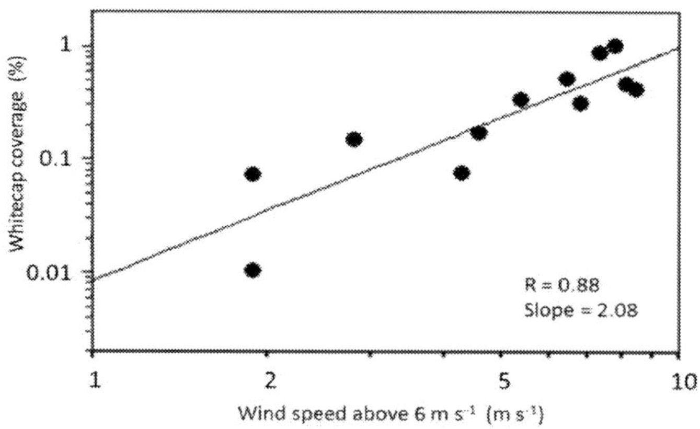

Figure 10.2. Whitecapping as a function of wind speed above a transitional velocity of 6 m s^{-1}, based on Figure 5 of Kondo et al. (1973).

In concept, whitecapping would destroy any thin boundary layers otherwise limiting air-sea exchange, e.g., of trace gases. A dependence of average exchange coefficients on the square of the wind speed would be an obvious result, as is indeed observed in the cases of the gases considered in the collection presented by Liss and Duce (1997). However, other workers have based their considerations on available kinetic energy, leading to an expectation that the dependence of W on wind speed should be according to the cubic power rather than the square. The matter appears to need more attention.

If the interpretation by Deacon (1979) is correct, then it would seem that the large areas of ocean affected by pollution and/or by algal and especially dinoflagellate "blooms" could serve as regions of reduced surface drag and therefore of enhanced wind speed. At the same time, these areas would display gas exchange characteristics different from the unaffected ocean, without any immediate indication of how that difference would reveal itself. The issue is of importance in the context of algal blooms, because it has been shown that occurrences of intense turbulence inhibit the growth of algae at the surface (Gibson and Thomas, 1995). Turbulent intermittency at sea is widely observed and it would be an unsurprising evolutionary consequence for such sensitive algae in surface water to have an innate protection against the consequences of breaking waves. The question obviously arises as to whether wave breaking is inhibited in oceanic areas of algal blooms. It is also now postulated that surface tension could well serve as a factor to explain the differences in transitional velocities — the wind speeds at which field studies indicate that the expectations based on aerodynamic smoothness transfer into a wind speed dependence with whitecapping.

Chamberlain (1983) points out that the Charnock expression (Equation 10.1) works for other surfaces. "It appears that the roughness created by modification of a mobile surface is numerically similar whether the surface is water, sand or snow." This observation is clearly hiding some truths that remain to be revealed, however when the observation about surface tension is added to the picture it emerges that the low transition wind speed evident in the historic open-ocean data might increase when surface tension rises above the level normally associated with sea water ($\approx$ 0.2 mN m^{-1}). Note that the surface tension of fresh water is substantially lower, c. 0.078 mN m^{-1} and therefore a slight difference should be found between the long-fetch drag coefficients of salt water and fresh. To check this out, results from a series of field measurements over Lake Michigan (USA, fresh water) were compared to results obtained using identical equipment over Bass Strait (between Victoria and Tasmania, in Australia, salt). The results are shown in Figure 10.1. Any differences are too small to be detected (q. v. Hicks, 1972).

Conclusion

Most of this planet is covered by salt water. Life depends on unsalted water. The association of these two is a major factor contributing to the viability of life. It deserves more attention.

Part II: Rethinking the Basics

Chapter 11

Stepping Forward

'An island of correlation in a sea of chaos' – Noel E. Ridere

So far, the discussion has emphasized the evolution of the micrometeorological specialty, as it has grown slowly from early studies of the atmosphere over specific surfaces to multi-faceted investigations incorporating its sibling specialties – soil science, agricultural meteorology, forest meteorology, urban meteorology and oceanography. There are many textbooks that describe each of these in detail. It is not the present purpose to repeat such detailed examinations, but instead to look for recent results that would reveal new aspects of the phenomenon of turbulent exchange.

The reason for the continued interest in the details of air-surface exchange and the relationships between fluxes and gradients in the SBL (and PBL) is the need to construct reliable predictions when observations are few. In the case of crops as considered above, demands for scarce water impose a need to get the evaporation rate quantified correctly, while allowing for both the photosynthetic demands of whatever crop is grown and the changes in soil moisture content. In the case of weather forecasting, understanding of the free atmosphere permits prediction of synoptic-scale velocity fields with considerable confidence but predictions of this velocity field aloft do not easily project themselves to the surface. It is near the surface that society and ecosystems are most affected. What is needed is a way to describe the linkage between the atmosphere and the surface 'seen' by the atmosphere.

A basic goal is to formulate effects in terms of their causes. If the intent is to describe the variation of the wind speed with height, then the determining variables are the wind speed aloft, this being the source of momentum, and drag at the surface. Specifying the wind aloft is an issue for synoptics. For the lower boundary condition, the no-slip criterion requires that there is no residual momentum in air in contact with the surface. Formulating the wind profile amounts to interpolating between the wind speed aloft and the no-slip criterion below.

As already reviewed above, the classical development of micrometeorology started with consideration of neutral conditions, building on the results of studies in wind tunnels and other controlled environments, largely resulting from studies of aeronautics. The velocity was always high and heat exchange played a negligible role. At first, there was a conscious search for real-world landscapes and circumstances such that the demanding conditions imposed by the aeronautical heritage could be extended to the SBL. The resulting studies were reassuringly successful and the real-world value of the von Karman constant was found to be the same as that found in laboratory situations (Sutton, 1952; $k = 0.41$). It was accepted that the quantities describing a surface, the roughness length (z_0) and the displacement height (d), were unchanging characteristics of the surface. In practice, early Australian field studies were sometimes abandoned because neutral situations (by which the site could be calibrated in terms of z_0 and d) were not encountered. In the search for truly neutral stratification, the controlling factors were that conditions should be unchanging with time, wind speeds should be high and heat exchange rates (involving latent as well as sensible heat) low. The transition period at dawn was avoided, since the thermal characteristics of the atmosphere are then changing most rapidly with time and the wind speeds are characteristically low.

In contemporary simulations, legacy flux-gradient relationships are widely accepted as 'proven' although based on limited data and often extrapolated far beyond their actual range of applicability. The warnings of those who developed these legacy formulations are often found to be inconvenient. Improving the situation has proved difficult. A common viewpoint is that there is no benefit in additional studies because the topic has already been well explored and the results are satisfactorily summarized in textbooks. This is despite the fact that new instruments have been developed, enabling more and better understanding to be generated at minimal cost. The discussions to follow will focus on new observations, often obtained in experimental programs peripheral to mainstream micrometeorology, such as in agrometeorological and forest meteorological experiments. These experiments have been conducted in situations often far from ideal and as such they can be taken to represent circumstances closer to the 'real world' than the classical studies of the distant past.

The present approach is to regard legacy understandings of relationships between fluxes and gradients as sound guidance and then to adopt a purely empirical view: What do modern measurements actually tell us? The considerations to follow are guided by experiences of the early micrometeorological field campaigns, at Kerang and Hay in Australia (q.v. Garratt and Hicks, 1990). In these field studies, modern data acquisition methods were not yet available and the instrumentation required continuous monitoring by experienced observers. Three independent flux systems were used, the 'Evapotrons' and later the 'Fluxatrons' as were introduced earlier. These were operated by three technical experts, working independently: the Scientist in Charge, Arch Dyer; a Meteorological Technician, Guntis Grauze; and an Experimental Officer, yours truly. (The three of us and our separate instruments were known as 'Cerberus.') Acceptable results were reported when there was no conflict among the three results. As the atmosphere was observed and the dials watched, there was a lot of discussion about what the atmosphere was actually doing. Working on a sheep farm (at Hay), two separate convective cells (small dust devils; 'willy willies') were observed to migrate along the sheep tracks until they combined into a single, larger cell when the tracks came together. On many occasions, a strong burst of convection provided a short burst of activity that was then averaged into the sensible heat flux reported for a half-hour period. (As Arch Dyer said – "Guys, we just saw the sensible heat flux for this run.") The concept of fluxes resulting from a continuous spectrum of turbulence was at odds with observations, especially for the daytime $\overline{w'T'}$. The nighttime case was even more intermittent. It is the intermittency of both unstable and stable surface boundary layers that dominates much of the discussion to follow.

The observations of the turbulent atmosphere made during the outback Australian Kerang and Hay experiments and later in the great plains and mountain environments of North America have colored the interpretation of all datasets addressed here. In many instances, these observations have been at odds with commonly accepted beliefs. In such cases the focus will be on understanding the observations with little desire to reinforce related dogma.

> "The eye, which is called the window of the soul, is the chief means whereby the understanding may most fully and abundantly appreciate the infinite works of nature"
> – Leonardo da Vinci (as in 'Notebooks, Leonardo da Vinci'— Irma A. Richter, 1980).

Chapter 12

Turbulence, Fluxes and Statistics

"Hidden in the sea of chaotic turbulence are the small islands of covariance."

At the time of early studies of micrometeorology, it was recognized that errors of measurement made it difficult to discern niceties of the interactions among variables that lay at the core of the physics. The pragmatic response was to defeat the resulting uncertainties by averaging, usually by constructing ensembles of measurements sorted according to some other property considered to be controlling – most commonly atmospheric stability. Plotting average effects against averages of assumed causes was a fundamental first attempt at extracting meaning from the scattered observations that were initially obtained. In distant retrospect, the resulting plots of (say) non-dimensionalized temperature gradients against stability do not reveal much about the processes involved. The causative factors are the fluxes and the gradients respond to them (or maybe vice versa). The stability (either ζ or *Ri*) is then a symptom, not a cause.

Turbulence is a feature of fluid flow that exerts its influence in many ways. Until instrumentation capable of recording fast fluctuations became available, laboratory studies of fluid dynamics focused on flow regimes (streamlines and their obstruction, often in conduits). Micrometeorological research has been similarly influenced by the recent developments of fast-response anemometry, about which more will be said later. For the present, note that the association between turbulence and flow patterns is not a simple problem. Fluid dynamics reliance on indicators such as the Reynolds number ($Re \equiv \bar{u}d'/v$, where $\bar{u}$ is fluid speed, d' is the relevant distance scale and v is viscosity) have obvious relevance to flow over natural surfaces and permit conditions for the generation of turbulence to be anticipated. But the uncertainties associated with the necessary extension of pipe-flow and flate-plate results to a natural landscape impose daunting insecurities — a natural surface cannot be well described as an array of geometric obstacles, such as are imagined by theoreticians. The closest approximation to ordering could well be the applicability of fractal descriptions in the case of some vegetation (e.g., ferns).

Turbulence is largely chaotic. It is only a small part of the turbulence observed at any place or time that can be associated with the eddy flux of momentum (or of any other quantity) and hence with the description of the wind profile. Experience says that there is a strong element of hysteresis involved — while turbulence can be generated quicky, it decays slowly. Aircraft contrails are an easily understood manifestation of this phenomenon, although at altitudes much above those of relevance here. In the lower atmosphere, the wind profile itself carries the signature of all of the regimes of turbulence experienced upwind. The part of this total turbulence regime that relates to the transfer of momentum is small. Typically, less than 20% of the turbulent kinetic energy can be associated with the momentum flux; this will be discussed further in the considerations of recent field programs, to follow. It is small wonder that description of the wind profile in terms of its defining quantities has been so challenging. It is likewise understandable that descriptions of the wind profile based on considerations of turbulence intensity have yielded such a wide variety of results. The empirical constant associated with the KEYPS equation (more appropriately, the O'KEYPS equation; see Li, 2021) has been variously quantified as anywhere in a range extending over an order of magnitude, a result contributing substantially to the failure of the relationship to gain favor within the Australian community. Dyer (1974) and Swinbank criticized the KEYPS formalism, with the latter famously rearranging the acronym and referring to the PESKY relationship.

Figure 12.1 illustrates the chaotic characteristic of turbulence in the SBL and the way that flux-carrying eddies are contained within the population of turbulent cells. The data shown derive from a single sonic anemometer, operated at about 4 m height above an arid and poorly vegetated surface (sagebrush) in Idaho, during June of 2014. The anemometer operated at 10 Hz, with subsequent derivation of 1 s averages that were archived in runs of 5 min duration. Thus, there were 300 data sets for every run. The data recording system reported the conventional micrometeorological variables, such as variances and covariances, but also identified the time associated with the detection of the minima in wind speed, temperature and vertical velocity for every run. Times of maxima were also recorded. Data have been sorted into two month-long time frames, 1000 to 1600 MST to represent the daytime, convective period, and 2200 to 0400 MST for the nighttime vase. (Mountain Standard Time, MST, is 7 hours delayed from UTC.) In all of the panels of Figure 12.1, the abscissa identifies where the maximum (or minimum) in the vertical velocity occurred. The panels identified as a1 to a4 (daytime) and for b1 to b4 (night) present the time-of-occurrences for the *w* and *T* data streams. Panels c1 to c4 (daytime) and d1 to d4 (night) show the same sets of results, for the *w*, *u* pairing. For the a-series of panels, a1 shows that the maxima in *T* occurred simultaneously with the maximum of *w* on a well-defined basis – the line of equality. Panel a2 shows that there were essentially no occasions when a minimum in *T* occurred at the same time as a maximum in *w*. Panels a3 and a4 show that there were no simultaneities with minima of *w*, for either maxima of *T* (a3) or minima of *T* (a4).

Panels c1 to c4 parallel those of the a-sequence, addressed above. For the case of momentum exchange, it is seen that minima of *u* correspond best with maxima of *w* (c2). One again, there is no evidence of simultaneity in either panels c3 or c4 (minima occurrences of *w* – downdrafts). There is slight evidence of an association in panel c1, but by far the dominant association is between updrafts and wind speed minima. In other words, the momentum flux is mostly due to the transport of lulls upwards than with high wind speeds downwards. Considering the evidence from panel a1, it must be concluded that the momentum flux in the unstable conditions considered here was dominated by the consequences of buoyancy. It is also clear, from panel c2, that the wind speed minima occurred after the occurrence of the vertical velocity maxima. The driving force was the vertical velocity associated with convection and the momentum flux was largely derivative.

The nighttime results indicate that the major simultaneity revealed by the velocity data is between maxima in *w* and minima in *u*, as seen in panel d2. However, in this case there is no reason to propose that buoyancy played a strong part and indeed there is no sign of a positive/positive association in panel b1. Panel b2 looks like panel d2 and similarity is implied. At no stage is there evidence of strong coupling involving minima (downdrafts) in *w*.

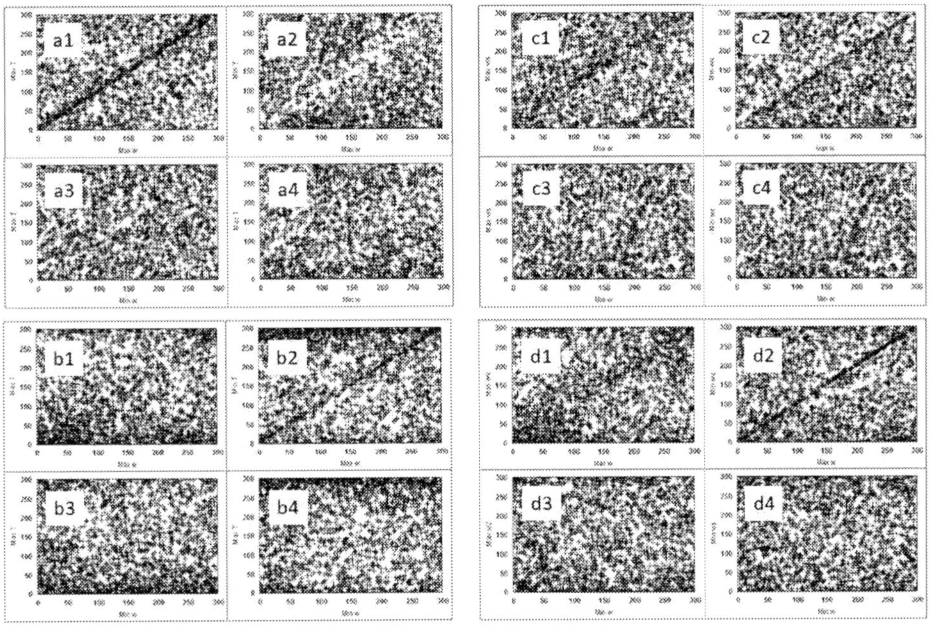

Figure 12.1. Scatter diagrams showing the association between the times of occurrence of vertical wind speed maxima (or minima) within 5-min runs and the times of maxima (or minima) in T (the panels to the left, a1 – a4 and b1 – b4) and wind speed (to the right, c1 – c4 and d1 – d4). In the four panels to the upper left, panel a1 relates to maximum w and maximum T, a2 is for maximum w and minimum T, a3 is for minimum w and maximum T and a4 is for minimum w and minimum T. The other three sets of results are similarly arranged. The upper pair of set of eight panels shows daytime results (1000 – 1600 local time); the lower set is for the central nighttime period (2200 – 0400). Axes are scaled according to the number of seconds past the beginning of each 5–min run.

Figure 12.1 tells only a small part of the overall story. However it is clear that the most ordering is associated with heat exchange in daytime and with momentum exchange at night. The uniformly distributed array of points plotted in all of the plots for minimum *w* is interpreted as good evidence of the ubiquitous chaotic characteristic associated with SBL turbulence, disrupted rarely by covariance associations.

The focus of micrometeorological research has not been solely on the physical processes contributing to turbulent exchange, but also on energy balance characteristics and especially the water loss of landscapes, especially those of farms. This focus continues to modern times. For example, Figure 12.2 shows average diurnal cycles of measured heat energy fluxes (derived using BREB methodology) for a selection of two-week periods during the year 2015, through the growing season of a field of maize in central Ohio. The averages plotted reveal some features of terrestrial air-surface exchange that were not readily apparent in early results. In early June, the data of Figure 12.2 indicate that one of the crops was growing faster than the other, as indicated by the dissimilarity in the L_wE and H cycles. Later in the growing season, the energy usage of the two crops equalized. The most striking and informative feature of the sequence of diagrams is the changing contribution of the growing vegetation. The loss of water by the transpiring vegetation is clearly a major consideration, maximizing when the crop was growing most rapidly — starting in June and extending into August. The plots show that the near-noon Bowen ratio $\beta \equiv H/L_WE$ minimized when the demands for energy by the growing plants caused L_wE to increase at the expense of H. This brings to mind the common and contrasting acceptance of the Priestley and Taylor (1972) conclusion that the large-area ratio of sensible to latent heat fluxes can be estimated on the basis of surface temperature alone (see Chapter I), clearly not relevant for crops like are considered here.

During the night, soil evaporation probably continued, so explaining the positive latent heat fluxes then observed by the eddy covariance systems. This serves as evidence that interpretation of fluxes measured above the surface is subject to influences other than those attributable to the vegetation alone and raises obvious questions about the applicability of conventional Bowen ratio methodology in other than periods of strong incoming radiation. In this context, the differences between sensible heat fluxes measured by eddy covariances and the results of BREB calculations indicate the difficulties associated with using the former in studies of small test plots. The requisite height of deployment of covariance systems imposes a need to consider the role of fetch. In the case now illustrated, the covariance system responded to air-surface exchange extending further upwind than the BREB systems, a matter that will be explored further in subsequent discussion.

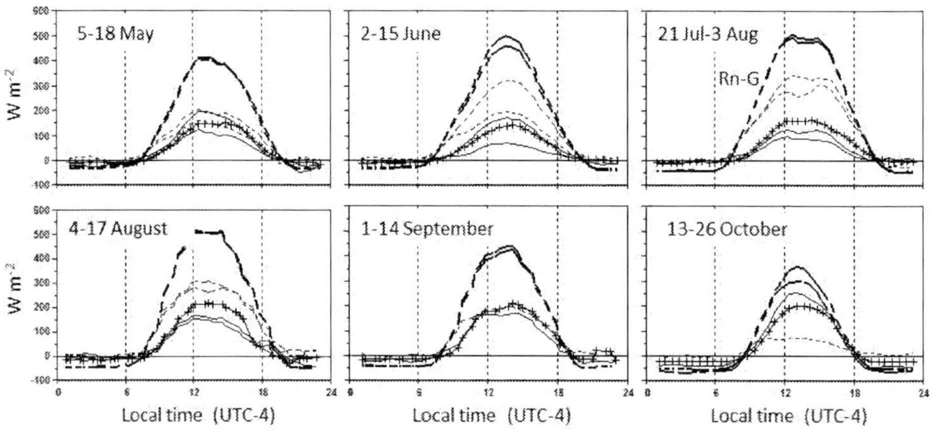

Figure 12.2. Averaged diurnal cycles of net radiation (Rn, the heavy black, uppermost curves), sensible heat flux (H, light solid curves)and the latent heat of evaporation (L_WE, dotted)as measured over a field of maize through the growing season of 2015. The site was at 40.606 °N, 82.674 °W, near Mansfield, Ohio. Observations over two adjacent fields are shown. The crosses depict eddy covariance results, made midway between the two plots. Averages are constructed over two-week periods.

Conclusion

Paraphrasing Geiger (1965) — the essence of air-surface exchange is statistical.

Chapter 13

Data Normalization and Dimensional Analysis

"Dimensional analysis is an art full of dangers"
— Saint-Guilhem (1962); translation by F. A. Gifford, (1967)

The process of data normalization has a definite purpose. Consider a variable of specific interest. Once the major contributor to its variation is known and formulated, the original variable can be normalized so that the effects of other contributing factors can then become more apparent. In the case of micrometeorology, a starting point is acceptance of the Fickian dependence of the momentum flux on the wind gradient:

$$\tau = -\rho \cdot \partial\bar{u}/\partial z \tag{13.1}$$

where the negative sign arises from the familiar micrometeorological convention that gradients increase upwards and fluxes are positive when upwards. This topic has already been explored in some detail, but in the context of present interest it is worth revisiting.

The surface drag τ is used to define a scaling velocity, the friction velocity

$$u_* = (-\tau/\rho)^{0.5}. \tag{13.2}$$

Experiments confirm that in neutral stratification and sufficiently above the zero plane (at height d) the SBL wind follows a logarithmic law, as predicted theoretically and hence defining two constants, a and b, as in

$$\bar{u}/u_* = a \cdot (\ln(Z) - b) \tag{13.3}$$

Here, Z is used to represent the height above the zero plane, d — $Z = (z - d)$ where z is the height above the ground. The value of a has been shown to be common to both aeronautical and meteorological applications as about 2.5 and this determines the reciprocal as the von Karman constant $k \sim 0.4$. (Earlier, it was pointed out that Sutton (1952) obtained $k = 0.41$, with which the results of the present author agree.) The finding that the same value worked for both aeronautical wind tunnel applications and also in the SBL generated considerable confidence in the extension of high-velocity engineering results to low-velocity atmospheric conditions.

Trivial manipulation yields the logarithmic wind profile

$$\bar{u} = (u_*/k) \cdot \ln(Z/z_0) \tag{13.4}$$

in which the familiar roughness length z_0 derives from the value of b in Eq. (13.3), The local wind gradient is then (obviously)

$$\partial\bar{u}/\partial z = u_*/kZ. \tag{13.5}$$

Field studies support the applicability of this relationship when wind speeds are high and there is negligible confounding influence by buoyancy, terrain, vegetation, etc. — i.e., in neutrally stratified conditions over micrometeorologically adequate sites. (What constitutes 'micrometeorologically adequate' will be discussed later),

The framework thus outlined forms the basis for investigating the role of atmospheric stability over land – resulting from solar heating of the surface in daytime and radiative cooling of it at night. All early thinking accepted that the role of stability should be examined in terms of the consequential departure of profiles from neutrality. The stability index is conveniently considered as being like the ratio of buoyant to mechanical production rates of turbulent energy, in two alternative ways. The gradient Richardson number is then defined as

$$Ri = (g/\bar{\theta})\cdot(\partial\bar{\theta}/\partial z)/(\partial\bar{u}/\partial z)^2. \tag{13.6}$$

and in terms of fluxes

$$\zeta = Z/L = -(g/\bar{\theta})kZ\cdot(\overline{w'T'})/u_*^3. \tag{13.7}$$

Following work by Lettau (1951), Priestley (1955) proposed a simple description of free convection by relating the sensible heat flux to the temperature gradient in the SBL on the assumption that the friction velocity is not a contributing variable. The results of subsequent field studies of the unstable SBL revealed that the relationship between fluxes and gradients closely approached the free convection expectation when the instability reached some readily-attainable level (as will be discussed later). An obvious corollary is that in these situations when u_* is not important, formulating exchange processes on the assumption that u_* remains a contributing variable is misleading, if not fundamentally incorrect.

In the same period as the publication of Priestley's analysis and the development of Monin-Obukhov similarity theory (MOST), a number of mathematicians reminded readers of the problems associated with dimensional analysis. For example, Saint-Guilhem (1962; translation by F. A. Gifford, 1967) concludes that "dimensional analysis is an art full of dangers" due to the often-overlooked requirement that one must first be convinced that the selected variables do indeed contribute. Hicks (1978) expanded on this, by showing that the sharing of variables leads to apparent correlations that are entirely artificial.

MOST expressed the influence of stability in terms of departures of profiles from neutral, so that a dimensionless wind profile element ϕ_m was introduced —

$$\phi_m \equiv (kZ/u_*)\cdot\partial\bar{u}/\partial z \tag{13.8}$$

as was seen earlier. The goal was then to express ϕ_m as a function of stability. For convenience and in accord with MOST, the most common stability index was $\zeta = Z/L$, as in Equation 13.7. In practice, ζ depends on z and on $\overline{w'u'}$ ϕ_m. There is an obvious likelihood of generating a visually convincing plot of ϕ_m versus ζ even if there is no physical association between these two quantities. Relating ϕ_m to ζ is, in essence, relating a measure of the relationship to a symptom of it.

The problem that arises when ϕ_m is considered as a function of ζ is not necessarily avoided if a different stability index is used. If the stability index is Ri, then the wind gradient $\partial\bar{u}/\partial z$ appears on both sides of the relationship. Hence, a plot of ϕ_m against Ri would be expected to have the appearance of a plot of x versus x^{-2} if all the other quantities were held constant. If an analysis of field data results in a best fit that shows that ϕ_m varies according to Ri^{-2}, then a first order conclusion would be that there is no evidence of a significant physical dependence among the variables of main concern. If the analysis makes use of ζ as the stability index, then it is u_* that is shared and a plot indicating that ϕ_m varies with ζ^{-3} would be a similarly discouraging result. In this case, of course, z is also a shared variable, with additional confounding consequences.

In today's literature, this phenomenon is often referred to as the shared variable syndrome. While the problem is well known, it is still sometimes ignored by authors who erroneously perceive a physical behavior when the facts are otherwise (see Andreas and Hicks, 2002; Baas et al. 2006). As an example, consider the case of turbulence, as characterized by the statistical quantity σ_w, the standard deviation of the vertical wind component derived from measurements at a single location through some convenient averaging period, typically in the range 15 to 60 min. At neutral, the association with eddy fluxes is simply expressed, since by definition we can write

$$u_* = (-\overline{w'u'})^{0.5} = (-R_{wu}\cdot\sigma_w\cdot\sigma_u)^{0.5} \tag{13.9}$$

by which the correlation coefficient between vertical and longitudinal fluctuations in the wind is defined: R_{wu}. Accepting the role of the friction velocity as the appropriate scaling parameter, we can then consider the association of the normalized turbulence property σ_w/u_* with some other property deemed to be influential, such as atmospheric stability. Following the dogma of the era, researchers in the 1960s and 1970s confidently plotted σ_w/u_* against ζ and reported that the relationship confirmed the predictions of dimensional analysis that

$$\sigma_w/u_* = A|\zeta|^{1/3} \tag{13.10}$$

with the constant A being about 1.3. The same basic dependence was found for the other velocity components, but with different values for the constant A. The values of A are as indicated in Figure 13.1, which is a modification of diagram frequently shown, representing the results of an analysis of 'data' collected during a fictitious study conducted at Conargo, New South Wales. Conargo is a place that actually exists, in southern New South Wales, but at the time of the spate of Australian studies of the 1960s the reference is mainly to recognize the Conargo pub, a convenient but isolated watering hole at the intersection of two highways (at the time, each no more than two lanes of poorly maintained macadam). The data ascribed to the fictitious Conargo experiment are random numbers arranged to have the same ranges as had been reported in the open literature.

The curves shown in Figure 13.1 have been published in the open literature. The agreement with the 'data' plotted indicates that these curves can be reproduced when there is no physical association among any of the contributing variables. The issue has now been acknowledged by the community at large. In the interim, it has been a satisfying pastime to randomize published data and to reproduce the authors' interpretations using the modified dataset.

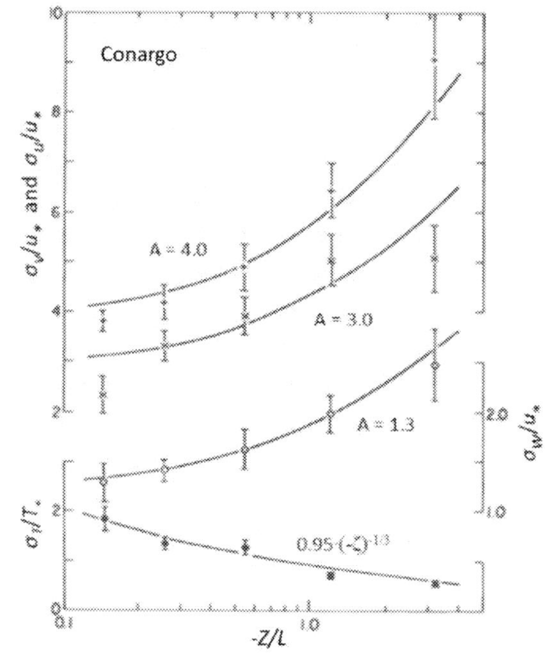

Figure 13.1. Results of analyses of the 'Conargo' data. Averages and standard errors are shown. The curves drawn represent the relationships accepted at the time of this work (the early 1970s). The basic "data" are random numbers arranged to have the same ranges as in published datasets. An early version of this diagram was published by Hicks (1981).

> When one expert published results that appeared vulnerable to the shared variable syndrome, a set of his data after randomization was provided to him along with an analysis that showed that his conclusions were solely due to the method of analysis. His response was memorable. "Thank you for your data. I will use them in my next report."

It remains to address a basic question: Do the velocity fluctuation statistics vary with height and/or stability? To clarify the discussion, Figure 13.2 presents results derived from a 2011 experiment in Ocotillo, Texas (see Hicks et al. 2014), in a manner not influenced by the shared variable syndrome. First, the results are plotted against a stability index based on gradients alone, not fluxes — the Richardson number (*Ri*) quantified using data from 3 m and 21 m. Values of the velocity statistics σ_u, σ_v and σ_w were measured at several intermediate levels. For illustration here, results from 3 m, 8.5 m, 14 m and 21 m are shown. Temperature statistics are also shown. For each 15-min run, data from each level of measurement have been averaged and these averages have been used to normalize the raw observations (so that the results can be shown most informatively). The resulting dataset was then ordered according to *Ri* and sorted into sequential groups, each containing 50 records, from which the averages now plotted were computed. Plotting against *Ri* compresses the stable stratification results considerably and ready interpretation of the stable data is not easy. However, it is clear that the ordering evident for the unstable case does not extend into the stable.

Conventional wisdom decrees an ordered increase in σ_u/u_*, σ_v/u_* and σ_w/u_* with increasing instability. Assuming that u_* is constant with height, it is therefore commonly assumed that all of σ_u, σ_v and σ_w increase with height. In contrast, Fig 13.2 shows that for σ_v any increase with height is trivial. For σ_u the increase is minor in comparison to that evident for σ_w. In standard micrometeorologically-acceptable situations (good fetch and stationary conditions), depictions like those identified in Figure 13.1 are therefore entirely a consequence of the way the variables are constructed and have little practical relevance. The case of σ_T is in accord with expectations, decreasing with height. In the real world of more complicated surroundings, consistent height variations of the turbulence quantities might well be real, due to changes in the surface upwind. Moreover, modern applications recognize that the ratios in question, as considered in Figure 13.1, depend on the depth of the atmospheric boundary (mixed) layer (h_m) in daytime (see Hanna, 1982; Hicks, 1985; Ryall and Maryon, 1997). The method used to normalize the ratios displayed in Figure 13.2 is intended to bypass such PBL considerations; the data shown do not preclude a dependence involving h_m. The stable case requires additional consideration, as will become evident in discussion to follow.

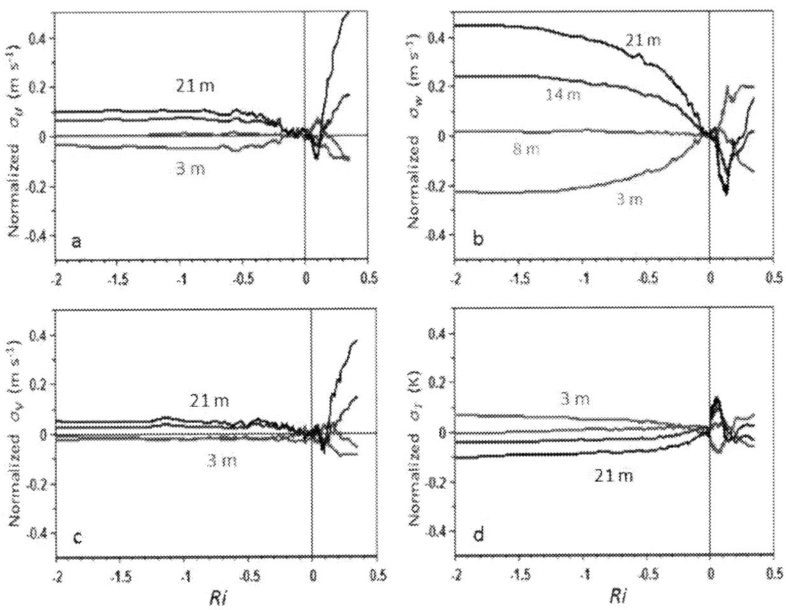

Figure 13.2. The change of turbulence with height and with stability, expressed as a function of the Richardson number Ri so as to avoid the shared variable syndrome when plotting against ζ. Heights of measurement are as shown in (b). (Based on Figure 10 of Hicks et al. 2015).

Conclusion

It is surely time to rethink.

Chapter 14

Time Stationarity in the SBL

"The things that you're li'ble to read in the Bible, they ain't necessarily so."
— Ira Gershwin.

Figure 14.1a is an example of how $\overline{w'T'}$ varies through the day, plotted using the time of sunrise as the origin and averaging data obtained over many months of study at the Ocotillo site. The plots are for several heights of measurement, with the uppermost at noon representing the 27 m height, the lowest 3 m and data from other levels falling in sequence. Plotted in Figure 14.1b are the average values of the rates of change of temperature with time, as indicated by independent platinum resistance thermometers. Several aspects of Figure 14.1b cause micrometeorological concern. First, the concept of time stationarity is clearly questionable, especially during the hours first following dawn, when the rate of change of air temperature with time averages nearly 2 K hr^{-1}, with particular days varying substantially from this average. Later in the day, the rate of cooling is slightly less than the morning rate of heating.

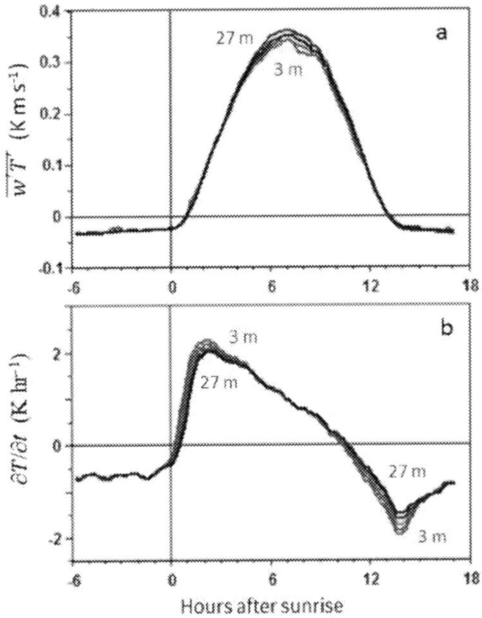

Figure 14.1. Averages derived from measurements at several heights, during the Ocotillo (Texas) experiment of 2011. The plots are of data collected at heights of 3, 8.5, 17.2, 21.0 and 27.4 m, with the highest and lowest as indicated. The curves between these extremes fall in order of the measurement height. (a) Average values of $\overline{w'T'}$ $\partial \bar{T}/\partial t$.

Micrometeorological convention decrees that its relationships apply in 'constant flux' conditions, when the fluxes are invariant with height. In that case, the plots shown in Figure

14.1a for different heights should overlie each other. The data suggest a variation if the sensible heat flux with height at midday, which is counter-intuitive and could indicate a fetch or instrumentation limitation. At this same time (about solar noon, about six hours after sunrise), the time rate of change of temperature evident in Figure 14.1b is constant with height and is trending towards zero as time progresses. During the six (or so) hours centered on solar noon and during the night, the air mass heats or cools as a single entity, up to the greatest height of measurement: 27 m.

Heating rates during the dawn and dusk periods are shown in more detail in Figure 14.2, where the heating is seen to continue uniformly following sunrise, starting nearest the ground and progressing upwards. The heating rate stabilizes (becomes constant with height) about two hours after it maximizes. Late in the afternoon the cooling anticipates sunset, starting nearest the ground. Likewise, after minimizing shortly after sunset the cooling rate stabilizes after a further one to two hours.

During the morning and late afternoon hours, the different instruments yield heating (or cooling) rates that vary with height. As expected, the lowest level always responds sooner — the source or sink of heat is the surface and the air closest to the surface responds to changes in the surface more quickly than air further above it. This feature is shown with more clarity in Figure 14.2, where it is seen that the lowest levels respond quickly to the input of radiative heat energy at sunrise, with the higher levels lagging behind and in height sequence.

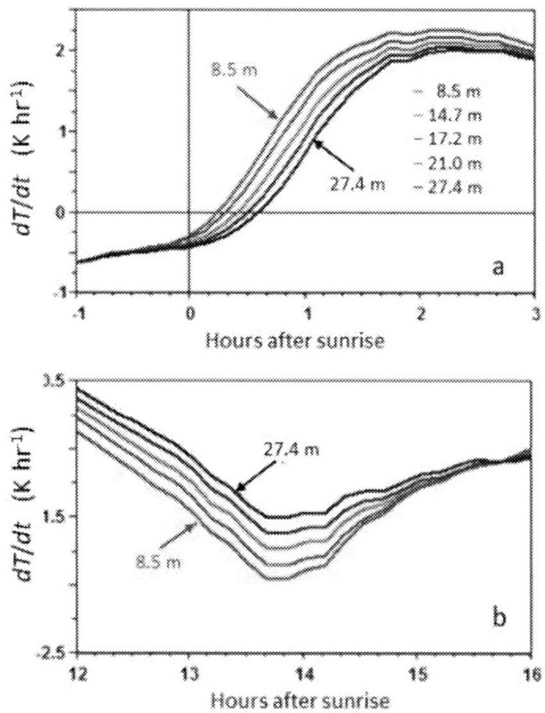

Figure 14.2. The rate of heating of air at several different levels, as a function of time through the transition periods of dawn and dusk. Ocotillo data are used (based on Hicks et al. 2014b).

The smoothness of the curves plotted in Figures 14.1 and 14.2 is a direct result of (a) the use of a very long data record and (b) the analytical method employed. The initial seven-month data record was of 15-min data, which were averaged after sorting according to the time after sunrise. The lines are drawn to connect averages based on sequential ensembles containing 100 data points. Sunrise times were derived using www.esrl.noaa.gov/gmd/grad/solcalc/sunrise.html.

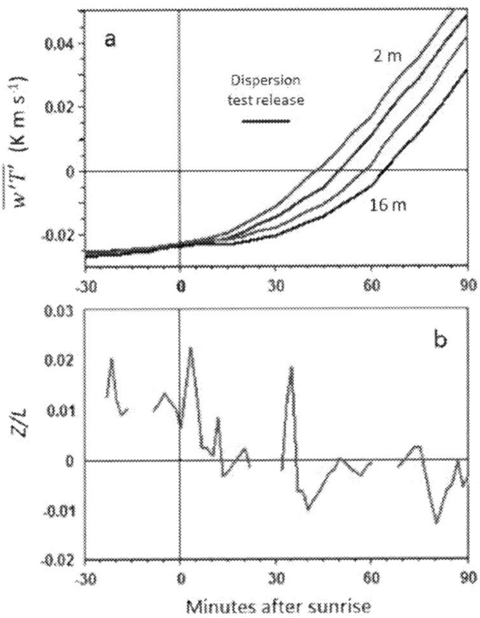

Figure 14.3. Results from the Jack Rabbit dispersion study conducted at the Dugway Proving Ground in Utah, in 2010 (Hanna et al. 2012, Hicks et al. 2013). In this case, data have been analyzed using a 5-min sampling period, so that the change after local sunrise can be seen better. The upper panel (a) shows trends in $\overline{w'T'}$

(b) shows the variations in stability indicated by the 2–m data. In (b), gaps in the record indicate periods of low u_* such that $\overline{w'u'}$ ^{2} s^{-2}. The horizontal line in (a) indicates the period during which a dispersion experiment was conducted.

The results summarized in Figures 14.1 and 14.2 are not unusual. Figure 14.3 illustrates similar behavior derived from multi-tower micrometeorological data collected as a component of a dispersion study conducted over the arid and sparsely vegetated land of Dugway Proving Ground in Utah (Hanna et al. 2012). This site is part of the now-dry lakebed of the ancient Lake Bonneville. Data are arranged to show how the heat fluxes changed with height during the period following dawn. In this case, results from three towers have been combined after covariance analysis using 5-min averaging times. A short averaging time has been used to yield finer time resolution near sunrise. Overall, the Utah observations confirm the Ocotillo results depicted in Figures 14.1 and 14.2. Sunrise is soon followed by initiation of convective sensible heat flux, first observed at the lowest levels and proceeding upwards with notable consistency. At any particular time during this warming period sensible heat fluxes decrease with height. At the lowest level of measurement (2 m in the Utah example of Figure 14.3a), heating is first detected about 10 min after sunrise. For a considerable time after sunrise, the assumption of

constancy of $\overline{w'T'}$
micrometeorological relationships is certainly questionable while this heating of the lowest layers of the atmosphere takes place.

The horizontal bar evident in Figure 14.3a indicates the time period during which a plume dispersion experiment took place (Fox and Storwold, 2011). The requirement for this particular dispersion study was for conditions most amenable to simulation, so that characteristics specific to the chemical employed (chlorine) could be investigated. The simplest atmospheric situation to simulate is that near neutral. A period of neutrality would obviously be rare and in practice there was indeed a large variation of stability during and following the period of the chlorine release. Figure 14.3b shows how atmospheric stability at 2-m height changed during this particular early morning period. While it has been a standard practice of dispersion experimenters to conduct studies in the post-dawn calm, when sensible heat fluxes are low, examination of the stability regime as indicated in Fig 14.3b reveals that the post-dawn period is among the most variable times of the day.

All of the datasets now available display the same features. During the few hours around the transition periods, especially following sunrise, conditions are not such that legacy micrometeorological constraints would always be satisfied. Fluxes of heat are not constant with height, nor are conditions stationary in time. It may or may not be that the departures are important. For the present, it is sufficient to note the conceptual difficulties involved and to accept that these matters need to be considered in the analysis of data and in the application of whatever formulations might result.

Conclusion

Careful! The atmosphere awakens slowly.

Chapter 15

Extending the Legacy Relationships

"For a good inferno, the more logs the better" — Noel E. Ridere

It is known that scaling the wind profile near the surface involves the logarithm of the height. It is also well known that atmospheric stability plays a role and that atmospheric stability changes with height. The MOST micrometeorological flux/gradient relationships provide an attractive solution to what is otherwise a confounding problem.

All known intensive micrometeorological studies of the relationship between fluxes and gradients in the SBL have largely confirmed the relationships derived in the first studies of such kind, although with minor deviations. As summarized by Dyer (1974) and as already addressed in Part I, these equations are, for unstable stratification,

$$\phi_H = (1 - 16\zeta)^{-0.5} \tag{15.1}$$

$$\phi_m = (1 - 16\zeta)^{-0.25} \tag{15.2}$$

and for stable conditions

$$\phi_H = \phi_m = 1 + 6\zeta. \tag{15.3}$$

There is uncertainty about the values (16 and 6) in these equations. This matter will not be discussed here, because examination of available new data (as will follow) indicates that there is considerable site-to-site variability and moreover that even the imposition of unity at neutral is questionable although probably valid on the average.

Now, many decades later, sonic anemometers allow the relationships to be addressed in more detail. Figure 8.2 showed results from the NOAA Ocotillo study in Texas (White et al. 2014). As has been emphasized in formal presentations of these data (Hicks et al. 2014b; Hicks et al. 2015b), the site was far from perfect. The agreement with conventional expectations evident in Figure 8.2 is therefore gratifying, since it implies a broader applicability of the early 'perfect site but imperfect instrumentation' results to an imperfect site with much better instrumentation. Note, however, that the nighttime data show some evidence of dropping below the log-linear expectations at high stability.

Figure 15.1 shows results from a similar analysis of data from a 61 m tower located on the reservation of Idaho National Laboratory (INL). Data from 4, 8 and 16 m are used here, to replicate the procedures of earlier micrometeorological studies. The consequences of site complexity are obvious in the differences between the averages derived for the two different height intervals: 4 – 8 m and 8 – 16 m. These differences amplify as the heights of consideration increase – a direct indication of site limitations such as will be discussed later. The lines represent the expectations, corresponding to Equations 15.1 and 15.2 for unstable conditions and to Equation 15.3 for stable; the results do not differ greatly from those of the Ocotillo study, nor from legacy expectations provided $|\zeta| < 1$. For unstable stratification, agreement between field observations and legacy expectations is common for the range of stabilities originally prescribed: $|\zeta| < 1$. However, the way in which the plotted variables are constructed generates insecurity about how well the contributing physics is described and consequently about the validity of its extrapolation.

Figure 15.1 indicates that as stability increases beyond the normal constraints (i.e., for $|\zeta| > 1$) ϕ_m tends to revert to near-neutral in the unstable case (Figure 15.1a) and that both ϕ_m and ϕ_H tend to asymptote to a value in the range of 7 to 10 in exceedingly strong stability (Figure 15.1b). This is not a surprise; similar conclusions have been reached elsewhere.

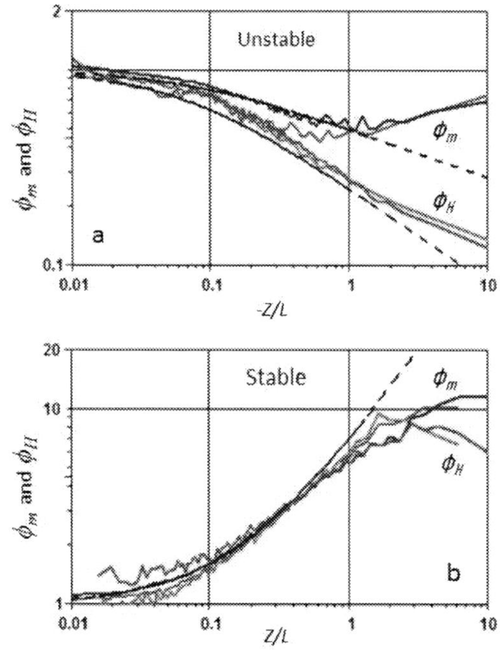

Figure 15.1. The dependence of the dimensionless gradients on stability, using INL data from 4, 8 and 16 m, showing the departure from classical expectations as $|z/L|$ exceeds unity. The dashed lines represent the legacy relationships, Equations 15.1, 15.2 and 15.3.

Figure 15.2 is an expanded view of the upper right quadrants of Figure 8.1 (recomputed so as to identify and exclude several outliers) but now with linear axes and showing every fiftieth point after ordering, not averages. The scatter in ϕ_m appears greater when viewed linearly and the scatter in ϕ_H even more so. Figure 15.2 indicates more obvious support for the assumption of a straight-line description for moderately stable conditions, say $\zeta < 0.5$. The line drawn corresponds to $\phi_{m,H} = 1 + 6\zeta$, as before. There is obvious curvature as stability increases, largely in line with the INL results shown in Figure 15.1. It is commonly assumed that in conditions of stable stratification the dimensionless gradients for heat and momentum depend on stability in the same way, buoyancy contributing mainly as a small restoring force due to the change in air temperature associated with a vertical displacement. The consequences should appear in comparisons of ϕ_H and ϕ_m, but are typically obscured by scatter introduced in analysis by the requisite algebraic manipulation of very small values.

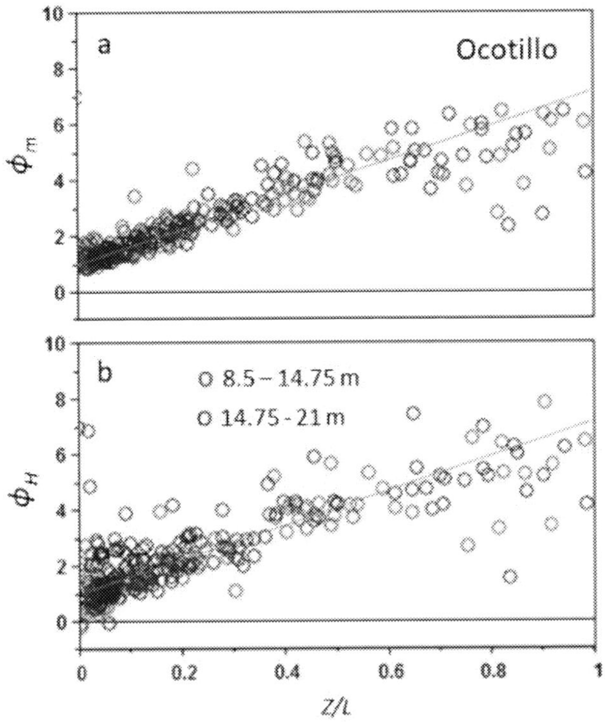

Figure 15.2. Another view of the variation with stability of the dimensionless wind gradient ϕ_m and temperature gradient ϕ_H as indicated by Ocotillo nighttime data (2200 to 0400 hrs). For this presentation, original data have been ordered according to $\zeta = Z/L$ and then every fiftieth point is plotted. In this way, the scatter of the raw data is more evident. (Based on Figure 7 of Hicks et al. 2014).

In unstable conditions, the repeated verification of the Dyer relationships provides an outwardly satisfying solution to the dominant problem of relating fluxes and gradients. But the essence of the usual question posed is to describe the fluxes in terms of gradients and the non-dimensionalizing procedure of MOST; this obscures any physical association that might be dominant. A plot of ϕ_m against ζ displays the association between two symptoms rather than the linkage between causes and consequences.

Conclusion

The *appearance* of veracity can be more impelling than veracity itself.

Chapter 16

Correlations and the Role of u_*

"Instead of thinking about u-star barred, we should be considering u-bar starred"
— W. C. Swinbank

As was discussed in Chapter 9, the familiar flux-gradient relationships arise from relating non-dimensionalized gradients to stability, imposing the specter of the shared variable syndrome. The development of these relationships followed from an initial assumption that the neutral wind profile provided a basis for the extension to situations that are not neutral. The key assumption is that the fluxes are related closely to the associated gradients. Now that measurements of the fluxes are possible, using eddy covariance, this assumption can be tested.

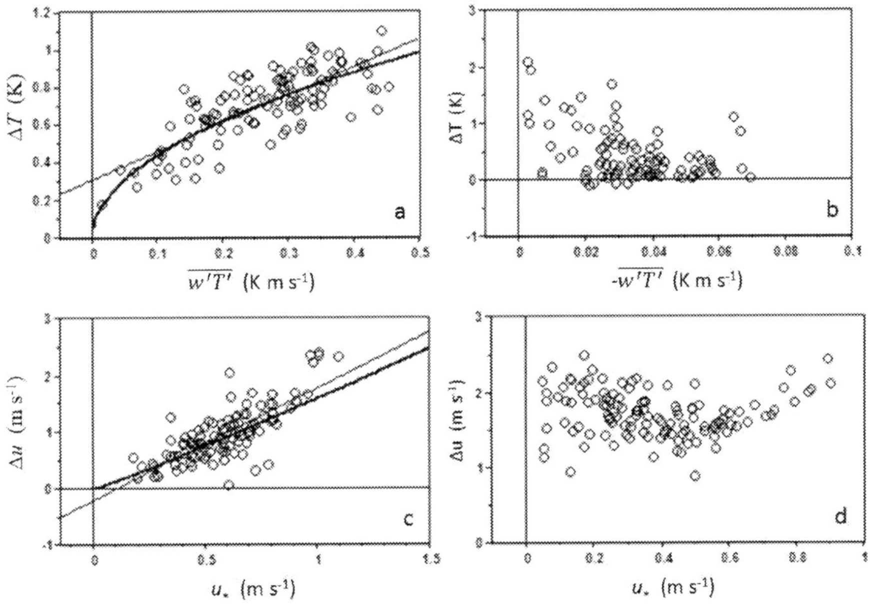

Figure 16.1. Using data from the Ocotillo (Texas) study, plots showing the relationships between eddy fluxes and the corresponding difference in state variables. Diagrams (a) and (b) address the case of the heat flux and temperature gradient. The temperature difference between 8.5 m and 21 m is used as a measure of the relevant gradient. The lower pair of diagrams (c) and (d) address the $\Delta\bar{u}$ and u_* association. Every hundredth point is plotted, after randomized ordering.

Figure 16.1 addresses the relationship between sensible heat fluxes (plotted as $\overline{w'T'}$ and temperature differences ($\Delta\bar{T}_a$) derived from measurements made during the Ocotillo campaign, at 8.5 m and 21 m. As previously, the original 15-minute data have been averaged into 100–run groups, after sequential ordering according to $\overline{w'T'}$
and (b) and to u_* for the lower pair. The diagrams to the left are for daytime conditions (1000

to 1600 local time), to the right for nighttime (2200 to 0400). Every hundredth data point is plotted, selected from the entire set of sequential 15–min averages. The lines drawn are the result of regressions based on the entire dataset, not just confined to the points plotted. The temperature differences shown here have not been converted to potential temperatures and hence the temperature differences do not converge to zero when $\overline{w'T'}$
zero offset corresponds to the conventional correction for the slope of the dry adiabat: 9.77 K km^{-1}).

The unstable cases of Figure 16.1 appear well ordered. In each of the unstable plots (16.1a and 16.1c), two regression lines are shown: a linear and a power law. In the case of the temperature data (Figure 16.1a) the preferred result is the power law, yielding an exponent of 0.51 (i.e., $\Delta\bar{T}_a = \hat{A}(\overline{w'T'})^{0.51}$ where $\hat{A}$ is a constant). However, this is not the case for the stable data. Figures 16.1b and 16.1d show little basis for accepting the hypothesis that the fluxes and gradients are closely associated at night. Hence, for the nighttime case there is no strong experimental basis for normalizing $\partial\bar{T}_a/\partial z$ by $\overline{w'T'}$, nor for normalizing $\partial\bar{u}/\partial z$ by u_*, except perhaps as u_* increases beyond about 0.5 m s^{-1} when the linear relationship accepted for near-neutral conditions is approached.

Figure 16.2 illustrates the correlations between the variables contributing to the eddy fluxes. Two distinct bands of data are shown, corresponding to R_{wu} and R_{wT}, the correlation coefficients associated with the transfers of momentum and heat, respectively. In the case of R_{wu}, there is little evidence that the correlation coefficients change with height. However, this is not the case for the R_{wT} data for the midday period. In addition, there appears to be a change in the turbulence regime occurring about two hours after sunrise, i.e., about one hour after the SBL switches from stable to unstable. Afterwards, and for most of the convective period, the correlation coefficients increase consistently with increasing height. It is of relevance to note that the switch from height-independent R_{wT} is at about the same time as a deviation is seen in the trend of R_{wu}, as indicated by the circled data in the diagram. Reference to Figure 14.2b shows that it is then that the rate of change with time of air temperature maximizes.

In passing, note that the correlation coefficients do not vary greatly through the evening hours and that the *w* and *u* fluctuations are more strongly coupled than the *w* and *T*. This was also evident in the simultaneity plots of Figure 12.1; it corresponds to an interpretation already presented and which will arise again — that in stable conditions the heat flux is derivative on the exchange of momentum, with minor modification due to the restraining effects of buoyancy. This is what conventional understanding would anticipate. There is no reason to modify convention, in this regard, but the verification using modern data is indeed satisfying.

The magnitudes of the various correlation coefficients are revealing. At night, only about 10% of the turbulent fluctuation variance is associated with the actual fluxes. (The variance attributed is quantified by the square of the correlation coefficient plotted.) In daytime, the association is stronger, peaking at 20% to 25% but varying with height. This and other general features illustrated here appear to be common aspects of data from terrestrial sites not overly subject to terrain complexity. This indicates a feature of the turbulence regime that compounds the experimental challenge in relating fluxes to gradients. It is only a small part of the overall field of turbulence that can be explained in terms of the fluxes. This result relates directly to the reliance in some mesoscale models on TKE as a route to extract meaningful values of the friction velocity. The matter will be explored further, below.

The circle in Figure 16.2 indicates a period during which the values of $|R_{wu}|$ decreased before subsequently returning to their earlier value. In discussion to follow, it will become apparent that the cause could be associated with onset of local katabatic flow.

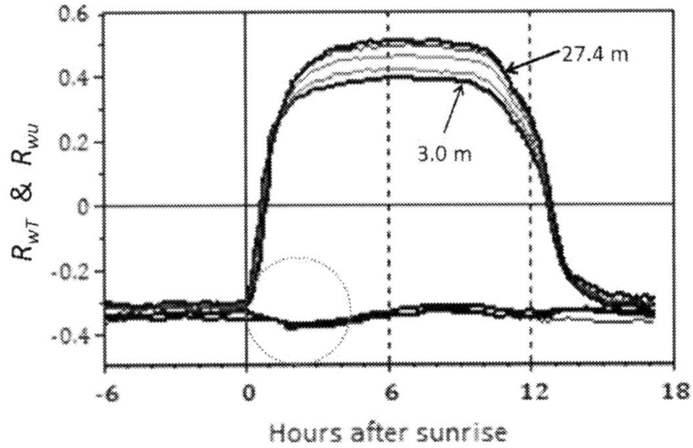

Figure 16.2. Average daily cycles of the correlation coefficients R_{wu} and R_{wT}, the first associated with the eddy flux of momentum, the second with the flux of sensible heat. Data are from the Ocotillo (Texas) study, using sonic anemometer measurements made at heights of 3.0, 8,5, 14.7, 21.0 and 27.4 m. 15-min evaluations of the correlation coefficients are used. The plots are of averages constructed in ensembles sorted according to time after sunrise. The heavy lines identify the lowest and greatest heights of measurement. Curves for other height fall between, in order.

Figure 16.3 shows details of the R_{wT} and R_{wu} behavior through dusk and dawn, in this case expressing the evening transition in terms of time after sunset. Consistent height variability is seen in both diagrams. In the morning case, R_{wT} decreases with height during the hours immediately following dawn, but switches to increasing with height after about an hour following the transfer from stable to unstable (when R_{wT} changes sign). This is an additional indicator of a fundamental change of some kind. It is at this same time that R_{uw} deviates from its nighttime baseline value. A similar (although much smaller) departure is evident at about the time that R_{wT} changes sign before sunset.

The role of u_* in formulations describing the heat flux is explored further in Figure 16.4, which builds upon the examinations of Ocotillo data illustrated in Figures 16.2 and 16.3. To avoid any contamination of the results due to a shared role of u_*, the plots are now against Ri (based on 8.5 m and 21 m data) instead of ζ. In this case the data have been ordered (according to Ri) and every fiftieth data point has then been plotted. The scatter is now revealed, but nevertheless several aspects of the overall behavior are obvious. In Figure 16.4a, the rapid decrease of R_{wT} in unstable conditions as neutral is approached is accompanied by a slow increase in $|R_{wu}|$, indicating that the dominance of the convective exchange process extends to quite near neutral. When shown on an expanded Ri scale (as in Figure 16.4b) it is seen that the inflection point is in the vicinity of $Ri = -0.03$, much as expected on considerations already discussed. The stable dataset of these two diagrams suggests that the rapid change of R_{wT} terminates when Ri approaches $Ri = 0.03$ (see Figure 16.4b), however as stability increases both correlations rapidly approach zero when $Ri \approx 0.5$ (in Figure 16.4a).

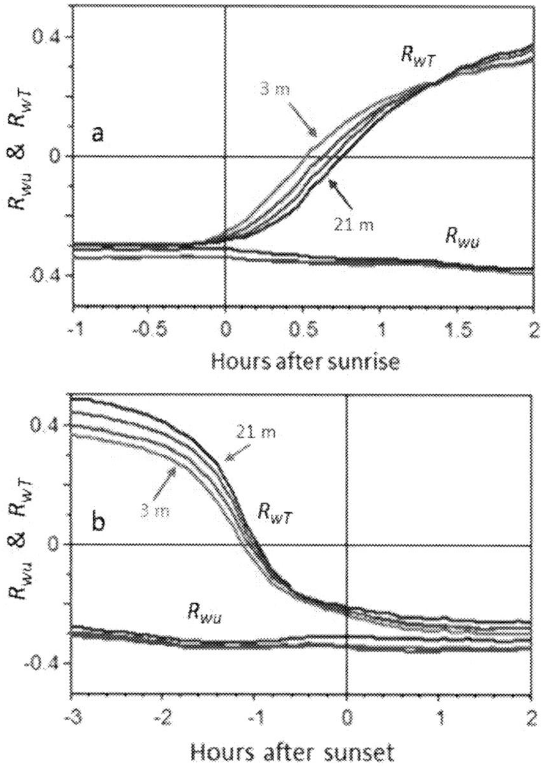

Figure 16.3. As in Figure 16.2, but with expanded time scale to show details through the morning and evening transition periods. The 27 m results are not shown — they often overlie the 21 m. (Based on Figure 4 of Hicks et al. 2015).

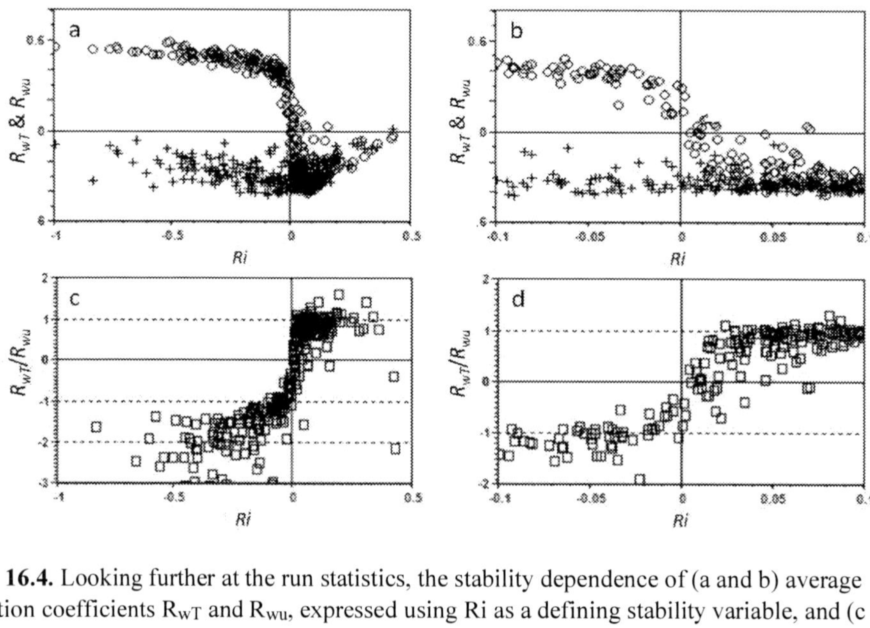

Figure 16.4. Looking further at the run statistics, the stability dependence of (a and b) average correlation coefficients R_{wT} and R_{wu}, expressed using Ri as a defining stability variable, and (c and d) of the ratios R_{wT}/R_{wu}. To avoid any influence of the shared variable syndrome, stability is quantified as Ri, determined using 8.5 and 21 m data.

The same data are used in Figure 16.4c and 16.4d to show how the two turbulence regimes are associated — the diagrams depict the variation of the average ratios R_{wT}/R_{wu}. The extent to which the absolute value of this ratio R_{wT}/R_{wu} grows to eventually exceed unity in unstable situations is highly informative: it enforces previous indications that increasing convection (with a flux of heat from the surface) moderates the resulting flux of momentum (to the surface). On the other hand, even though the correlations R_{wT} and R_{wu} both decrease as conditions become more stable (the right-hand side of Figure 16.4a), the results in Figures 16.4c and 16.4d indicate that the two fluxes vary hand-in-hand in stable conditions such that Ri > about 0.03, outliers excepted.

Figure 16.5 shows a selection of results derived from a partial correlation analysis. Simultaneous 15-min vertically averaged values of $\overline{w'u'}$ and $\overline{w'T'}$ covariances measured at all heights from 8.5 m to 21 m) and the bulk differences $\Delta\bar{u}$ and $\Delta\bar{T}_a$ (defined as between 8.5 m and 21 m) are used. (Close inspection of the data indicate some discrepancies in the 3-m and 27.4-m observations; these measurements have been omitted. The focus is on the role of u_*. To this end, raw data have (as before) been sorted according to stability (ζ) and sequential groups of fifty cases constructed. Correlation coefficients and partial correlation coefficients have then been computed within each group and the results plotted against the corresponding average stability. In the figure, the correlation coefficients are identified in terms of the relevant micrometeorological variables: $\overline{w'u'}$, $\overline{w'T'}$ $\partial\bar{u}/\partial z$ and $\partial\bar{T}_a/\partial z$. The results shown in Figure 16.5a indicate that in unstable conditions for $\zeta < -0.1$ there is negligible remaining correlation between u_* and $\partial\bar{u}/\partial z$ after the effects of $\partial\bar{T}_a/\partial z$ are accounted for. In Figure 16.5c the situation is clarified, with the same range of instability yielding a consistent relationship between $\partial\bar{u}/\partial z$ and $\partial\bar{T}_a/\partial z$ such that the correlation coefficient between them is in the range -0.3 to -0.4. Extending this discussion to the heat flux case, Figure 16.5d shows that the association between the two gradients is about the same ($R_{12.3} \approx -0.3$) when the heat flux is held constant. However, in Figure 16.5b it is clear that the correlation between $\partial\bar{u}/\partial z$ and $\partial\bar{T}_a/\partial z$ is unaffected by variations in $\overline{w'T'}$ provided $\zeta < -0.05$ (or close). As instability increases, the correlation coefficient between $\overline{w'T'}$ and $\partial\bar{T}_a/\partial z$ decreases uniformly, with no apparent complication corresponding to a contribution of $\partial\bar{u}/\partial z$.

It appears that u_* is not a contributing variable in considerations of the dependence of $\overline{w'T'}$ on $\partial\bar{T}_a/\partial z$ in unstable conditions except for instability very close to neutral. Figure 16.5b reveals that there is the expected strong correlation between $\overline{w'T'}$ $\partial\bar{T}_a/\partial z$ over almost the entire range of instability, but that there is no change after the effects of $\partial\bar{u}/\partial z$ are accounted for. That is, in conditions more unstable than $\zeta \approx -0.05$, $\partial\bar{u}/\partial z$ (like u_*) is not a variable that influences the heat flux or the temperature gradient. The consistent decrease in the magnitude of R_{wT} as instability increases corresponds well to the observed characteristics of the turbulence regime. As instability increases, fluxes are increasingly associated with bursts of turbulence with decreasing frequency. Since $\overline{w'T'}$ represents an average dominated by the times during these bursts, while $\partial\bar{T}_a/\partial z$ is influenced by the time between them, the correlation between these two quantities mist be expected to diminish with increasing instability. This does not negate the dominance of free convection as the driving process.

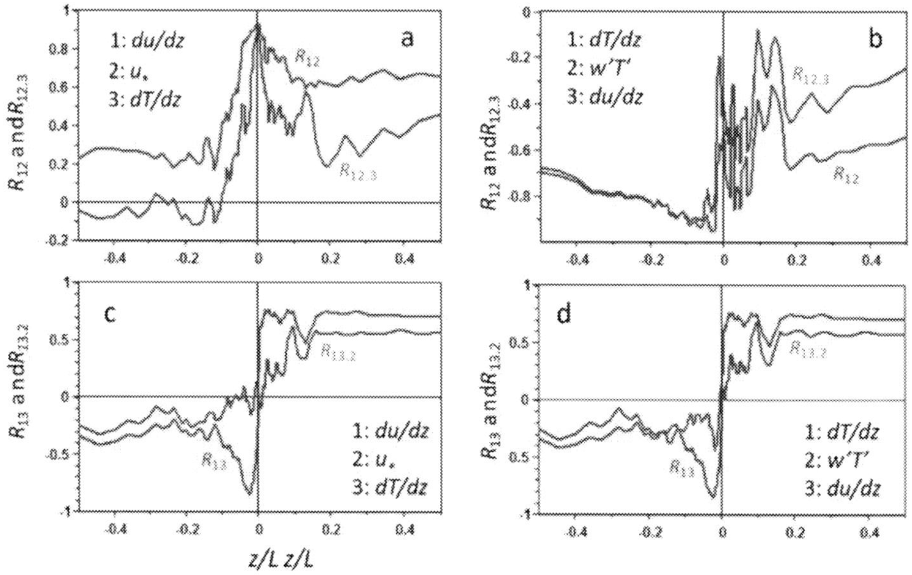

Figure 16.5. The variations of selected correlation coefficients with stability, from the Ocotillo data. To the left, the intent is to test the relationship between u_* and the gradients of wind speed and temperature. To the right, the test is of the $\overline{w'T'}$
total correlation coefficients, R_{12} and the partial, $R_{12.3}$.

The peculiarity of the near-neutral situation is evidenced in Figure 16.5a, where confidence in the hypothesis that u_* is a central variable decreases rapidly with increasing instability. The change in $R_{12.3}$ indicates that for $\zeta < -0.1$ there is no statistical basis for associating $\partial\bar{u}/\partial z$ with u_*. (The specification of the limiting $\zeta \approx -0.1$ is poor. Inspection of the diagrams shows that values of –0.05 or –0.03 could be justified and would be more compatible with the original analyses summarized by Priestley, 1959.) This is in complete contrast to a common modeling rationale — that the key to describing the SBL is to derive u_*, after which all else follows.

On the stable side, a major feature is that $\partial\bar{T}/\partial z$ and $\partial\bar{u}/\partial z$ appear to be strongly correlated through most of the stable regime. There is no difference between the plots of the total correlation coefficient in Figures 16.4c and 16.4d, but the departures of the $R_{12.3}$ values from the R_{12} indicate that there is little reason to relate the wind and temperature gradients near neutral. In both (c) and (d) cases, smoothed curves pass through the origin.

Figures 16.4 and 16.5 lead to a consistent conclusion. The analysis so far confirms that for most daytime situations the formulation of the relationship between the sensible heat flux and the temperature gradient does not require inclusion of u_* as a contributing variable. The transfer of heat from the surface into the air is governed by the temperature regimes of the surface and the air; within the airstream it does not matter how quickly the air is moving across the surface. Convective eddies are carried with the air. Figure 16.6 provides clarification, showing the direct association between the dominating eddy fluxes $\overline{w'T'}$ and $\overline{w'u'}$
and stability (b). In Figure 16.6a, data have been sorted according to time of day after sunrise. Data analyzed separately from measurements made at all five heights of the Ocotillo tower are shown (3.0, 8.5, 14.7, 21.0, 27.4 m). Results from the different levels of measurement track

well, except for the 3-m case which is suspected to be affected by local terrain slope (as was noted above).

In Figure 16.6a it is seen that the flux-to-flux association at night tends to be stronger than in the daytime. This feature becomes even more striking in Figure 16.6b, where the dependence on stability is shown. In Figure 16.6b, data from all of the measurement heights illustrated in Figure 16.6a have been combined and sorted by stability (ζ), before grouping into ensembles of fifty 15-min periods. The grouping does not permit illustration of behavior in extremely unstable conditions because there are relatively few such observations. However, the message of Figure 16.6b is clear. It is only in a narrow band of stabilities defining the transition zone near neutral that the covariances $\overline{w'T'}$ and $\overline{w'u'}$

Outside this band, in stable conditions the momentum flux controls. In unstable conditions it is the heat flux.

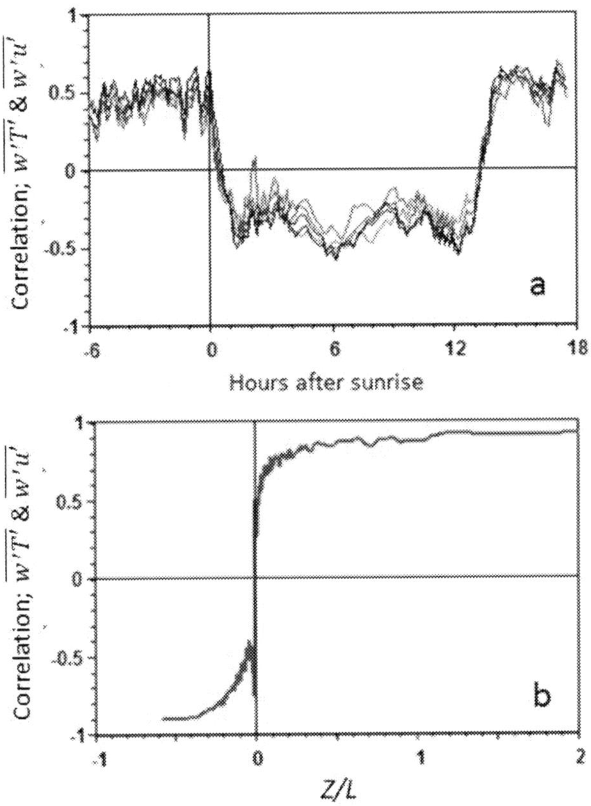

Figure 16.6. The ensemble variation of the correlation coefficient between the fluxes of heat and momentum as a function of (a) time of day and (b) atmospheric stability. In (a), relevant heights of measurement vary from 3 m to 21 m, with the lowest height yielding the least significant correlation.

A note of caution arises. The two eddy fluxes involved in the discussion above were derived from the same instruments — sonic anemometers. The gradients of wind speed were also derived from these anemometers and so there is some commonality that could introduce artificial and unwarranted correlations (maybe positive and maybe negative). However, the temperature gradients were measured independently by aspirated platinum resistance thermometers. Hence, there is a lesser opportunity for measurement 'cross-talk' in consideration of $\overline{w'T'}$ $\partial \bar{T}_a/\partial z$ than for the case of $\overline{w'u'}$ $\partial \bar{u}/\partial z$.

The major conclusion from the discussion above is that the range of justified applicability of the MOST flux-gradient relationships is small. Within this range, the legacy flux-gradient relationships certainly provide a meaningful and well-founded method for transitioning from one stability regime to another, through the neutral transition. Figure 16.7 emphasizes this conclusion. Here, data from an Alabama field study of 2014 are used to quantify the proportion of variance of the observed temperature difference accounted for by changes in $\overline{w'T'}$ and $\overline{w'u'}$ Also shown is the corresponding proportion of the variance of the wind speed gradient. The values of this total variance attribution are calculated as

$$R_{1.23}^2 = R_{12.3}^2(1 - R_{13}^2) + R_{13}^2 \qquad (16.1)$$

where $R_{12.3}$ is the familiar partial correlation between variables 1 and 2 when variable 3 is held constant. Clearly, the total amount of variance explained in variable 1 due to the influence of variables 2 and 3 ($R_{1.23}^2$) must be the same as that derived when variables 2 and 3 are switched, i.e., $R_{1.23}^2 = R_{1.32}^2$, a requirement that provides a convenient test of corresponding data analyses.

In the unstable regime outside the central transition range (approximately $-0.1 < \zeta < 0.1$), Figure 16.7 shows that the temperature gradient is well accounted for, but not the wind gradient. Within this transition zone, the wind gradient is best described. (The situation for the temperature gradient is indeterminate, because the heat flux is near zero in the vicinity of neutral.) On the stable side, the temperature gradient is not well described by any combination of the prevailing local eddy fluxes (c.f. Figure 16.5).

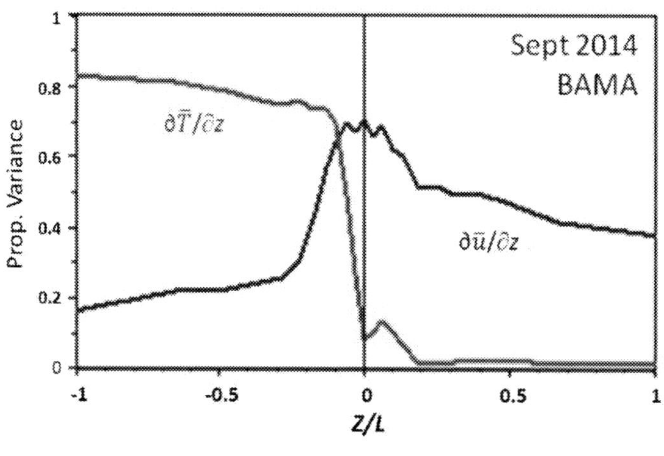

Figure 16.7. The proportion of the variance in measurements of temperature and wind speed gradients explained by the contributing micrometeorological covariances: $\overline{w'T'}$ and $\overline{w'u'}$ evaluations of $R_{1.23}^2$ as in Equation 16.1, with X_1 being either of $\partial \bar{T}_a/\partial z)$ or $\partial \bar{u}/\partial z$ (as identified) and with $X_2 = \overline{w'T'}$ and $X_3 = \overline{w'u'}$. In this case, data are from an experiment over grazed pasture, in Alabama.

Similar behavior has been seen in all datasets that have been examined in the analyses presented here. Figure 16.8 shows results from the Idaho National Laboratory site, on a plain affected by the nearby Rocky Mountains. Figure 16.8a parallels Figure 16.7, but there are considerable differences between the two. There is no obvious explanation of these differences, but note that they share the demonstration that the wind speed gradient is well described near neutral and that the temperature gradient is poorly explained in stable conditions. Both Figures 16.7a and 16.8a indicate that the temperature gradient is best accounted for in unstable situations, but the depiction of results from two different height intervals in Figure 16.8a indicates a height dependence of the association that remains to be explored.

The matter of height dependence is also evident in the width of the near-neutral peak seen in Figure 16.8a. If the same data were plotted against $1/L$ instead of against $\zeta = Z/L$, then the various curves of Figure 16.8a would collapse, as is evident in Figure 16.8b. In this depiction it is evident that the only remaining height dependence is for the case of the unstable temperature gradient.

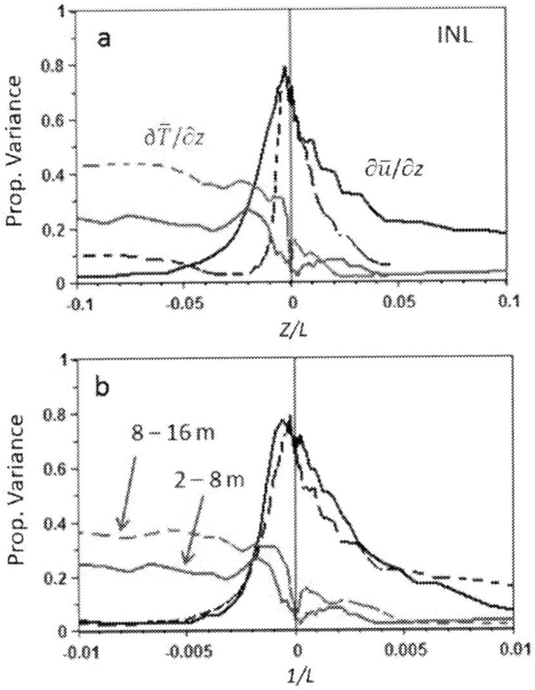

Figure 16.8. (a) As in Figure 16.7, but using data from the INL site in Idaho, for height intervals: 2 – 8 m (solid lines) and 8 – 16 m (dashed). (b) The same data as are plotted in (a), but using 1/L as a stability index rather than ζ= Z/L.

The discussion above leads to conclusions that can be summarized succinctly — when daytime convection is vigorous, buoyancy dominates and momentum is transferred largely as a consequence of it. In the opposite circumstances, when stability is strongest, momentum exchange dominates and heat transfer is then a consequence. It is in the intermediate transitional conditions, near neutral, that the transfer is apportioned between the heat and momentum causative factors and it is then that MOST is most applicable. Definition of the limits of MOST applicability depend on the interpretation of data now available. MOST is certainly applicable over the range $-0.03 < Z/L < 0.03$, but different presentations of the data (as in the figures shown here) indicate that this range of applicability might alternatively be considered as $-0.1 < Z/L < 0.1$.

A purist might object to all of the diagrams shown here because of the nature of the sites. It is acknowledged that each of these sites is subject to the influence of processes outside the range of variables considered in conventional micrometeorological (SBL) field studies, which are invariably and intentionally constrained to avoid the influence of spatial inhomogeneities and temporal changes. In contrast, the present data are from sites beyond the constraints of micrometeorological purism, selected to provide a step towards the real terrestrial world such as must be described in working numerical simulations. However, the present sites are not such that the descriptor 'complex terrain' would seem appropriate. For those with related curiosity, the surroundings of each location can be examined using readily available satellite imagery, using the coordinates listed in Appendix 3.

Conclusion

In daytime, buoyancy dominates; at night, momentum exchange. The role of the friction velocity requires careful consideration.

Chapter 17

The Unstable Case and Free Convection

"Why spend so much time on relating ϕ_H to z/L?"
— a question posed to the present author by C. H. B. Priestley (c. 1964).

It has already been emphasized that the conventional development of flux-gradient relationships near the surface assumed the neutral case as a benchmark and quantified departures from it as functions of stability, either Ri or ζ. Chapter 16 concentrated on how the velocity and temperature variables are correlated, with the eddy fluxes being a consequence of this correlation. A major conclusion resulting from the discussion of Chapter 16 is that for a major part of the diurnal cycle at the sites being studied, u_* should not be considered as a causative variable. However, for unstable conditions, there is evidence that the sensible heat flux conforms to free convection expectations. In early work, Dyer (1965) concluded that 'The formula $H = 24\Delta\bar{\theta}^{3/2}$ based on Priestley's regime with sensible heat H in mW cm^{-2}, and potential temperature difference $\Delta\bar{\theta}$ between 1 and 4 metres in °C provides a practical basis for the evaluation of the heat flux from profile measurements.' Dyer saw no need for u_* to enter the relationship.

Prandtl (1932) and a number of later researchers explored the situation when the friction velocity is not a key component of the convective exchange process. Free convection is described differently by different authorities, but in practice its definition is simple. It occurs when the surface is heated enough that convection dominates vertical turbulent exchange. The familiar method for judging whether the wind speed affects the convection mechanism is to compare the rate of production of turbulent energy by buoyancy to the corresponding quantity associated with surface drag. That is, to compare $(g/\theta)\cdot\overline{w'T'}$ to $u_*^2(\partial\bar{u}/\partial z)$. This is little more than relying on ζ as a determining quantity.

The corresponding physical picture is that buoyant elements are generated at or near the surface and subsequently move with the flow field as these buoyant entities ascend. Priestley (1955) considered the situation of extreme instability, in which convection dominates the transfer processes in the SBL and in which u_* can therefore be eliminated from the list of relevant variables. Dimensional analysis led him to expect a relationship of the form

$$\overline{w'T'} = H^*(g/\theta)^{1/2}(\partial\bar{\theta}/\partial z)^{3/2}Z^2 \qquad (17.1)$$

where H^* is a constant to be determined empirically. Minor manipulation of Equation 17.1 leads to a form more familiar to those with a MOST background:

$$\phi_H = (k/H^*)\cdot(-\zeta)^{-1/3}. \qquad (17.2)$$

Priestley (1959) proposed that Equation 17.1 is a good representation of the SBL for instabilities such that $\zeta < -0.03$, with $H^* \approx 1.3$. This is obviously at odds with the interpretation that is often heard — that free convection is a limit approached as the wind speed tends to zero or ζ becomes highly negative. In fact, the derivation of Equation 17.1 assumes only that u_* can be eliminated from the list of relevant variables.

Back in 1963 and 1964, when Arch Dyer was contemplating the results of the first daytime field studies intended to reveal the dependence of ϕ_H on ζ, there was an acceptance that ϕ_H must approach unity as stability trends towards neutral but also that the processes being studied are fundamentally statistical. Modern data show that the distribution of ϕ_H measured at any particular ζ is indeed close to log-normal, with average values in agreement with either Equation 15.1 or Equation 17.2, depending on how a researcher views the issue. There is simple logic that applies for the case of momentum exchange. Within any narrow band of instabilities, the dimensionless wind gradient ϕ_m must be expected to vary between two limits — one corresponding to the logarithmic profile associated with the situation of negligible heat flux and the other to free convection. If it is assumed that the wind gradient is fully determined by the diffusivity associated with the freely convective mechanism during the part of the time that free convection is the dominant feature and by the neutral log-law when it is not, then within any band of instabilities centered on a stability ζ (as determined by the relevant fluxes and the height of measurement) we must expect to find experimental evaluations of ϕ_m to lie between ϕ_H and unity, with (once again) an approximately log-normal distribution. Hence, if the heat exchange is the dominant contributing mechanism in unstable conditions, with statistical variability such that at instability ζ the normalized temperature gradient is ϕ_H, the normalized momentum gradient must lie between neutral ($\phi_m = 1$) and ϕ_H, with an expected log-normal distribution. The geometric mean of the distribution of ϕ_m at this same stability must then be such that

$$\phi_m = \phi_H{}^{0.5} \tag{17.3}$$

This simple prediction agrees well with observations but entails no fundamental consideration of the relevant physics, save that the heat transfer controls the momentum exchange in most unstable stratifications. Once again, it seems that the essence of micrometeorology is largely statistics.

It is often claimed that free convection can be identified with highly unstable conditions, e.g., $\zeta < -0.3$. In practice, observations like those examined here continue to support the conclusion reached by Priestley (1959) that free convection is a good description of the relationship between sensible heat fluxes and temperature gradients beyond $\zeta \approx -0.03$, an order of magnitude closer to neutral. Figure 17.1 shows results from the Ocotillo field campaign of 2011 (Hicks et al. 2014), which show that the free convection regime starts somewhere in the range $-0.05 < \zeta < -0.03$. To construct the plot, the data have been ordered according to ζ and then every fiftieth point has been plotted. (There has been no grouping or averaging of the values now shown.) The density of plotted points indicates the distribution of stabilities encountered during the study. The regression lines are based on the entire dataset. It should be noted that the Ocotillo site is far from perfect, but its results mirror those found elsewhere, although often with increased scatter. The two regression lines shown for ϕ_H in Fig 17.1a have slopes of –0.325 and –0.378 for the lower and upper height intervals, respectively. The free convection expectation is –0.333. In Figure 17.1b the corresponding slopes are –0.166 and –0.149. The statistical/heuristic expectation is –0.167. The dashed lines correspond to the theoretical free convection expectation; the dotted line in Figure 17.1b identifies the dependence expected for ϕ_H.

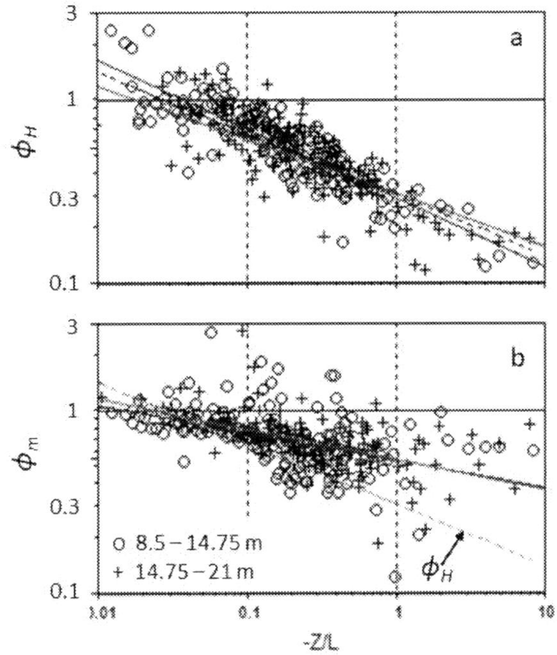

Figure 17.1. Plots of the dimensionless temperature and momentum profiles (ϕ_H and ϕ_m) against –Z/L, using data from the Ocotillo (Texas) field study of 2011. Height intervals are as shown. The straight lines show regression best fits. The dashed lines show the dependence predicted by the Priestley free convection formula, Equation 17.2. (Based on Figure 8 of Hicks et al. 2014).

The discussion above illustrates the need to consider average eddy fluxes as entities with statistical characteristics. The concept of an optimal sampling time requires that attention be paid to the inherent ability to relate measurements at any location to the area surrounding it and to the length of time necessary to specify the average flux at a single location to some specified accuracy. These two considerations will be explored in the discussion to follow.

It is concluded here that the free convection description of turbulence in the fully-buoyant situation explains much of what is observed of the unstable flux-gradient relationships provided ζ is less than –0.03 and that the role of the friction velocity is then largely imaginative. In this regard, MOST serves as a convenient and familiar method for smoothing between free convection expectations and neutral for the range from neutral to about ζ = -0.03. In similar vein, Webb (1970) points out that his examination of then-available data indicates that the log-linear profile extends from the stable regime through neutral to the same instability, ζ = –0.03.

Conclusion

In the unstable SBL, free convection often provides adequate formulation.

Chapter 18

Nighttime and Intermittency

"Chaos was the law of nature. Order was the dream of man" — Henry Adams.

A feature of conventional micrometeorological observations is that the results for stable stratification are highly scattered. Howell and Sun (1999) attribute the increasing scatter in ϕ_m as stability increases to intermittency. Every one of the datasets now being considered reveals some aspect of nighttime enhanced flux variability and intermittency. The complexity of the matter is obvious in conventional plots of ϕ_m against ζ or Ri, as in Figure 18.1a. Here, the plot is based on data from the Idaho field site, using measurements from 2 m and 8 m. The fine lines correspond to ± one standard error bounds. The smooth line shows the conventional expectations and these fall within the error band until $\zeta \approx 0.2$. Beyond this point, the results appear to indicate an asymptotic approach to some maximal value, likely $\phi_m = 6$ or 7 in this case. The leveling of the ϕ_m curve for ζ beyond about 0.5 is a common finding, most often leading to an eventual value of 9 or 10. The present value is considerably lower, perhaps a consequence of the topography of the Idaho field site (30 km east of the foothills of the Rocky Mountains).

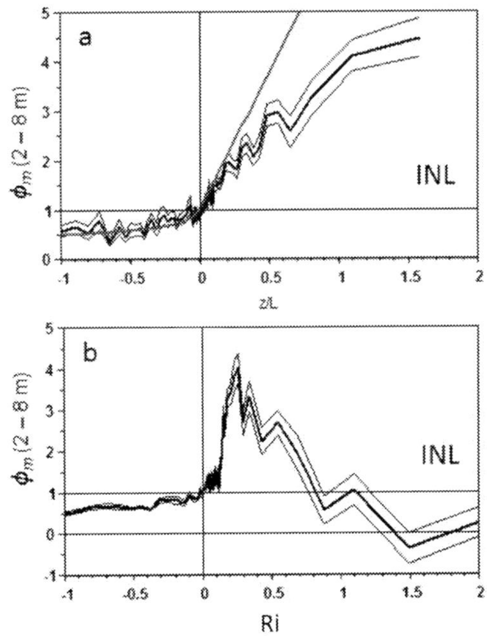

Figure 18.1. Evaluations of the dimensionless wind gradient, ϕ_m, derived using INL (Idaho) data (2 m and 8 m), showing average values and bounds corresponding to ± one standard error. (a) when plotted against $\zeta = Z/L$ and (b) against the Richardson number derived using data from the same two levels of measurement.

The departure of ϕ_m and ϕ_H from legacy expectations as stability increases has been reported extensively. However, Figure 18.1b tells the story differently — that ϕ_m tends to zero as Ri increases, after maximizing when Ri = 0.25. The reason is obvious. Reduced to its essentials, Figure 18.1b is a plot of $\partial\bar{u}/\partial z$ against $(\partial\bar{u}/\partial z)^{-2}$, modified by the presence of other variables not associated with $\partial\bar{u}/\partial z$. The plot is therefore like one of X versus X^{-2} and the behavior evident in the very stable part of the diagram is then as expected.

Data examined earlier (Chapter 16) indicate that temperature gradients are not well associated with sensible heat fluxes in stable conditions beyond the near-neutral transition regime. Moreover, wind gradients are then not closely linked to momentum fluxes. Hence, some of the fundamental assumptions underpinning presentations like those of Figure 18.1 break down in strongly stable conditions.

The concept of intermittency, along with the constraints imposed by averaging and ensembling, generate doubt about the physical relevance of any deterministic formulation of the ϕ functions beyond $\zeta \approx 0.5$ (or whatever value best fits some particular set of observations). As already emphasized, the central questions relate to the relationships between fluxes and gradients and addressing these questions using quantities like ϕ_m and ζ (or Ri) in stable conditions serves to obscure the relationships rather than reveal them.

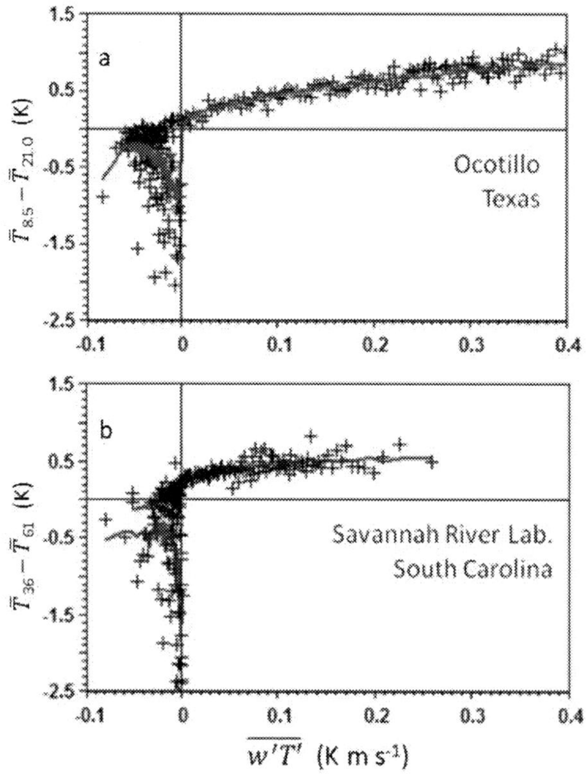

Figure 18.2. Plots of temperature differences against sensible heat flux covariances measured in two field studies – over sparse vegetation in Texas and over a mature pine forest at the Savannah River National Laboratory field site in South Carolina.

Data obtained in two completely different field campaigns are plotted in Figure 18.2. These are raw data; the vertical axis represents actual, not potential, temperature differences. Lines are drawn to join averages, each containing fifty values after ordering by $\overline{w'T'}$. The lines in the lower left quadrant link averages for which both the $\overline{w'T'}$
temperature differences both indicate stable. All available data contribute to the ordered sequence of averages to the right of the zero abscissa, without any such constraint. The lack of a clear-cut relationship in stable conditions ($\overline{w'T'} < 0$) is obvious. In contrast, the unstable situation is well ordered.

Figure 18.3 shows results an experimental study of alternative agricultural practices conducted in Zimbabwe, in 2013 (O'Dell et al. 2015). In Figure 18.3a the data indicate much the same behavior in stable conditions as the two datasets illustrated in Figure 18.2. In Figure 18.3 a day/night distinction has been applied to avoid transition periods. Daytime data are from 1000 to 1600 LST, nighttime from 2200 to 0400. The size of the dataset is such that plotting each point is not revealing. To show the spread of the data and their scatter, every fiftieth point is plotted after ordering according to the abscissa.

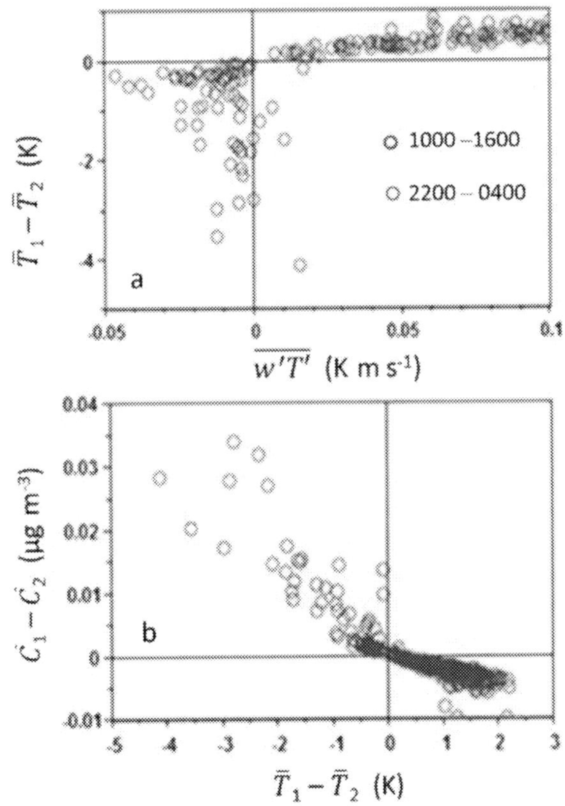

Figure 18.3. Data from the Zimbabwe experiment of 2013, showing (a) the ordered behavior of $\Delta\bar{\theta}$ versus $\overline{w'T'}$ in daytime (1000 – 1600), in comparison to nighttime (2200 – 0400). (b) the association between $\Delta\bar{\theta}$ and the corresponding difference of CO_2 concentration — $\Delta\bar{C}$. In these plots, stable conditions are to the left, unstable to the right.

The Zimbabwe dataset includes CO_2 gradient observations, which are used in Figure 18.3b to demonstrate the way in which CO_2 gradients tend to follow the temperature gradients, even at night. The eddy fluxes are also well associated, as is seen in Figure 18.4. In summary, the set of fluxes contains variables that are well correlated and so does the set of gradients. Yet these two sets are largely independent. A seemingly unavoidable conclusion is that in stable conditions, the same processes contribute to the fluxes but that these processes have little influence on the corresponding gradients.

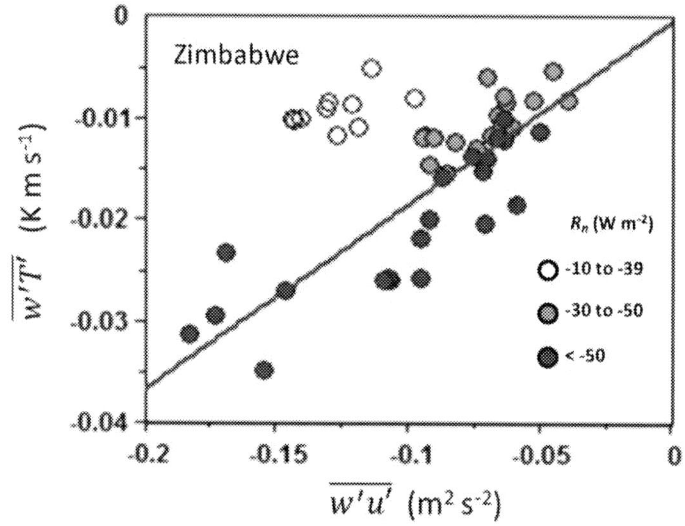

Figure 18.4. For nighttime conditions (2200 to 0400 hrs, local time) the association of the sensible heat and momentum covariances computed as 30–min averages. Data are identified according to the prevailing net radiation in three bands as shown: -10 to -30 W m^{-2}, -30 to -50 W m^{-2}, and <–50 W m^{-2}). As before, every fiftieth point is plotted after ordering by $\overline{w'u'}$. The line drawn is the result of a regression using all observations for R_n < -30 W m^{-2}.

It is admitted that scatter is high in Figure 18.4. The line drawn is the result of a linear regression for all data for which R_n < -30 W m^{-2}, not limited to the data points plotted. Figure 18.5 shows results from the same procedure, applied to data from the Idaho National Laboratory field site. One line in Figure 18.5 corresponds to the regression line drawn in Figure 18.4, for all data for which R_n < –30 W m^{-2}. A second regression line is shown, for R_n < –50 W m^{-2}. The statistical uncertainties associated with these plots is such that there is no basis for concluding that the results differ from a proportionality. The slopes of the lines are about 0.2 K s m^{-1} for both the data of Figure 18.4 and Figure 18.5, even though these refer to greatly different surroundings.

The conceptual picture that arises is that at night and mainly under clear skies the prevailing stratification (associated with temperature gradients) is intermittently penetrated by turbulent events that are the flux-carrying entities. The observation of such nocturnal turbulent intermittency is common, apparently regardless of the surface itself. For example, Klipp and Mahrt (2004) raise the issue of intermittency affecting legacy nighttime flux-gradient relationships, concluding that the expected relationships appear to be well described when the effects of intermittency are removed. The attribution of intermittency to specific causes remains contentious. It is known, for example, that episodes of unusually high turbulence can occur over land, generating the honami of micrometeorological lore (Inoue, 1955, a, b). The 'cat's paws' observed over still waters are a similar phenomenon. These behaviors are attributed to features of air-surface exchange in the SBL involving both the atmosphere and the nature of the surface itself. However, the alternative PBL cause is a better explanation for many terrestrial observations.

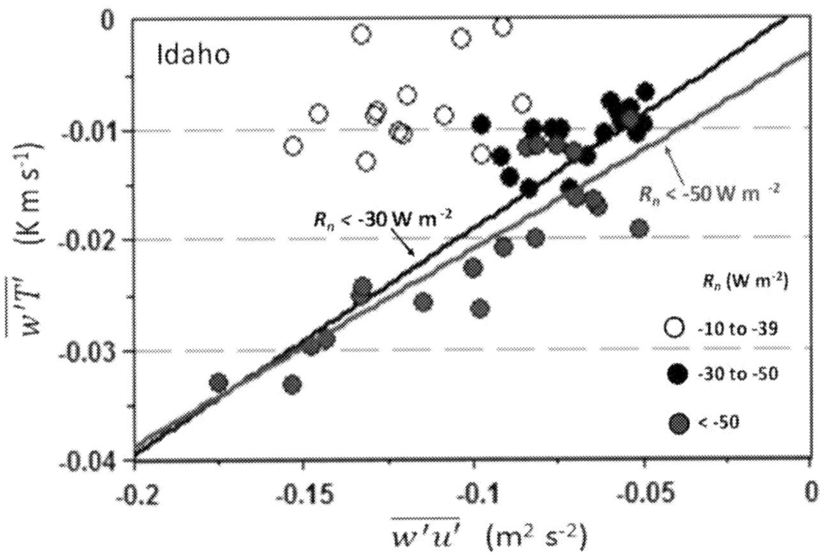

Figure 18.5. An independent set of data reflecting the nighttime association between eddy fluxes of heat and momentum in clear skies, at night. The analysis is as in Figure 18.4, but this time the results are from an arid area – Idaho National Laboratory. Symbols are as in Figure 18.4. Lines relate to regression results based on all data within the R_n bands identified.

To provide some further justification, consider data obtained in an agricultural study of growing maize conducted by the University of Tennessee in Ohio (40.606 °N, 82.674 °W; see Odell et al. 2018). Details and results of the experiment will be considered in more detail later. Figure 18.6 shows results for two five-day time periods, one in the Spring of 2014 (Julian days 126 to 131, after tilling but before planting of maize) and the other in the Autumn (days 240 to 245, when the crop was mature). The turbulence data on which the present discussion is based were obtained about 3 m above the top of a maize canopy. The data stream is of five-minute averages and of the maxima and minima observed within each such period. The uppermost diagrams (a) and (b) show that meteorological conditions were complex – the net radiation record was clearly affected by frequent cloud and the sensible heat flux covariances gave similar indications. However, the net radiation at night was frequently such that the syndrome addressed above could have been anticipated. Note that the vertical scales for (a) and (b) are linear. Those for (c), (d), (e) and (f) are logarithmic. The time scale refers to the midnight beginning of each 24-hr day.

The $\overline{w'T'}$

flux were present on all nights illustrated. However, nocturnal intermittency is most evident in the central pair of diagrams (c) and (d); these show the differences between maximum and minimum values of vertical velocities and concentrations of CO_2. The biggest swings in CO_2 concentration occur at night, but the biggest swings in the vertical velocity are in daytime.

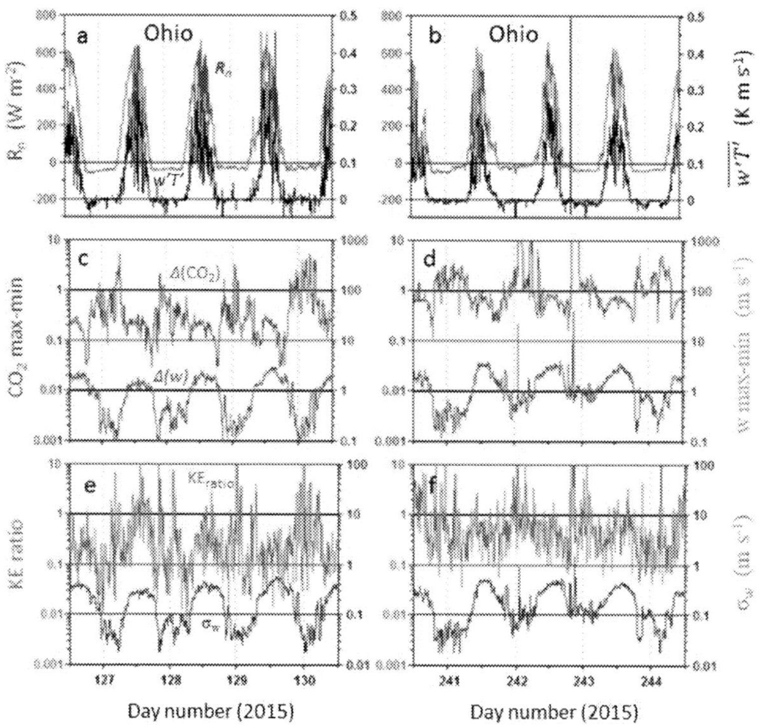

Figure 18.6. Time series of a selection of SBL properties observed within 5–min periods at the Ohio site of the University of Tennessee, in 2015. Differences (Δ) refer to the spread between recorded maximum and minimum values observed during each period. CO_2 concentrations are in mg m^{-3}.

These observations are compatible with the understanding of the SBL now presented — convection dominates the daytime scene, with large updrafts occurring irregularly. Turbulence intermittency controls the nighttime flux regime. At night, the soil efflux of CO_2 and the consequent pooling in the lowest levels (due to the gravity and the pervasive role of stability) lead to high CO_2 concentrations which are greatly reduced when affected by migrating packets of turbulence. At night, large convective updrafts are not present and the swings in w are therefore far smaller. This last feature is also evident in the σ_w data (Figs 18.6e and 18.6f). Also shown in these figures are measurements of the kinetic energy ratio: $KE_r = (\sigma_u^2 + \sigma_v^2 + \sigma_w^2)/\bar{u}^2$. A value exceeding unity means that there is more turbulent kinetic energy present than could be extracted from the average wind. Again, the evidence of these diagrams is compatible with an assumption that a nighttime source of intermittent turbulence episodes is aloft.

Conclusion

Intermittency is a normal feature. At night, radiative heat loss and intermittency appear to go hand in hand.

Chapter 19

The Morning Transition

"All bubbles have a way of bursting or deflating in the end" — Barry Gibb.

Sunrise imposes a moving thermal shock to the surface, to which the soil and the atmosphere slowly respond, along with any vegetation or other materials on the surface. It is commonly assumed that the rising sun heats some parts of a landscape preferentially, resulting in hotspots from which convective cells of increasing intensity can be generated, subsequently moving downwind. Geiger (1965) described observations of dust plumes and eddies visualized using water vapor as a tracer. These showed that when "an upward whirl is initiated through some outside agency, the adjacent layers are drawn into the movement and the phenomenon proceeds, borne along by the wind, affecting new layers one after another, and thus gaining strength." The details of how heating of a gaseous fluid can cause convective cells to be generated has been the subject of extensive laboratory and theoretical studies, which need not be discussed here but which will become central concepts in later considerations of vegetative canopies.

Figure 19.1 shows how intermittency can affect the SBL in the period soon after dawn. The data were derived during a field study at the Dugway Proving Ground of the U.S. Army (q.v. Hanna et al. 2012). The field site was an extensive arid surface – the bed of the historic and now dry Lake Bonneville in Utah. (The Great Salt Lake is all that now remains of the large inland sea that preceded it and the surrounding old lakebed now serves as, for example, a favored venue for land speed records – the Bonneville Salt Flats.) Details of the experiments have been presented by Fox and Stowold (2011). The circumstances were somewhat reminiscent of the "outdoor laboratory" micrometeorological experiments of the distant past. However, such conditions cannot be accepted as either representative or common, unless the focus of attention is on flat areas like those of the arid North American west.

The covariances plotted are one-minute block averages of 10 Hz data. Such short-term covariances $\widetilde{w'T'}$ and $\widetilde{w'u'}$ are rarely seen; it is longer-time averages of covariances that are usually of main interest, conventionally using time steps in the range 15 min to 60 min. (Note the modification of convention here. The squiggle overbar indicates an incomplete covariance.) The behaviors revealed by the present analysis are largely obscured when such longer averages are considered. The time sequences shown here give indications of the periodicity associated with increasing sensible heat flux as time processes. Similar figures will follow. In every case, a major feature is the presence of short-term episodic turbulence much as expected by theoreticians, except in circumstances when the situation is influenced greatly by background advected turbulence.

The sequence illustrated in Figure 19.1 shows that there are two dominant determinants of the spectra: the pulse repetition frequency and the frequency of the causative fluctuations within the packets of turbulence. The repercussions of this dichotomy of frequency regimes are evident in the classical power spectra of SBL turbulence such as are found in many textbooks. In near-neutral conditions, the power spectrum appears well ordered. As conditions depart from neutrality into increasing instability, the influence of intermittent convection becomes more evident, injecting a low-frequency perturbation of the higher-frequency spectrum otherwise expected (as discussed, for example, by Kaimal, et al. 1976; Garratt, 1992).

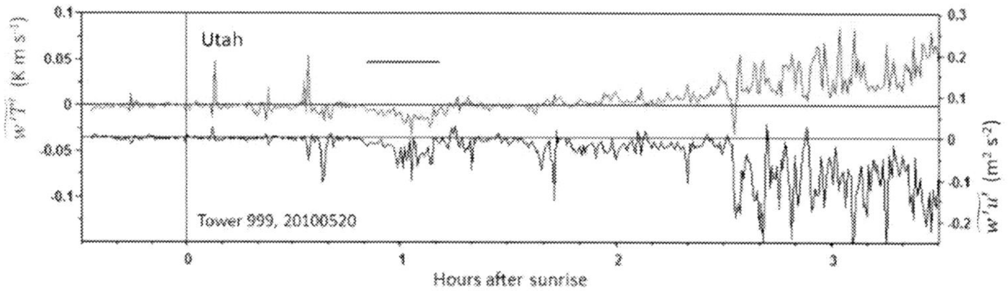

Figure 19.1. An example of how short-term covariances $\widetilde{w'T'}$ and $\widetilde{w'u'}$ vary with time following sunrise (at time zero on the horizontal axis). Data are from a field study conducted at the Dugway Proving Ground (40.206 °N, 113.266 °W) in 2010 (the Jack Rabbit experiment, see Fox and Storwold, 2011, Hanna et al. 2012).

A common view of the onset of heat-generated turbulence following dawn is the process described by Lettau and Davidson (1957), involving the initial generation of buoyant cells at some frequency determined by the situation and the subsequent frequency doubling of these turbulent episodes. Studies of the onset of convective turbulence in laboratory controlled experiments (e.g., in Couette flow) have led to a revised understanding, involving the generation of intermittent turbulence events of gradually increasing intensity and frequency as the energy input into the system increases. From the conventional micrometeorological perspective, it is the time average of these intermittent episodes that controls the stratification and other bulk qualities of the surface air mass. However, the laboratory experiments and the mathematical investigations of the onset phenomenon agree that the generation of small-scale convective cells is essentially chaotic, with a dominant log-normal probability density function. This has obvious relevance to such considerations as the optimal period of sampling to derive accurate and relevant quantifications of the sensible heat flux, the momentum flux and any atmospheric exchange of similar kind.

There is extensive literature relating to the onset of convection and it is not the present intention to review this. This is an evolving science and the interpretations given here are certainly susceptible to subsequent modifications. In large, these recent mathematical and laboratory studies differ somewhat from the Prandtl frequency-doubling mechanism of micrometeorological convention (see Schroeder, 1991). However, the influence of post-dawn intermittency (often regular intermittency) on the associations of SBL gradients with time-averaged surface fluxes remains inadequately studied.

Theoretical studies of the initiation of convective turbulence indicate that the origin of packets of turbulence can be purely chaotic, regardless of the underlying surface. It can be concluded that in relatively calm conditions the morning transition often involves a series of turbulent occurrences, increasing slowly in magnitude and in frequency and eventually evolving into a continuum of turbulence. Some vestiges of the initial intermittency can be retained throughout the daytime period. At the opposite extreme, with higher winds, a sufficiently turbulent nocturnal atmosphere will permit the background turbulence spectrum to persist through the morning transition, with no apparent role for the intermittency that would occur if conditions were calmer.

Figure 19.2 shows that the morning evolution of the SBL is not always as clear-cut as is indicated in Figure 19.1. Data from a University of Tennessee field site in Loudon, Tennessee are shown, for two days of June 2017. The crop was rapidly-growing maize. On the first of these days (3 June 2017) skies were clear. On the second (5 July 2017), there was cloud cover. To show the post-dawn intermittency clearly, values of covariances less than 0.001 (in the cases of both $\widetilde{w'T'}$ and $\widetilde{w'u'}$, in SI units) have been omitted.

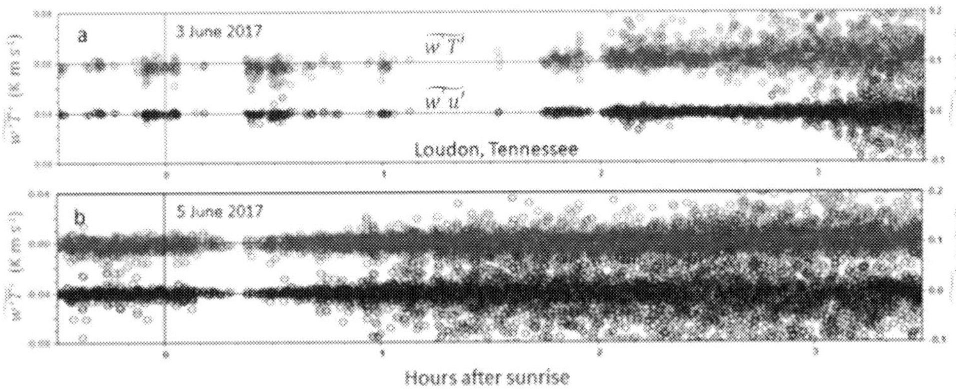

Figure 19.2. Time sequences following dawn of short-term quantifications of $\widetilde{w'T'}$ and $\widetilde{w'u'}$ for two days in early June 2017, observed at the UT Loudon site in Tennessee. 3 June data were obtained under clear skies, 5 June under complete cloud cover. Low values of covariances have been omitted.

Data obtained at the same location in August, under clear skies and over mature maize, provide a somewhat different perspective. In Figure 19.3, there is evidence of a somewhat regular intermittency, in this case with about 40 minute period, after sunrise and before the onset of chaotic turbulence starting about three hours later. Moreover, it might be imagined that there is some indication of about half this 40-min periodicity as the chaotic increase initiates. Deriving a smooth sequence of 15–min or 30–min averages through the period depicted would be difficult (to say nothing of being inappropriate since the plotted values are actually covariances computed over periods too short to permit flux computation without omitting part of the flux-carrying spectrum of turbulence). The variability of stability, as normally computed, would be exceedingly large, regardless of the averaging time interval employed. Also shown in Figure 19.3 is the corresponding sequence of TKE measurements, once again based on averages from three anemometers. The post-dawn intermittency is more evident in the TKE trace than in either the $\widetilde{w'T'}$ or the $\widetilde{w'u'}$.

Figures 19.1, 19.2a and 19.3 present examples of what happens in clear skies. The characteristics when clouds are present are seen in the 5 June data of Figure 19.2b. Visual reports of cloud cover are not common in contemporary SBL experimentation, but a useful indicator is the variability of measured net radiation, as is plotted in Figure 19.4 for the Loudon, Tennessee site of present focus. A smooth curve describing R_n is an indicator of clear skies, with few fair-weather cumuli. The passage of some of these rare cumulus clouds caused the spikes and other short-term variability seen in the diagram.

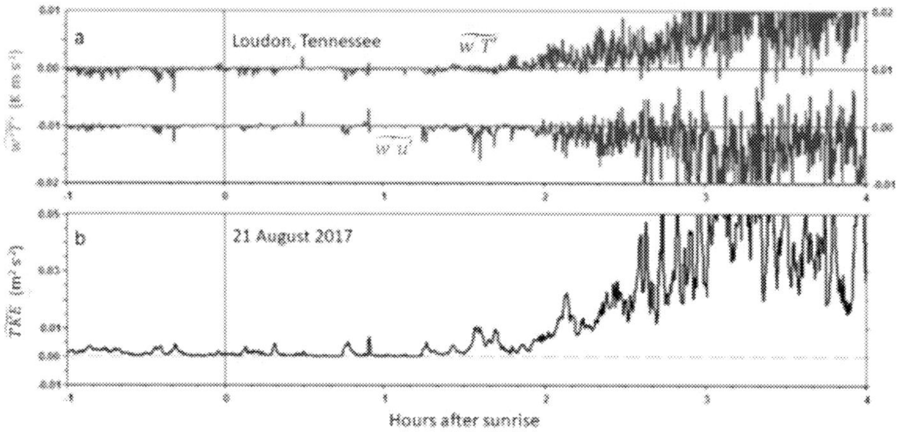

Figure 19.3. The upper panel shows time series of short-term variations in the heat flux and momentum covariances computed (in the same way as in Figures 19.1 and 19.2), for 21 August 2017. Data from three independent but closely located sonic anemometers are used. All anemometers were erected about 2 m above a field of mature maize. The lower panel shows the variations in TKE through the same period. Data from the Loudon site are used here.

The cloudy sky situation requires additional attention. Two of the days depicted in Figure 19.4 appear to have been minimally affected by cloud cover (5 and 6 August), two others indicate a strong influence of cloud (7 and the morning of 8 August). The first two days yield R_n trends that are smooth; the other periods are dominated by scatter that appears chaotic. The wind speed data (also shown in Figure 19.4) show the expected trends during the periods of inferred clear sky. For example, the evening of 5 August was clear, with high negative net radiation and light winds extending well into the following day.

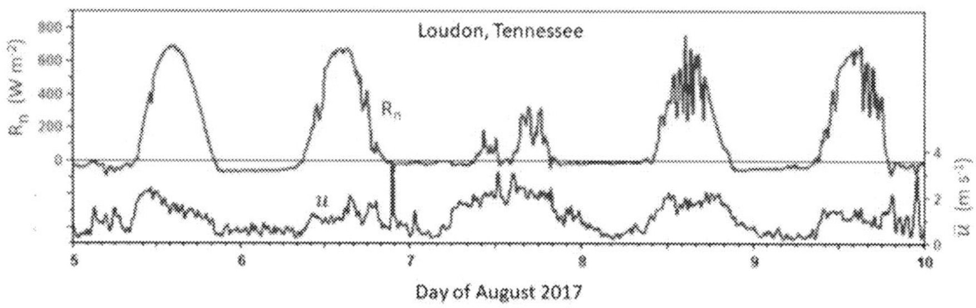

Figure 19.4. A time series of observations of net radiation, starting at midnight on 5 August 2017 and extending for 5 days. The effects of cloud cover are apparent as noise. Also shown are wind speed data at a height of about 3 m, above the crop (senescent maize).

Figure 19.5 shows time sequences for the mornings of 6 and 7 August, 2017, following the same format as for Figure 19.1. Figure 19.4 shows that the first of these days was mainly clear (the net radiation curve showed little scatter), whereas the sunrise of the 7th was influenced by heavy cloud cover with considerable scatter in R_n. The differences of the turbulence fine structure are considerable. There could be some intermittency affecting the time sequence of Figure 19.5b, but if so it is well disguised. In general, the many cases now available indicate that the intermittency following sunrise is not always present but is likely on clear sky mornings and mostly with near-calm conditions.

The behavior of short-term covariances such as are depicted in the various figures here is a well known factor of air-surface exchange, originally explored by the first developers of eddy correlation methods (such as Scrase, 1930; Deacon, 1959). In later years, the matter has been studied extensively using spectral analysis of various kinds, such as the use of lag-correlograms, Fourier analyses and wavelets. Analyses of such kind have been addressed extensively elsewhere (see any of the many micrometeorological textbooks) and it is not the purpose here to repeat these reviews. Instead, it is appropriate to point out that such classical analytical methods require the extraction of a sample of continuous data from a longer period of similar characteristics – i.e., satisfying the need for spectral homogeneity. The records displayed here indicate that the transition period following dawn does not satisfy this requirement, perhaps with the exception of situations with substantial cloud cover.

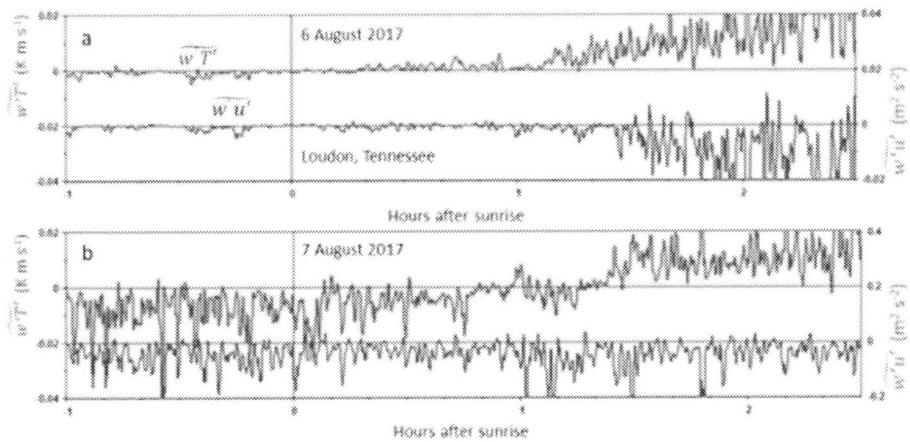

Figure 19.5. Time sequences of fine-scale flux structure following sunrise on 6 and 7 August, 2017, at the Loudon (Tennessee) field site of the University of Tennessee.

It is a common expectation that the spectral characteristics associated with the fluxes of such properties as water vapor and carbon dioxide are well described by results obtained for sensible heat. All such turbulent exchange is facilitated by the same causative factors – the exchanges of momentum and heat. If the intent is to derive accurate fluxes from analysis of fast-response data, as in the conventional eddy-flux methodology, then the duration of the time averages becomes a central consideration. If, for example, an averaging period of ten minutes were used in the post-sunrise period illustrated in Figure 19.1, then successive averages might or might not witness the passage of a packet of turbulence. A reported value of near zero might be followed by a measurement considerably different, with the next back to near zero. As the averaging time increases, the chances of encountering a consistent number of turbulence packets increases and the apparent run-to-run scatter of the resulting time sequence decreases. However, increasing the run averaging time has a severe drawback: it leads to violation of a basic requirement of micrometeorology, time stationarity. It is for this reason that many of the analyses presented here favor the use of a 15–min run length. A question arises that is beyond the present scope — How does the concept of stability apply when the atmosphere is chaotically intermittent?

There is a contributing role of plant physiology that can become a confounding issue. The field studies contributing the data plotted in Figures 19.2 to 19.5 offer some relevant insight. In Figure 19.6, an example is given of the behavior of short-term water vapor and carbon dioxide covariances (measured close to the top of the maize crop) through much of 21 August 2016. In the immediate post-dawn period, the heat and water vapor fluxes appear to increase in tandem. The CO_2 covariance shows a different behavior. This is attributable to the well-known exhalation of CO_2 from soil, resulting from microbial activity. In the stable conditions prevailing at night, CO_2 emissions can be accumulated in the atmospheric layers immediately adjacent to the surface. The resulting CO_2 pooling will be accessed by convective turbulence once solar heating is adequate, resulting in positive CO_2 fluxes and covariances early in the post-dawn period. It is only after this accumulation of CO_2 is diminished that the consequences of above-surface plant physiology will become apparent, and the atmospheric fluxes of CO_2 turn negative (i.e., towards the crop).

In Figure 19.6, the effects of a complete solar eclipse are evident, starting at about 6.5 hours after sunrise. The data obtained during the eclipse give a clearer view of the role of the surface biology. When the solar eclipse caused the incoming radiation to drop to zero, the sensible heat flux reversed in sign while the evaporation rate and CO_2 exhalation continued. As the eclipse ended, all of the heat, evaporation and CO_2 covariances responded quickly, but as night approached the heat flux dropped to near zero while the evaporation and CO_2 exchange continued.

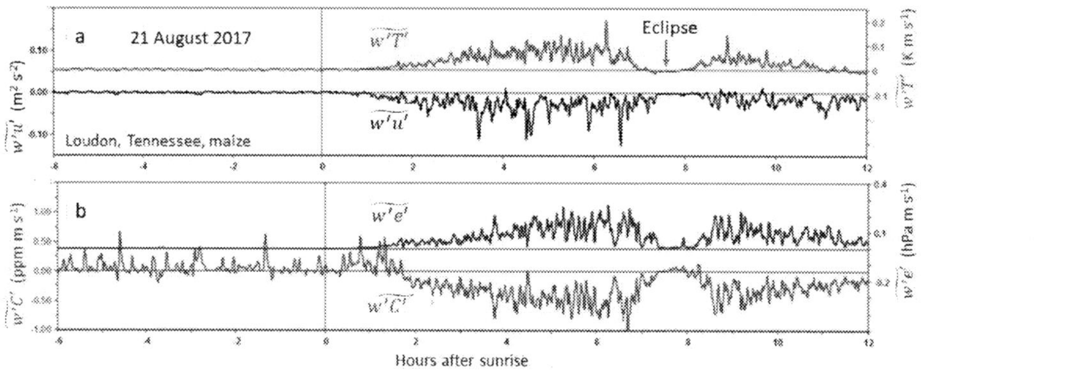

Figure 19.6. For 21 August, 2017, time trends of covariance fine structure through the daylight hours at the Loudon site. The effects of the total solar eclipse are evident, starting about 6.5 hours after sunrise. Plots are shown of the partial covariances $\overline{w'T'}$, $\overline{w'u'}$, $\overline{w'e'}$ and $\overline{w'C'}$, all calculated from raw measurements.

The physical interpretation of the energy balance at dawn is straightforward. Immediately after sunrise, solar radiation will impact the vegetation or other obstacles that might be present at a location of interest. If there is enough water available, the foliage will utilize the energy first to promote photosynthesis and transpiration. The first convective response to sunrise might therefore be evident in a trace of $\overline{w'e'}$ or $\overline{w'C'}$ where C represents the concentration of CO_2. The first response to sunrise evident in the covariances plotted in Figure 19.6b is indeed the burst of CO_2 upwards soon after dawn, presumably originating in the pool of surface CO_2 that accumulated during the night and in advance of the initiation of both photosynthesis and convection.

> During one of the series of Sangamon (1976) studies of the associations between the SBL and the PBL, a kytoon was used to explore the turbulence and mean structure around the low–level jet in the pre-dawn hours. After an extended period of probing, the kytoon, without warning, became unstable and proceeded to describe loops of steadily increasing amplitude until it impacted the surface, leaving behind a trail of expensive germanium and silicon circuitry and astonished observers who now believed in stable stratification intermittency.

In the case of the intermittency illustrated in the figures presented here, the initial trigger mechanisms remain to be identified well enough to satisfy critics. Early work by Lorenz (1963) introduced chaos to the meteorological community and subsequent studies in fluid mechanics and mathematics have led to the development of theoretical treatments of the onset of chaotic behavior. A common understanding is that solar radiation heats leaves and other elements of the surface, creating hot spots that then serve as sources of convective cells. Experience from watching a pan of water starting to boil is a familiar analog, however the kitchenware observation is influenced by a phase change and resulting buoyancy that is not relevant in the SBL context. In other disciplines, attention to the topic (e.g., by Feigenbaum, 1979, 1980; Collet et al. 1981; Weng and Lau, 1994) has led to an alternative view — that diffusive transfer of heat from the a warmed surface increases the entropy of the surrounding air and that chaotic motions within this air will eventually lead to the production of buoyant parcels. In other words, convection starts within the air, not at the surface bounding it. In a study of the turbulent interaction between different fluids, Moore et al. (1983) conclude that the repetition rate of bursts of turbulence is somewhat predictable, and that the amplitude of the bursts satisfies a universal rule if the overall energy input to the system remains the same. In this sense, the key features of the SBL turbulence frequency spectrum would be expected to be sequences of pulses when fluxes are small, evolving into a continuum of turbulence when the heat flux gets large enough. However, the visual observations that frequented early field studies, such as of dust devils and convective behavior in general, indicated that when the heat fluxes are highest, packets of turbulence migrate with the wind, retaining their identity for distances that have yet to be quantified.

Conclusion

Intermittency is common in daytime. The doors for future research are wide open.

Chapter 20

The Evening Transition

"Habit is stronger than logic" — Santayana.

The data collected at the Loudon site in 2017 contain clear examples of how fluxes vary with time following sunset. Four different flux quantities are shown in Figure 20.1: $\overline{w'T'}$ and $\overline{w'u'}$ in the upper panel and the resulting mass fluxes of water E and of CO_2 (F_{CO2}) in the lower. The covariances plotted in Figure 20.1a have been derived using a 100–sec digital filter, plotted every minute. The intermittency seen earlier (in the preceding Chapter) is clearly evident. Figure 20.1b shows results from independent instrumentation, based on the use of 1–min block averages. Clearly, it does not matter which of many different methods is used to display the physics inherent in the mechanisms depicted, provided the analytical time step is short enough. The two diagrams of Figure 20.1 illustrate that there are indeed differences in the visualizations derived from different analytical methods. In particular, the 1–min block averaging results indicate a slow but consistent ramp between the obvious intermittent bursts and this is not evident in the other traces. However, the fluxes of water vapor and carbon dioxide were consistently upwards (positive) during the period shown, while $\overline{w'T'}$
This is in line with expectations — that water and CO_2 continued to be emitted from the surface even though sensible heat flux was downwards. Most important in the present discussion is, however, the convincing similarity between the four signals plotted. The interpretation already discussed is clearly relevant. There is little "action" for much of the time; the average fluxes result from periodic bursts of turbulence.

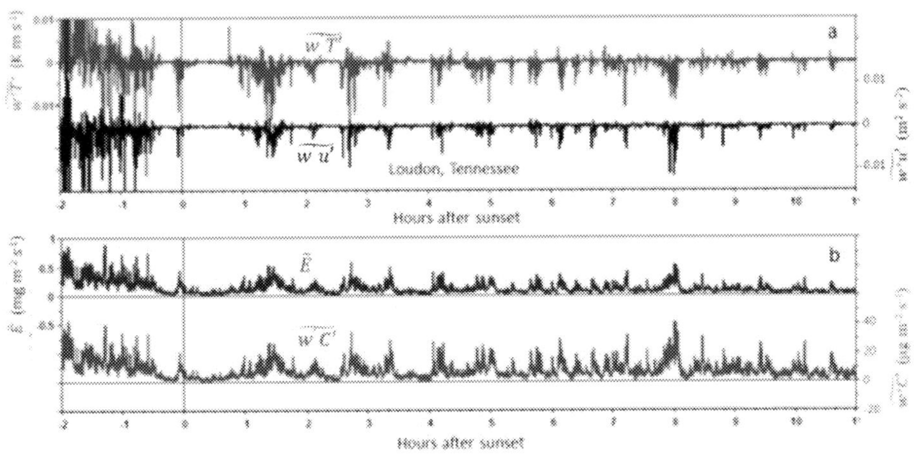

Figure 20.1. Data collected at the Loudon site, showing the period through sunset and extending to nearly dawn.

The observations represented in Figure 20.1 were obtained when cloud cover was present, to an extent not quantifiable but indicated by the net radiation records now presented as Figure 20.2. Two periods are illustrated, the first (a) corresponding to Figure 20.1, showing the signature of cloud cover which caused net radiation to be limited to about –20 W m^{-2} over the night. On this occasion, Figure 20.1 shows that intermittency continued throughout the night. In contrast, the period emphasized in Figure 20.2b was initially of clear skies, with net radiation consistently about –50 W m^{-2}. About 2.5 hours after sunset, the first of two short-term reductions in outgoing net radiation occurred.

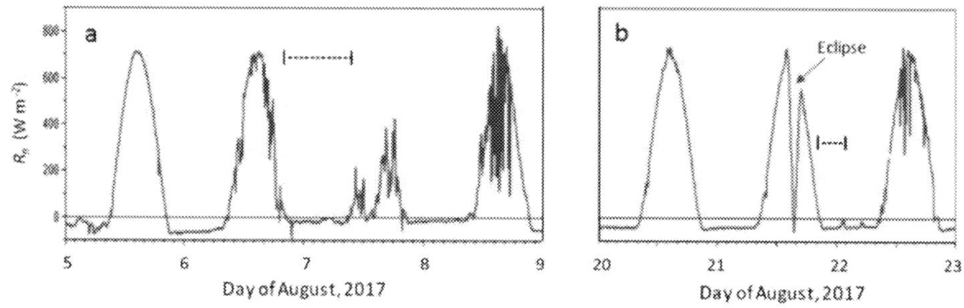

Figure 20.2. Diurnal cycles of net radiation corresponding to the periods of observation considered here — with bars indicating (in (a)) the period illustrated in Figure 20.1 and (in (b)) that of Figure 20.3.

Figure 20.3 shows how the initial intermittency following dusk evolved into chaotic turbulence as the evening progressed during the second period of present interest, as identified in Figure 20.2b. The transition occurred at about 2.5 hours after sunset, corresponding to the obvious change in net radiation at that time.

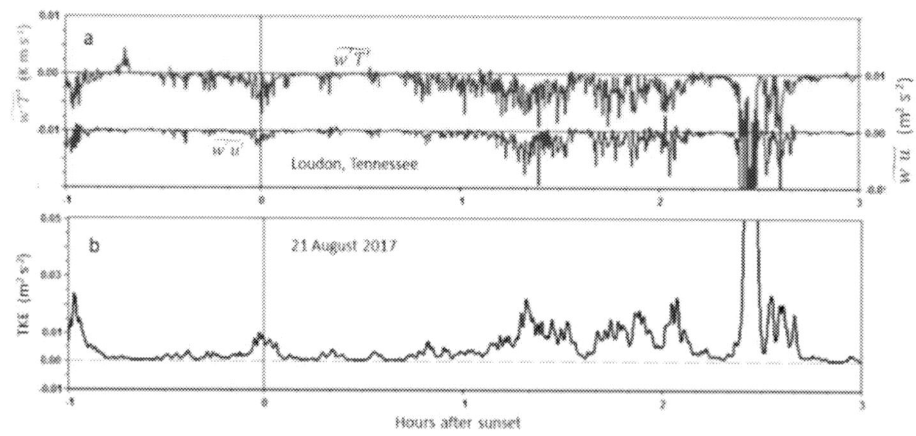

Figure 20.3. Using observations obtained on 21 August, 2017, at the Loudon (Tennessee) site of the University of Tennessee, (a) the variation of partial covariances as time progressed through the sunset and (b) for comparison, the variation of TKE.

Examination of the data provides support for the criterion identified earlier, that $R_n < -30$ W m^{-2} serves as a tentative boundary beyond which quasi-regular intermittency reigns and eddy fluxes become well correlated regardless of profiles. This conclusion serves as a starting point for many of the considerations to follow. The physical explanation for these phenomena appears to be as described above but remains to be explored in detail. Examination of the several data sets considered here reveals a commonality that relates well to the clear sky observation. Clear skies appear to be the portent of nighttime intermittency, much as must be expected on the simplest of all possible grounds: PBL stratification is strongest when there is no cloud blanket interrupting the radiative loss of heat from the surface. Moreover, the variations of TKE shown in Figure 20.3b mirror the variations seen in Figure 20.3a, whereas in daytime TKE measurements show little influence of the local fluxes.

There are many ways in which a data stream can be analyzed in order to reveal intermittency. Salmond (2005) considered the ramp structure evident in time records of ozone concentrations near the surface and introduced wavelet analysis as a valuable analytical method. Hicks et al. (2015) examined records of CO_2 concentrations at night at the research site in Zimbabwe, Africa, as already mentioned, using rapid departures from previous concentrations as an indication of intermittency. In practice, selection of the best methodology in some specific situation is likely to be influenced by both the circumstances of the study and the instrumentation available. The present preference is to avoid spectral analysis because all such methods require a representative sample of adequate duration and address a continuum of uniform turbulence. The very concept of 'representative' appears at odds with the present focus on intermittency — which can be considered to involve episodes that are not representative. Webb (1982) noted that a detailed examination of the log-linear wind profile revealed periods in highly stable conditions when the potential temperature gradient fell to nearly zero for periods of several minutes. He warned that analysis of the stable SBL must be extended to include factors external to the SBL itself, a foresight that has definite contemporary relevance.

Whereas some studies indicate that the breakdown and generation of turbulence is associated with a critical Richardson number $Ri = 0.25$, other work indicates that such a stability criterion could vary over a wide range, extending as far as $Ri = 1$ (Swinbank 1952b; Businger, 1969, 1973; Miles, 1986). Field data have repeatedly shown that SBL values of Ri at night can often exceed unity, e.g., during the Wangara experiment (Hicks, 1976).

The many possible causes of nocturnal turbulence intermittency were reviewed by Mahrt (1981), emphasizing the common occurrence of low-level jets (LLJs; see Bonner, 1968) as a major contributing factor but also pointing out the geostrophic and topographic features can contribute. Among the processes now relevant is the inertial oscillation initially observed at Porton (as reported by Sutton at the 1951 'International Symposium on Atmospheric Turbulence in the Boundary Layer' — Hewson, 1952). The fundamental characteristics (specifically of bursts of turbulence occurring with a recurrence period of 20 to more than 60 minutes) were discussed by the attendees at the symposium and summarized by Lettau (1952). Understanding of the overall behavior was refined and modeled by Blackadar (1957). In essence, the onset of stability following sunset diminishes surface drag and releases the wind aloft from at least part of the surface drag that had previously been restraining it. The wind aloft can then accelerate, causing low level jets to form. When the consequent wind shear aloft rises sufficiently, turbulence is generated. This turbulence serves to decrease the overall resistivity between the surface and the air above it. In turn, the jet is slowed and the generation of turbulence is curtailed. The process tends to be cyclic and intermittent packets of turbulence at the surface are a consequence. This has also been demonstrated in many field studies and is now well reported (see, e.g., Poulos et al. 2002; Mahrt and Vickers, 2002; Mahrt 2010). Sun et al. (2002, 2012). It must be mentioned that the mechanism described above is but one of several that were detected and explored during the Cooperative Atmosphere-Surface Exchange Study (CASES) experiment of 1999 (Poulos et al. 2002), as will be discussed later. Other mechanisms include circulations associated with density currents, solitary waves generated orographically, and the interaction among katabatic flows.

Extensive PBL studies conducted as part of the multi-year experiment conducted in Sangamon County, Illinois, in the 1970s relied on the use of tethered kytoons and tracking of pilot balloons (Hicks et al. 1981). The nocturnal low level jet was found to be a regularly recurring feature of the PBL and was tentatively attributed to the large-scale slope associated with the Great Plains, downwards towards the Mississippi River. Subsequently, Mahrt (1981) demonstrated that a low-level nocturnal jet is a characteristic feature of the Tennessee River Valley (in which the Loudon site now considered is located) as much as it is for the Great Plains of the American midwest. Brook (1985) showed that the same is true for the region of the Koorin study, in the Australian Northern Territory. It appears that the existence of a nocturnal low-level jet is as ubiquitous over the Australian outback as it is over the central USA. It has recently been shown that such jets also occur over the area east of the Appalachian Mountains in the eastern USA. Banta (2008) used data from two independent field programs to show that scaling of profiles according to the height and intensity of the low-level jet is a profitable approach. He advocates that the low-level jet must be considered a key component of the nocturnal SBL. These low-level jets are indisputably among the causes of intermittency observed at the surface.

Figure 20.4. A first step in tracking pilot balloons is to locate the balloon. (Photograph from the Sangamon study of 1977, showing Paul Frenzen, Dick Hart and yours truly with the binoculars).

The high correlation between nighttime fluxes evident in Figure 16.6b is now explained. Outside the stability band around neutral, any 15–min or 30–min run could contain a turbulence bursting event, perhaps none and maybe more than one. The time sequence of any flux average as conventionally reported will be similar to any other, all being affected by the same periods of turbulent intermittency and promoted vertical exchange as the others. A high correlation between pairs of these sequences is indeed observed. It has already been emphasized that this super-stable case arises in clear sky situations, when the stratification at the surface is the strongest (Figures 18.4 and 18.5).

The present use of net radiation as an indicator of intermittency is considered no more than an interim measure. In practice, the reliance on net radiation follows from its association with cloud cover at night and the pragmatic association of both with the stability regimes in which the causes of nocturnal intermittency can flourish. The midwest of the United States is known to be affected by low level jets at heights typically about 300 m above ground level and these are currently postulated to be a central factor causing the SBL nocturnal intermittency now being considered.

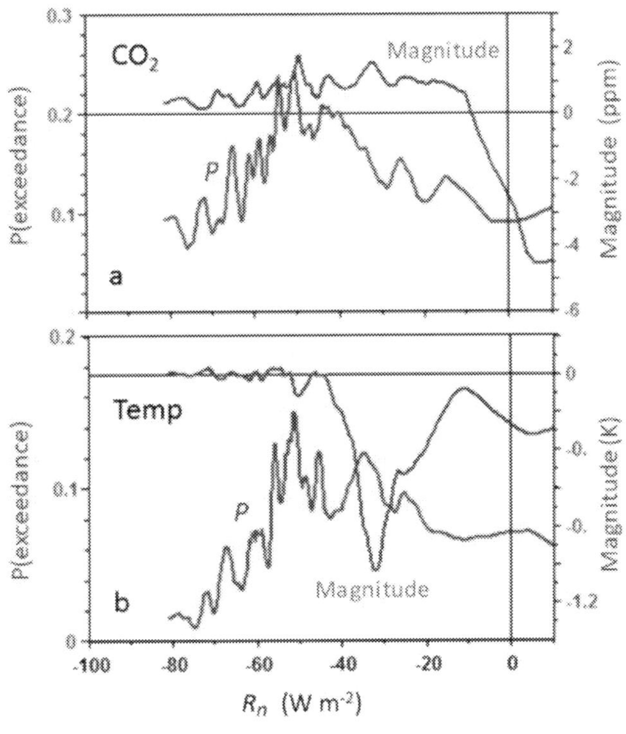

Figure 20.5. Using five-minute data, the probability (P) of finding a departure of more than two standard deviations from the 25–min period in which a measurement is nested, for (a) CO_2 concentrations and (b) air temperature, each as measured at about 2 m height above a fallow surface in Ohio in 2015. Also shown is the average magnitude of the exceedances. Results are shown as functions of measured net radiation.

Data obtained during a study of crop growth in central Ohio in 2015 (already used above) provide an opportunity to search for evidence of LLJ-related SBL nocturnal turbulence episodes. Figure 20.5 illustrates the results obtained. Figure 20.5a focuses on CO_2 data, showing the probability that any five-minute average of CO_2 concentrations (measured at about 2 m height) will be more than two standard deviations away from the average of the preceding five such five-minute averages. The results are shown as the probability of encountering such an exceedance in an hour, as a function of the measured net radiation. The results indicate a broad regime of high probability extending from about –30 to –60 W m^{-2}. Also shown are average values of the magnitude of such intermittences, showing that the nighttime cases of main interest here are consistently positive. This is as expected. The surface was fallow, with nighttime exudation from the soil. CO_2 concentrations in air in immediate contact with the soil would be expected to rise until interrupted by intermittent turbulence and hence the concentrations during a period of intermittency would be expected to be greater than the average, as observed. Note that this reasoning is not transferable to the near-neutral and unstable situations since the concept of slow pooling of the gas in the surface layer is then inappropriate.

Figure 20.5b addresses the temperature observations in an identical fashion. In this case, the nighttime probability peak is less pronounced. However, the exceedances are now negative. The behavior shown is again compatible with expectations. An enhanced diffusivity will improve coupling between the surface and the height of measurement, causing CO_2 concentrations to increase and temperatures to decrease as air from above is mixed with air at the surface.

The conclusion resulting from the considerations presented here can be simply expressed. At night, there is a finite probability that surface flux-gradient relationships will be disrupted by episodes of relatively strong turbulence arising from irregular processes external to the local SBL and hence not represented by the conventional SBL formulations. The situation in daytime is fundamentally different, because the causative factors generating the intermittency that is frequently observed are due to surface factors and are therefore taken into account in the standard flux-gradient descriptions.

Conclusion

Much as in daytime, nocturnal eddy fluxes often result from intermittent episodes rather than continual turbulent mixing.

Chapter 21

The Ocean Case — Some Common Issues

"It is all a matter of scale" — Will Pendergrass

The consequences of changing climate dominate global environmental thinking. The most important causes of these changes are certainly greenhouse gases and their persistence in the atmosphere. In practice, the systems of the earth struggle to maintain an equilibrium, or to generate a new one, and fluxes between these systems arise as mechanisms to restore such equilibria. There are two extreme aspects of this equilibrium picture. From the climate perspective, it is the long-term trend towards new equilibria generated by imposed changes in the controlling factors (most specifically, greenhouse gases that are of major concern. However from the immediate societal viewpoint it is the repercussions of shorter-term episodes that are most critical. Among these episodes of major concern are tornados, typhoons, hurricanes and tsunamis. In even the very recent past, there was insufficient understanding to simulate the details of such events, from origins to consequences. Today, understanding is growing as new observational methods become available.

The surface of the earth is largely ocean and depending on the time and space scales of concern the presence of these expanses of salty water must enter (if not dominate) consideration. While it is true that terrestrial micrometeorological experimentation is difficult to transfer to an open ocean situation, it is no longer true that observations of extreme oceanic conditions are unavailable. The advance is due to two technologies not available to researchers of the not-so-distant past — satellites and aircraft. The whitecapping discussion of Chapter 10 is of direct relevance to the use of satellite imagery, since the quantification of whitecap coverage can be used as a measure of the prevailing wind speed. However there are few in-situ measurements of wind speed with which to compare these satellite imagery inferences when the sea is very rough. It has been said (by an unknown but obviously experienced observer) "Once the average state of the ocean has arrived, any sane observer has long since gone home." As was emphasized above (Chapter 10) there is accumulating evidence that changes in surface films and surface tension will modify whitecap coverage. The considerations involved are necessarily micrometeorological, even though the identification of an actual "sea surface" might be impractical if not impossible.

To satisfy the micrometeorological experimental requirements developed for terrestrial applications, early open-ocean studies relied on the use of towers, either oil platforms (Hicks and Dyer, 1970; Smith et al. 1992) or similar installations in deep water or dedicated and instrumented towers in comparatively shallow, along the lines of the deployments yielding the data plotted in Figure 10.1. The Floating Instrument Platform (FLIP, q.v. Rudnik, 1964) provided a unique spar-buoy capability for investigating the open ocean, notably extending understanding from moderate wind speeds into fairly rough seas (e.g., Smith et al. 1987). Erecting instrumented towers near the bows of ships has been an attractive approach (e.g., Fairall et al. 1997; Miller et al. 2010) that has proved productive but is of limited utility when platform motion becomes severe. In most of these tower- or buoy-based operations, streamline deformation is an issue to be considered and dealt with.

To study air-sea exchange in high winds, airborne systems have become popular. Aircraft eddy covariance systems have been available for decades (e.g., Lenschow, 1986). In most cases, small aircraft fitted with fast-response anemometers detect velocity components which, when adjusted for aircraft platform motions, can then be used to derive measurements of the covariances as conventionally perceived, although with a requisite change from time-averaging at a single location to line-averaging along a specified path. The height at which covariances are computed is largely determined by aircraft safety considerations. Fluxes are measured at heights usually well above the reach of most tower installations.

The opportunity to make use of an experimental aircraft (a Rutan Long EZ; see Crawford et al. 1990) to yield covariances resulted in a series of studies over the ocean. After starting with studies over land, the focus shifted to coastal zones and eventually to areas more representative of open ocean. Measurements were made at heights between 10 and 20 m. These studies succeeded in demonstrating the expected exponential adjustment from a flux regime upwind that was characteristic of a vegetated land surface to a water-dominated regime representative of flow over the ocean, involving a distance constant of about 8 km. The diurnal cycles of measured covariances changed markedly as the distance from the shore increased. In a subsequent field program (Vandemark et al. 2001) the low-flying aircraft detected the presence of consistent variations in fluxes with distance scales of the order of 2 km, reminiscent of the intermittency occurring over land but attributed by the authors to roll vortices. These roll vortices were identified as contributing causes of the roughness of the sea surfaces, emphasizing the coupling that exists when the boundary is itself fluid. Further evidence that the atmospheric turbulence regime cannot be dissociated from the characteristics of the surface arose from the Coupled Boundary Layers and Air-Sea Transfer study, CBLAST (Edson et al. 2007). Even in light winds, momentum flux measurements indicated that the covariance extended through the surface, so that conventional quantification of a drag coefficient that assumes zero wind speed at the surface is misleading although probably not greatly concerning.

Black et al. (2007) report on aircraft measurements of air-sea exchange made by aircraft exploring the eyes of hurricanes. In this case, the conceptual picture is complicated, since the center itself could be relatively calm with a well-defined sea surface whereas the surroundings would likely be so disturbed that the surface is not easily identified. Quantifying such classical properties as the drag coefficient and the Dalton and Stanton numbers (as by Black et al. 2007) leaves some key questions unanswered. Here, the Dalton (Da) and Stanton numbers (St) are the energy equivalents to the familiar drag coefficient, quantified with reference to the 10 m meteorological conditions: $Da \equiv \overline{w'q'}/(\bar{u}_{10}\cdot(\bar{q}_0 - \bar{q}_{10})$ and $St \equiv \overline{w'T'}/(\bar{u}_{10}\cdot(\bar{T}_0 - \bar{T}_{10})$. In passing, it might be noted that 20 m waves are to be expected with a spray zone extending to a considerably greater height.

Conclusion

The atmospheric climate responds to changes in the oceans. Given my 'druthers', I would move many terrestrial research stations to strategically located islands (e.g., Kerguelen, Macquarie, South Georgia, Tristan de Cunha, Motu Mana) and fund them adequately.

Part III: Towards Complexity

Chapter 22

Revisiting Turbulence Statistics

"Every dogma has its day" — Anthony Burgess.

Conventional considerations of turbulence statistics in the SBL inevitably utilize u_* as the appropriate scaling velocity. The relevance of u_* in unstable stratification has been questioned repeatedly. Some insight can be gained by noting that the familiar ratio σ_w/u_* can be expressed as a function of the turbulence shape function σ_w/σ_u and the correlation coefficient between u and w:

$$\sigma_w/u_* = (\sigma_w/\sigma_u)^{0.5} \cdot |R_{uw}|^{-0.5} \qquad (22.1)$$

In practice, R_{uw} is a relatively constant property of the velocity field. Variations in σ_w/u_* are therefore mainly due to the injection of new vertical velocities such as when buoyancy becomes a major factor – i.e., in unstable stratification. Likewise, σ_u/u_* can be expanded to yield

$$\sigma_u/u_* = (\sigma_w/\sigma_u)^{-0.5} \cdot |R_{uw}|^{-0.5} \qquad (22.2)$$

Thus, the statistical turbulence shape ratios arise as central considerations. Figure 22.1 shows how σ_w/σ_u (a) and σ_v/σ_u (b) vary with stability. The data used are from the Ocotillo experiment (White et al. 2013; Hicks et al. 2014b).

Figure 22.1a focuses on the σ_w and σ_u comparison. The stability index employed is the Richardson number, Ri_B, based on the 8.5 m and 21.0 m observations. This particular stability index is selected to avoid any cross-talk among the variables included in the axes of the graphs. The variation with stability is as anticipated for unstable conditions, showing that for each height the ratio remains approximately constant and at a value attained slowly but consistently as instability increases. However, the plateau value changes markedly with height until the highest level of measurement is reached. One interpretation of Figure 22.1a is, therefore, that the direct influence of convective inflow is constrained to levels below the upper heights depicted, i.e., below some level in the 20 m to 30 m height range. The stable data also reveal a consistent variation with height of measurement, in the opposite sense than for the unstable case but this time with no obvious indication of an equivalent to the height constraint inferred for the unstable case. It is in the near-neutral region that there appears to be the greatest variation of σ_w/σ_u, where the average ratio maximizes.

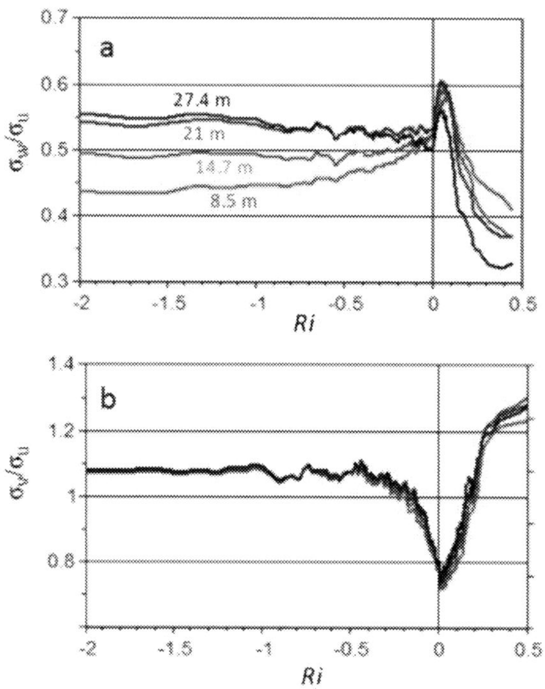

Figure 22.1. The dependence on stability of two turbulence shape ratios σ_w/σ_u (a) and σ_v/σ_u (b), derived from the Ocotillo (Texas) field study and plotted against Ri (8.5 m to 21 m). The lines show results for the heights as shown. (Based on Figure 9 of Hicks et al. 2015.)

Whereas the ratio σ_w/σ_u maximizes at what appears to be the slightly stable side of neutral in Figure 22.1a, in Figure 22.1b the ratio σ_v/σ_u is then seen to minimize. (Note that the departure from true neutral is likely no more than a result of the present lack of inclusion of humidity and evaporation in the evaluation of *Ri*.) High wind speeds characterize these near-neutral conditions, for which it is not unexpected that the contributing eddies will be stretched in the along-wind direction. An intriguing feature of Figure 22.1b is that the ratio σ_v/σ_u exceeds unity in unstable conditions. Earlier, Figure 13.2 failed to reveal this as clearly, because of the normalization then used. It is postulated here that the reason for the longitudinal shrinking (or cross-wind expansion) of eddies could lie in the nature of the convective cells that together define the velocity standard deviations. Consider the strongly convective case, with buoyant eddies moving freely with the wind. Well above the surface, it would be no surprise to find that these eddies would be circular in cross-section, on the average if not individually. However, as the surface is approached the wind will be pushing these eddies against the constraint imposed by friction, while at the same time buoyancy will be drawing air from the surroundings. At the lowest levels, the appearance of an eddy shape compressed in the longitudinal direction would be an expected feature, as has been verified by, for example, Wilczak and Tillman (1980). However, it also remains possible that the exceedance of σ_v over σ_u evident for unstable situations in Fig 22.1b could be due to instrumental factors, such as have been examined by Kochendorfer et al. (2012, 2013). Firm conclusions must await future studies. For the present, it is adequate to recognize that eddies quickly approach a constant cross-sectional geometry as stratification becomes more unstable, reaching a plateau beyond about $Ri = -0.04$.

In stable stratification, cross-wind exceedances appears even more pronounced. In very stable conditions meander would be a reasonable explanation, with σ_v being affected by it while σ_u and even more so σ_w decreasing as stratification starts to dominate. It is only in the near-neutral and fundamentally strong wind regime that σ_u and σ_v depart in the way depicted in many widely accepted formulations (such as have been satirized in Figure 13.1).

The utility of plotting the shape ratios as functions of *Ri* in Figure 22.1 is demonstrated by the four plots shown as Figure 22.2. These make use of the more conventional approach — using ζ as the stability measure. The dataset is the same for all plots shown as Figures 22.1 and 22.2. In Figure 22.2a, the plots are referred to a bulk stability property — ζ for a height of 10 m above the zero plane. This is a similar approach to that employed in Figure 22.1a, where all data are plotted as a function of the single bulk stability property *Ri*, derived from observations at 8.5 m and 21 m. In neither diagram is there allowance for the stability to vary with height. Nevertheless, the story is the same. There is an increase with height for unstable data, until the height approaches about 30 m. The scatter in Figure 22.2a is much greater than in Figure 22.1a — a consequence of the way in which the friction velocity and its component parts enter into the two quantities being considered. In reality, σ_u is a shared variable affecting all of the plots of Figure 22.2, in such a way that variability introduced by scatter in the measurements is amplified.

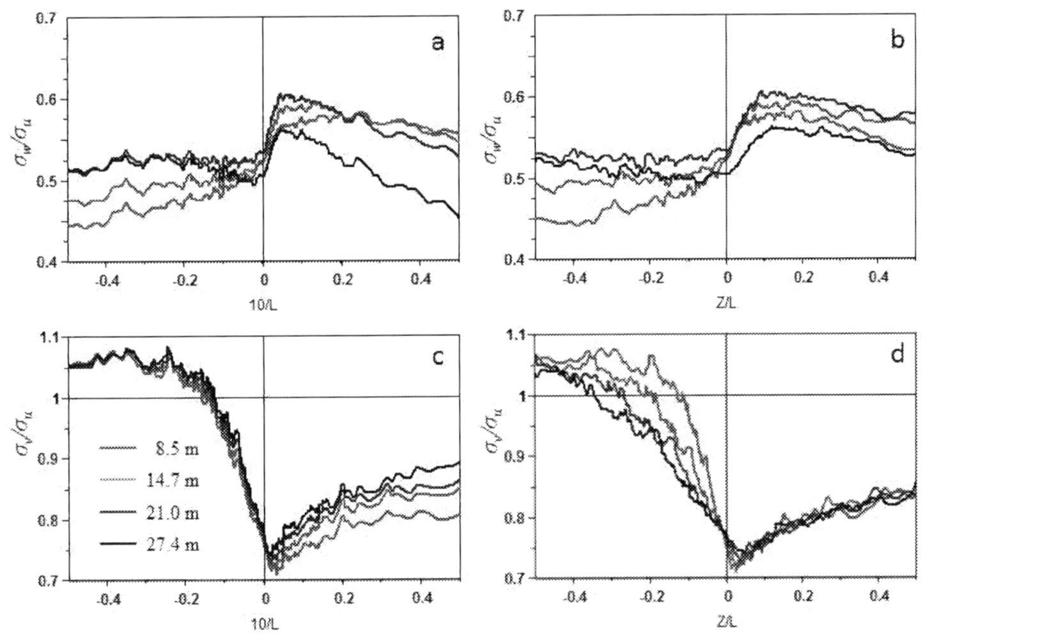

Figure 22.2. The data of Figure 22.1 expressed as functions of ζ = Z/L, using a single measure of stability to represent the entire profile, 10/L for the two diagrams to the left (a) and (c), and using conventional height-varying quantifications of ζ for the two diagrams to the right (b) and (d).

The diagrams to the right of Figure 22.2 depict the same data but show the dependence on conventional stability, varying with height. There is no readily apparent and convincing basis for detailed consideration of the differences between Figures 22.2a and 22.2b. Neither depiction causes the variations with height to collapse, hence both height and instability independently influence σ_w/σ_u. However, the comparison of Figures 22.2c and 22.2d shows clearly that the use of a bulk, height-independent ordering according to $10/L$ consolidates the transverse shape ratio dependence into a single curve, such that to $\sigma_v/\sigma_u \approx 0.7$ at neutral but asymptotes to ≈ 1.05 as instability increases. When examined in terms of z/L, the stable observations collapse into a single curve when plotted against z/L; the stability formulation accommodates the height dependence evident in Figure 22.2c.

An intriguing peculiarity of covariance statistics is that the often-mentioned quantity σ_θ/θ_* has little to do with temperature fluctuations. Expanding the covariances leads to

$$\sigma_\theta/\theta_* = (-|R_{wu}|^{1/2}/R_{wT})/(\sigma_w/\sigma_u)^{1/2} \tag{22.8}$$

The sole contribution of temperature information is via the correlation coefficient R_{wT}. Variations of R_{wT} and R_{wu} were examined earlier, but it is now apparent that a controlling consideration is, once again, the σ_w/σ_u shape ratio.

Conclusion

Neither Ri nor ζ is a perfect variable for exploring details of turbulence. In practice, use them both and interpret them together.

Chapter 23

Footprints and Fetch

"If a question needs to be asked, always question the answer" — Noel E. Ridere

If micrometeorological data are obtained over a surface differing from its surroundings, a question immediately arises: are these measurements indeed representative of the new surface? This question has been the concern of all experimentalists attempting to address the fundamental issues of micrometeorology and relates directly to the constraint that measurements must be made over spatially uniform surroundings.

Conventional micrometeorological wisdom holds that a new surface boundary layer is generated slowly as air moves from one surface situation to another. The differences in the two situations may be due to topography or obstructions, or to changes in surface roughness, temperature, or humidity. The effect is to generate a layer of air above the new surface, which is in equilibrium with it. Wind tunnel tests demonstrate that the rate of growth of the new boundary layer is greatest in the immediate downwind vicinity of the edge. Building upon this understanding, Garratt (1992) and Stull (1988) present what has become a familiar visualization of what happens when the wind strikes a surface of different roughness. A new surface layer develops, in equilibrium with the new surface and through which fluxes of heat and momentum (and presumably of other quantities as well) are constant with height. Within this layer, the flux/gradient relationships are those applicable to the new surface. Figure 23.1 is a familiar schematic representation of the situation now visualized, with measurements being made at a height Z_h (above the relevant zero plane) at a distance X downwind of the edge, within a growing equilibrated boundary layer of depth δ. Of course, it must be remembered that nowhere is nature as simple as is indicated by Figure 23.1.

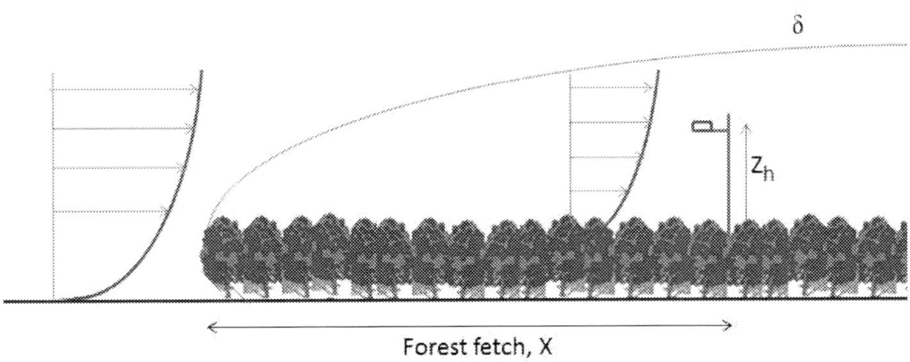

Figure 23.1. A schematic representation of the generation of an equilibrated boundary layer downwind of the leading edge of a forest canopy. The text discusses measurements at a height Z_h above the zero plane, at a distance X downwind of the forest edge, such that the measurements are within a constant flux layer of depth δ in equilibrium with the new surface. In the close vicinity of the treetops, a new roughness sub-layer develops, usually taken to be determined by the roughness characteristics of the canopy.

Obviously, there is no simple answer to the questions involved. There is no definitive level of demarcation corresponding to the height δ, instead there is a gradual and intermittent transition from mixing characteristics of the conditions upwind to those of the new surface. The situation can be addressed in several different ways, ranging from detailed simulation of specific situations to pragmatic avoidance by considering only data that are sufficiently close to the new surface. The operational questions are then what do 'sufficient' and 'effective' mean? Central in this line of thinking is the concept of the roughness sub-layer— the layer closest to the surface within which the consequences of individual surface sources and sinks remain evident.

In the case of the study of surface stress and the surface heat balance, theoretical analyses have led to a general acceptance that the depth of the equilibrated surface boundary layer (δ) increases at a rate which depends on several meteorological factors, especially including the roughness characteristics of the surfaces involved, their surface temperatures and atmospheric stability (*q.v.* Elliott, 1958; Dyer, 1967). Field experiments have supported the basic predictions of these theories and experimentalists have adopted as a practical expediency that micrometeorological studies should not extend above the level $Z_h \approx 0.005X$, where X is the uniform upwind fetch. The corresponding fetch/height ratio (200) is often quoted as a necessary requirement for micrometeorological research, although this is sometimes relaxed to a value of 100 and is acknowledged to be useful as guidance for daytime only. Irrespective of the details, when exchanges in the vertical are being considered it is necessary that full account is taken of changes in the horizontal and the easiest way to do this is to ensure that the flow being studied is in complete equilibrium with the surface below. Constancy of fluxes with height is an indicator of a satisfactory fetch, although as has already been seen this matter is confounded whenever the boundary layer is changing with time (as in the early post-sunrise hours).

There have been many field studies addressing the rate of adjustment of the SBL to a change in the surface. For example, Munro and Oke (1975) report that the adjustment from a comparatively smooth surface to a forest depends on the distance downwind of the edge raised to about the 4/5 power. Moreover, they caution against the application of the conventional fetch/height ratios when the edge of interest is as extreme as that of a forest downwind of pasture. McNaughton (1989) reviewed the downwind effects of both forest edges and shelter belts, emphasizing the fundamentally different micrometeorological consequences between transitions from smooth to rough and from rough to smooth. Once again, the assumption of some constant but preferred fetch to height ratio appears to be challenged.

A second key consideration relates to the estimation of the upwind 'footprint' of any measurement made above the surface, as illustrated in Figure 23.1. The footprint (as introduced by Shuepp et al. 1990) is viewed as the surface area most associated with the flux that is measured (q.v. Baldocchi, 1997). The problem can be viewed as one of simple geometry, as follows. The rate of mixing in the vertical is well described by the standard property σ_w. The rate of transport in the horizontal is $\bar{u}$. Over any sampling period, the area upwind that is associated with the relevant footprint is therefore determined by the ratio $\sigma_w/\bar{u}$ More detailed considerations result in a number of refinements (e.g., Rannik et al. 2000; Kormann and Meixner, 2001; Kljun et al. 2004), including attention to the cross-wind aspects of the upwind footprint area.

The property $\sigma_w/\bar{u}$ is an appropriate first-order indicator:

$$\sigma_w/\bar{u} = (\sigma_w/u_*)\cdot C_f \tag{23.1}$$

where C_f is the familiar friction coefficient. As already demonstrated, in daytime σ_w/u_* is a property of the field of turbulence and depends primarily on the height of measurement and secondarily on atmospheric stability. On the other hand, C_f depends on the surface roughness. Hence, Equation 23.1 breaks the problem into two pieces, one primarily depending on height and atmospheric stability, the other a function of the nature of the surface. Some of the simulations recently proposed formulate footprint-location relationships in terms of σ_w/u_* alone, overlooking the role of C_f.

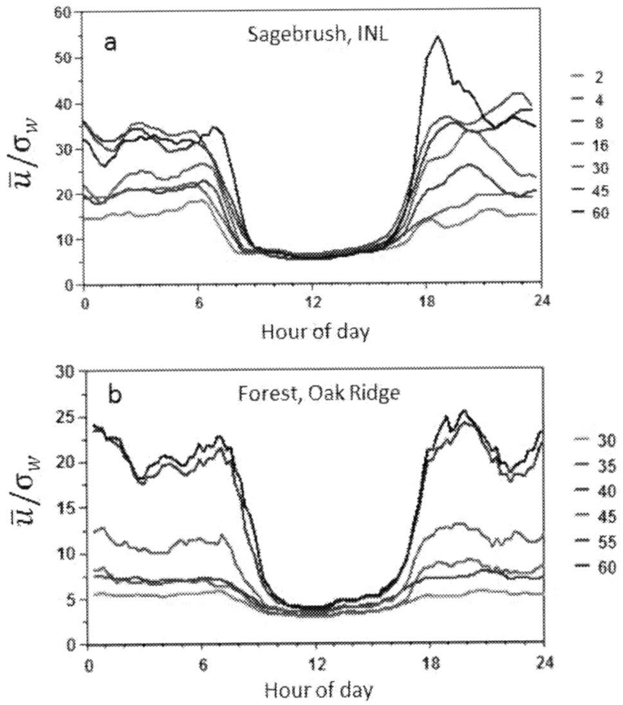

Figure 23.2. The average daily cycles of footprint ratios ($\bar{u}/\sigma_W$) derived from tall tower measurements above an arid sagebrush (a) and above a mixed deciduous forest (b). Heights of measurement (in meters above the ground surface) are as shown.

In the present case it is the inverse of the quantity expressed in Equation 23.1 that relates most directly to the footprint distance upwind of observations made at height Z_m. An appropriate 'footprint ratio' can be defined as $X_{ftprt} = Z_m \cdot \bar{u}/\sigma_w$. Figure 23.2 shows results from two 60 m towers, one operated over the sparsely vegetated experimental area at Idaho National Laboratory and the other over a mixed deciduous forest in Oak Ridge, Tennessee. The former data were collected as a part of the Sagebrush program conducted by the Field Research Division of the NOAA Air Resources Laboratory (ARL; 48.57N, 112.94W; see Finn et al. 2016), the latter was generated at the end of 2017 at the CHESS forest site of the NOAA/ARL Atmospheric Turbulence and Diffusion Division (35.93N; 84.33W; see https://www.atdd.noaa.gov/surface-layer-meteorology). In both cases, sonic anemometers were deployed at a number of heights above the surface. The heights listed in Figure 23.2 refer to the height above the ground surface. In contemplating the meaning of these results allowance should be made for the greatly different zero plane displacements of the two locations, estimated to have been zero in the Sagebrush case and about 27 m for the CHESS forest.

The Ocotillo dataset provides additional evidence. Figure 23.3 shows plots of the footprint ratio $\bar{u}/\sigma_W$ against stability, quantified as $\zeta = Z/L$. It is seen that the footprint ratio increases uniformly as neutrality is approached from the unstable side. (As pointed out previously, results for the lowest level, 3 m, appear somewhat anomalous.) This height uniformity extends into conditions of stable stratification, until about $\zeta = 1$. In more stable conditions, $\bar{u}/\sigma_W$ increases with height but appears to trend towards a constant as stability increases.

The data considered so far indicate that the unstable, daytime value of the footprint ratio is approximately constant, regardless of the height above the surface and of the kind of surface, The daytime value derived on the basis of Figure 23.3 is about the same as the average indicated by data from the two sites illustrated in Figure 23.2 – in the range 4 to 8. The stable, nighttime data are substantially different. In Figure 23.2 the ratio attains a value of about 20 to 30 at the greatest heights measured, apparently with little remaining height dependence above about 30 m above the zero plane. At lower levels there is a consistent variation with height. In this regard, the upper level data reflected in Figure 23.2 appear to be indicative of an intermediate between the lowest heights of measurement and a possible assymptotic height of about 30 m above *d*.

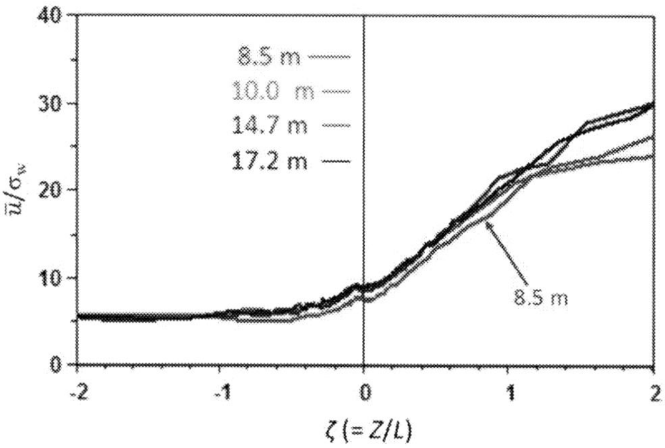

Figure 23.3. The dependence of the footprint ratio on stability. Data are from the Ocotillo experiment.

The footprint ratio considerations presented here are clearly little more than extended heurism. More detailed considerations have been presented in the past few decades, yielding more complex models to relate specific measurement situations (mainly over forests) to the originating surface or canopy areas (e.g., Leclere and Thurtell, 1990; Schmid, 1994; Baldocchi, 1997; Kljun et al. 2004). The expanded examinations of the footprint issue include allowance for the simple reality of profiles — in contrast to the simplistic considerations above, the ratio $\bar{u}/\sigma_w$ should not be constant with height.

Experimental evaluations of model predictions are scarce. Some guidance concerning the footprint ratio as defined here can be derived from results obtained in studies of dispersion. The inverse of the footprint ratio is familiar to dispersion meteorologists; it is closely related to the familiar standard deviation of the angle of elevation of the instantaneous wind vector, σ_ϕ Averages of these two quantities (σ_ϕ $\sigma_w/\overline{u}$) are not the same. Relating one average to the other is a trivial exercise in concept but a mathematical challenge, introducing terms that depend on stability. In practice, observations support the equivalency, as shown in Figure 23.4 where the first-order association that

$$\sigma_\phi \approx \text{Artan}(\sigma_w/\overline{u}) \tag{23.2}$$

is tested for central unstable and stable periods. To show that related results are not confined to the datasets already considered, the results shown in Figure 23.4 were obtained over a coniferous forest in South Carolina (Hicks et al. 2016). The points plotted represent 15-min averages at a measurement height of 61 m (39 m above the estimated neutral zero plane displacement, about 22 m) and are constrained to show the central daytime (1000 to 1600 hrs) and nighttime (2200 to 0400) periods. In these plots, high wind speed data are nearest the origin. The data support the expected near-linear relationship in strong winds, but with increasing scatter as wind speed decreases. The departure from linearity is in different directions for the daytime and nighttime cases, in accord with stability-related expectations.

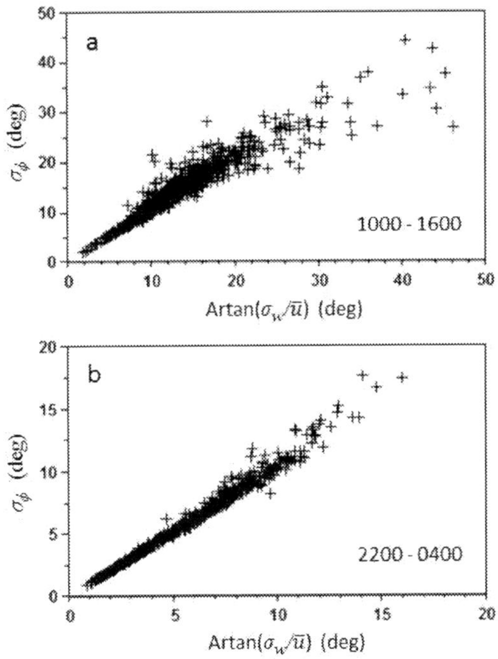

Figure 23.4. The relationship between the standard deviation of the wind angle of elevation (as is familiar in considerations of dispersion) and the angle associated with the inverse of the footprint ratio as defined here. Data are for the 61 m level of tower D of the Savannah River National Laboratory (see Figure 23.5).

Recent discussion of flux footprints has largely been in support of the carbon dioxide flux measurement programs now active across much of the global landmass. The focus of these programs is mainly on sequestration — the net uptake of atmospheric carbon by vegetation and its storage in the biomass and the soil. Most of the measurement locations are in forested terrain, where the carbon dioxide fluxes have been erroneously expected to be greatest and easiest to quantify. (In practice, the rapid growth periods of annual and deciduous plants yield CO_2 fluxes an order of magnitude greater than forests. The impact on the global atmospheric CO_2 regime of North American maize agriculture is probably considerably greater than that of North American forests.) Most measurement locations are equipped with eddy flux instrumentation at a single height, well above the anticipated zero plane.

Footprints and fetch are conceptually different, in that the former relates to where the surface most accounts for the fluxes observed at some height downwind, whereas the latter is intended to describe the upwind distance over which surface homogeneity can be assumed to account for all of the flux that is measured, i.e., so that data obtained are not unduly affected by upwind areas different from the surroundings of the measurement location. In both cases, classical micrometeorological guidelines recognize the substantial difference between daytime and nighttime conditions. A fetch/height ratio of 100 or 200 is commonly advocated for daytime and at least double this for nighttime, much as would be expected on the basis of the results plotted in Figure 23.3. Relating the discussion of footprint ratio above to the fetch/height expectations of convention is straightforward in concept but analytically demanding.

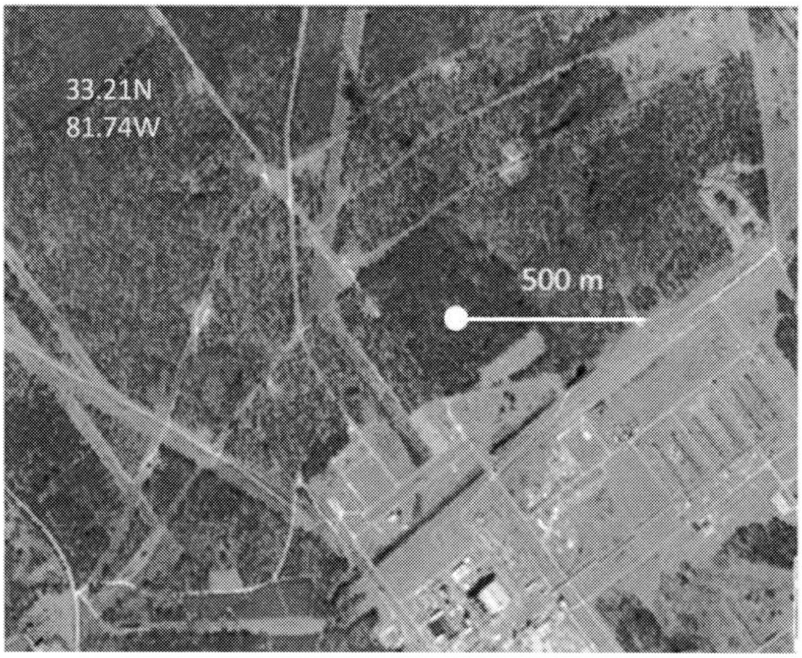

Figure 23.5. A view of the surroundings of the research tower 'D' on the reservation of Savannah River National Laboratory, in South Carolina.

Observations made at a study of forest micrometeorology at the pine plantation of the Savannah River National Laboratory in 2012 provide some useful guidance. In support of dispersion initialization, a network of 61–m towers across the SRNL reservation (q,v. Weber et al. 2012) provides wind field information, typically at a height of 61 m but with additional observations at different heights as required by particular circumstances. One location, tower

'D,' located as shown in Figure 23.5, was instrumented in 2012 to provide three-dimensional sonic anemometer data at heights of 36 m and 61 m. Details of the experimental program are given by Hicks et al. (2016).

The observations leading to Figure 23.4 have been considered in considerable detail elsewhere (Hicks et al. 2016). The aerial view of the site (Figure 23.5) shows that the forested area was bordered to the south by an extensive paved area. The measurements made permit regressions (among data in groups, ordered according to wind direction) yielding evaluations of the proportion of the observed 15–min TKE that is attributable to the observed sensible heat and momentum covariances $\overline{w'T'}$ and $\overline{w'u'}$ *TKE* is the classical turbulent kinetic energy computed from sonic anemometer data using

$$(\sigma_u^2 + \sigma_v^2 + \sigma_w^2) = 2 \cdot TKE/\rho. \qquad (23.3)$$

Figure 23.6 shows how this proportion varied with wind direction for the two levels of measurement (36 m and 61 m) and for the central daytime (1000 to 1600 hrs LST) and nighttime (2200 to 0400) periods.

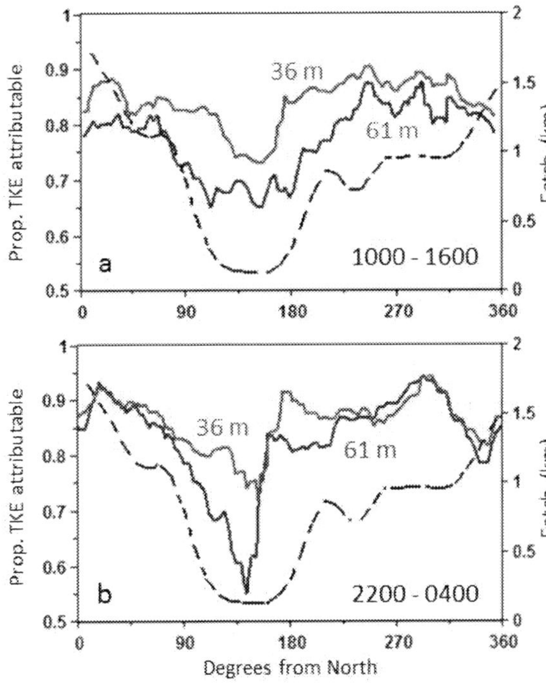

Figure 23.6. The proportion of observed TKE attributable to $\overline{w'T'}$ and $\overline{w'u'}$, for the two heights of observation of the Savannah River National Laboratory D tower, located as shown in Figure 23.5. The dashed line shows the measured fetch, derived from Figure 23.5. The zero-plane displacement is estimated to have been about 22 m.

The distance to the edge of the forest is also plotted in Figure 23.6 — a measure of the upwind fetch that is sometimes difficult to identify. The most striking feature is the close correspondence of the shortest fetch with the minimum in the proportion of *TKE* attributable to the forest. This is not a surprise. However, it is also clear that (1) the effects of the limited upwind fetch are most evident in the nighttime data, (2) that at night the 61–m observations (about 39 m above the zero plane, q.v. Weber et al. 2012) are far more affected by the memory of the upwind area than the 36–m results (about 14 m above *d*) and (3) that in daytime the effects of the fetch limitation are less severe but are spread across a broader sector. The correspondence between the turbulence data plots of Figure 23.6 and the fetch estimates is revealing, in that it confirms that the situation with the most demand for extensive fetch is the higher level of measurement at night.

The proportions of observed variance attributable to local flux measurements (as plotted in Figure 23.6) do not approximate unity under any of the circumstances experienced. An assumption that all of the observed velocity variance is due to local surface fluxes must be incorrect. Three possibilities arise immediately. First, the fluxes used in the present analysis might not be representative of the entire surroundings. Second, other causative factors might presently be ignored. Third, errors of measurement might be a major contributing factor. Reality likely includes contributions of all of these. However, the plots of Figure 23.6 confirm that association of the *TKE* with the forest is most appropriate for winds from outside the limited-fetch sector and that the contribution of other factors is larger at night than in daytime.

The quantity plotted in Figure 23.6 is the proportion of the measured total velocity variance that can be associated with the fluxes measured above the forest. Obviously, multiplying this proportion by the measured velocity variance represents the amount of variance the forest contributed, the forest attributable *TKE*. The data also permit quantification of the rate at which *TKE* (and hence the velocity variance) is being generated as air flows across the forest, presently approximated as

$$\varepsilon \approx u_*^3/kZ + (g/\theta)\cdot\overline{w'T'}. \qquad (23.4)$$

A more precise form of this relationship would include allowance for the role of stability, for example, but this complexity is avoided here as a matter of unwelcome complexity and unwarranted detail. The ratio of the forest-attributable velocity variance to ε indicates the time scale associated with the generation of the observed level of turbulence by the forest canopy. Multiplying this by the observed wind speed yields a distance scale Δx_f — indicative of the actual distance associated with adjustment of the field of turbulence to a change in the upwind surface. In simple concept, this corresponds to the footprint distance discussed above. Referring this last quantity to the height of measurement above the zero plane identifies an estimate of the appropriate footprint distance to height ratio.

In Figure 23.7 the footprint/height ratio $\Delta x_f/Z$ (= $\Delta x_f/(z - d)$) is plotted, following the form of Figure 23.6. It is evident that this ratio does not change much with height in daytime, generally hovering in the range 5 to 15 with a likely overall best fit over all wind directions of about 10. In comparison, the simplistic considerations presented earlier (involving $\bar{u}/\sigma_w$) would indicate a near-neutral value of about 5 for the equivalent footprint/height ratio, increasing substantially for the nighttime situation. Figure 23.7b shows (for the stable case) the expected increase in this ratio, but also shows a consistent and statistically significant difference between the results from the two heights considered. The present dataset does not warrant further examination from this perspective; the limits of credulity might already have been surpassed. The considerations presented here and the plethora of simulations recently presented to address the footprint issue indicate a need for further experimental study, especially because all of the simulations (to my knowledge) address the issue as if the source area defined as the 'footprint' were at the height of the zero plane, which may or may not be a relevant association. The matter of fetch adequacy is likewise in need of further attention. In the meantime, the safest option is to select sites so that the standard micrometeorological constraints are satisfied and to consider departures as opportunities for relevant research.

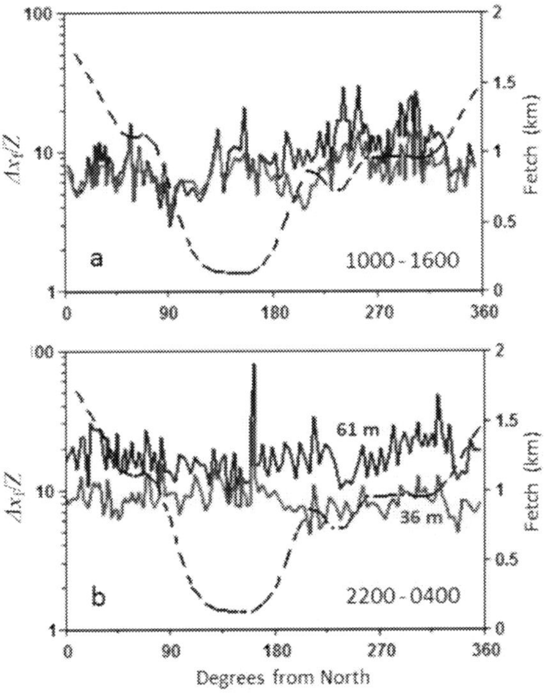

Figure 23.7. As in Figure 23.6 but showing the ratio of estimated footprint distance to height of measurement above the zero plane. As before, the dashed line shows the estimated fetch.

Conclusion

Conventional fetch guidelines are based on daytime understanding. Footprint considerations continue to evolve, however always beware of the nighttime case.

Chapter 24

Representativeness

"If a single viewpoint could represent a complicated mix, then democracy might work"
— Noel E. Ridere

A conventional test of model performance is to compare model predictions with observations, with appropriate consideration of how the statistical uncertainties involved diminish as the size of the sample increases. Clearly, the optimum situation would be to compare the predictions for model grid cells of minimum size with reliable observations made within similarly specified experimental areas. The matter of the spatial and temporal scales of turbulence has been examined in several field studies, as are conveniently summarized by Hanna and Chang (1992). Most of these studies focus on the characteristic scales of turbulence being transported across an instrumented grid cell (usually of carefully selected micrometeorological acceptability). These considerations do not immediately address the issue of current relevance — the matter of representativeness.

Micrometeorological field studies are most often over areas carefully selected to be uniform, but it is recognized that this does not impose a requirement that all observations made over this area must be the same. All of the variables of relevance are affected by turbulence, the consequences of which are evident in the variability associated with any measurement over such a uniform area. Clearly, if the standard deviation of wind speed at some particular location is σ_u, then this must be taken into account when considering how well this location represents the area as a whole. The practical consequence of such deliberations is usually to examine results after grouping them into relevant ensembles, so permitting the statistical uncertainties associated with any specific measurement to be reduced enough to satisfy some particular analytical requirement or (more often) the criticisms of some statistically uneducated reviewer.

Consider an area with a network of measurement systems, e.g., of wind speed. Then, consider a numerical model that requires the specification of a single measurement descriptive of a grid cell. Suppose there are n identical anemometers distributed across the area constituting such a grid cell. Over any particular sampling period, these anemometers yield observations which determine an average wind speed $<U>$. The standard deviation among them is quantifiable — σ_U. (Note that this is now a standard deviation in space, not in time as was the case in earlier discussions.) It is obvious that at any time, some of the anemometers will be reporting values higher than the spatial average and some lower. As an obvious corollary, it follows that no single anemometer can yield a measurement exactly equal to the spatial average. Nevertheless, the concept of a 'representative meteorological measurement' pervades the science, with much attention to specifying how such a measurement should be made. In reality the matter of representativeness could better be asked as "How can a single measurement be made to provide a best estimate of the spatial average?"

For the case of a 'perfect' micrometeorological site, a fundamental assumption is that an ensemble assembled from spatial network data will yield the same average as would an ensemble of data collected by a single station but sorted to represent the same sampling conditions. Typically, the sorting is based on stability, so that a result regarding (say) the flux-gradient relationship derived from a single tower in some particular landscape is then representative of what the result would be if the tower was located elsewhere within the same landscape. This principle of ergodicity underlies the basic assumptions of air-surface exchange and of micrometeorology in general, but it is rarely voiced (however, see Katul et al. 2004).

Data from the mesonet operated at the Idaho National Laboratory reservation (q.v. Clawson et al. 2007) offer an opportunity for a real-world test. The central part of the array is characterized by a flat surface sparsely vegetated by sagebrush. In this central area (of about 20 km x 20 km) there are eight mesonet sites, as indicated by the solid white points in Figure 24.1. These locations report 5–min average wind speeds and directions measured at 15–m height, plus temperature, humidity and solar radiation data. Outside this central area there are many other locations as will be considered later, stretching across the Snake River plain. During the Sagebrush dispersion experiment of 2015 (Finn et al. 2016), three additional locations were instrumented within this central area, using sonic anemometers at a height of 3.2 m. This small sub-network was within the central grid area.

Figure 24.1. The central array of the mesonet operated by the Field Research Division of the NOAA Air Resources Laboratory, within the reservation of Idaho National Laboratory. Attention here is concentrated on the 20 km x 20 km area in which eight reporting stations (white circles, with observations at 15–m height) are located. Note three white squares with central dots among the 8-member array; these were where measurements of eddy fluxes were made at 3.2–m height. The other points shown represent additional mesonet stations.

For purposes of the present discussion, data from one day have been selected: 21 August 2017. For every reporting time, the 5–min averages were used to derive a spatial average and a spatial standard deviation. Figure 24.2 shows how the spatially averaged wind speed ($\overline{U}$, from eight stations) varied through the day, with light black lines indicating the upper and lower statistical bounds: ± one standard error around the averages. These bounds illustrate the levels of uncertainty associated with the determination of the averages that are plotted. Also shown in the diagram is the variation of the spatial standard deviation derived from the same data, in this case showing the level of uncertainty associated with any single measurement.

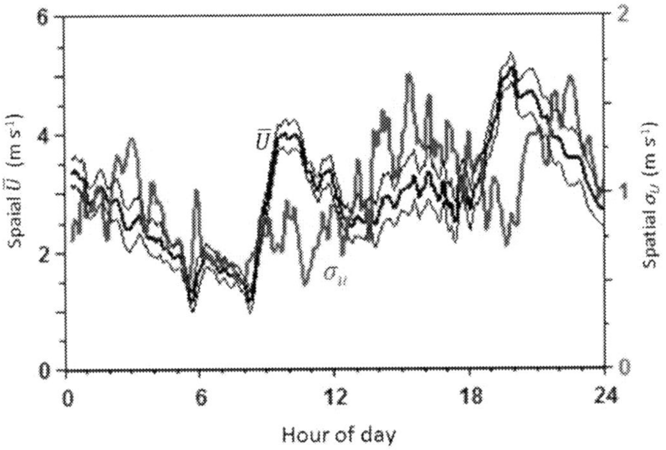

Figure 24.2. Average wind speeds based on eight measurements with identical systems distributed across a 20 km x 20 km area of Idaho National Laboratory. Averages and ± one standard error bounds are plotted in black. The single curve shows the corresponding standard deviations, as relate to any component of the eight-member array. The data are for 21 August 2017 and effects of the wind speed lull associated with the full partial solar eclipse on that date (late in the morning) are clearly evident.

The Idaho observations support the expectation that a model prediction of the average wind across the 400 km^2 domain of interest here should not be expected to yield a close equivalence with observations from any single site or even an array of sites. If there is only one anemometer to be used as a reference point, then a level of agreement better than about 1 m s^{-1} (and varying with time of day from about 0.5 to 1.5 m s^{-1}) should not be expected even if the model is well behaved. Elsewhere, it is shown that the lateral wind standard deviation does not vary greatly with height (q.v. Figure 13.2). Therefore, if the intent is to satisfy wind energy industry requirements (for which hourly-averaged wind speed predictions at a hub height of typically 60 m are most suitable) the ability to satisfy industrial desires for a precise prediction is confounded by the statistical behavior of the atmosphere even for the site considered here. The error margin considered so far relates to 5–min averages. A first order extrapolation of this result would indicate that the standard deviation affecting hourly averages would be about 30% of the range indicated in Figure 24.2 for the 5–min case (corresponding to $(5/60)^{0.5}$). This result applies to a 'perfect' site in unchanging conditions. Applying this line of reasoning to some particular situation would involve even more heurism or advanced modeling (probably leading to comparable results with similar credibility).

In practice, spatial arrays of measurement stations like that used here are rare. Most often available data are from a single measurement location. The question then arises as to how the matters of model predictability and spatial representativeness can be accounted for when data from only a single location are available. The ergodic nature of the lowest part of the atmosphere provides a path forward. The basis of conventional perfect-site studies is essentially that it does not matter where measurements are made in the selected research area. Provided the familiar site selection requirements are satisfied, the observations made at any location will have the same mean flow and turbulence characteristics as those made elsewhere in the same circumstances. The principle of ergodicity proposes that over a good site the variations in space of particular variables are like those in time. I.e., any single tower samples the spatial variations in the variable of interest as the atmosphere passes.

It is noted that some commercial providers claim to be able to predict the wind speed at some particular height with impressive accuracy. Their claim cannot relate to a single specific time interval or to a specific place. Instead, the claim can only be made with relevance to the averages reported on the basis of ensembling. That is, claims of a high-precision forecasts can be justified only if related to the prediction of ensemble averages. It is suspected that the purveyors of models with claims of highly accurate predictions also sell bridges in Brooklyn.

Some recent tests have revealed that the anticipation of ergodicity is appropriate unless there are changes that influence surface homogeneity and/or time stationarity. Even for an urban environment the ergodic assumption appears to be justified (Hicks et al. 2013). Using measurements made above the roofs of buildings distributed across Washington, DC and its surroundings, it was shown that the spatial variability of measurements averaged over a particular 15–min interval was about the same as the temporal variability experienced in sequential 15–min averages as measured at a single station within the same landscape.

The matter relates to the preceding discussion of footprints. If there are surface irregularities within the distance ordained by the footprint ratio, then clearly the concept of situating a measurement system to avoid the consequences of this upwind disturbance results in a need to view wind direction as a contributing factor. The fact that the disturbance in question might not be recognized by the model assumptions elevates the issue to one of first-order importance in model evaluations. It is for this reason that comparisons of model predictions with field observations are often considered after sorting according to wind direction. Any direction-specific instances of non-conformity can then be addressed.

In the study of dispersion, the issue of representativeness and of inherent uncertainty has received considerable attention. Stein and Wyngaard (2001) report an inherent (i.e., unbeatable) uncertainty of about 50% for a 1 hr prediction of dispersion, with a corresponding need for considerable ensemble averaging if the intent is to verify model predictions. Within the dispersion community, the term 'aleatoric' is used to describe the level of uncertainty that will remain unresolvable by deterministic models, regardless of their complexity. The parallel issue of how to organize data so that models can be properly initiated remains to be addressed. Given that it is well known that small departures in initial conditions can lead to wide perturbations in predictions (q.v. Lorentz, 1993), it follows that the ability of predictions to follow the natural progression of the atmosphere is intimately affected by the way in which the initialization process is carried out. Not only must measurements be refined to represent some uniform height above the appropriate zero plane but the statistical uncertainty associated with representing a grid-cell average by a single measurement must be addressed. The procedures involved in contemporary initialization programs are beyond the scope of the present treatment (i.e., the author knows nothing about them).

Conclusion

It is difficult to ensure that a single instrument reports precisely what it intends to measure at its time and place. It is far more difficult to extrapolate its observations to another time or place.

Chapter 25

The Peculiar Situation — Neutrality

"Custom adapts itself to expediency" — Tacitus c. AD 115.

Before revisiting the matter of neutrality it is appropriate to emphasize some of the conclusions reached so far, concerning legacy understanding of fluxes and profiles in the SBL.

1. It has been shown that the daytime terrestrial SBL flux-gradient relationship for sensible heat can be formulated in terms of free convection provided $\zeta < -0.03$ (or maybe –0.05). The value of H^* derived by Priestley is confirmed by modern data: $H^* \approx 1.3$.
2. In unstable conditions, classical descriptions of dimensionless gradients as functions of stability arise as smooth lines drawn through the consequences of free convection, after inserting the friction velocity to satisfy the requirement for a scaling velocity.
3. In unstable conditions, the momentum exchange varies as the most likely result between two extremes, the value corresponding to ϕ_H and the value associated with neutral. The most likely result is that $\phi_m = \phi_H^{0.5}$. In this regard, it is recognized that in strongly unstable conditions the turbulence regime is dominated by heat exchange and that the momentum exchange is modified by the intermittency imposed by convection.
4. In exceedingly unstable conditions, beyond (say) $\zeta = -2$, the mechanism causing almost all turbulent exchange is buoyancy. In this case, the dimensionless wind gradient ϕ_m tends to revert towards its neutral expectation.
5. At night, in stable conditions, gradients and fluxes are largely decoupled and the familiar associations between them (when expressed in dimensionless form) are mostly a consequence of the shared variable syndrome.
6. As stability increases, the turbulence causing the exchange becomes increasingly intermittent, with the consequence that all fluxes become well correlated in time.

This summary indicates that the factors causing exchange between the terrestrial surface and the atmosphere are dominated by heat exchange in daytime and momentum exchange at night. It is the intermediate case that appears to present the greatest difficulties. In this regard, it is appreciated that the present treatment is directly opposite to that of convention. Historical development has assumed that changes from neutrality form a reliable basis for describing the SBL in terms of stability. However, neutral stratification is a rare and usually transient circumstance. Often, what is postulated to be neutral is really a mix of stable and unstable short-term contributions selectively combined so that the average virtual heat flux is near zero. The requirement for a uniform field of turbulence through the selected period of neutral stratification is then violated.

The quiescent periods near dawn and dusk are especially vulnerable to violations of true neutrality. As has been repeatedly emphasized in this text, both the morning and evening transition periods are often characterized by the passage of bursts of turbulence, increasingly energetic as surface heating increases following sunrise and of gradually decreasing intensity as the SBL relaxes during the evening period. In both cases, and if conditions are right, there could be additional bursts due to the influence of PBL factors, such as upwind orography or any low-level jet that might occur. All in all, a search for true neutral conditions appears to necessitate considerations dominated by the wind speed. Extensive field experience led to early practical guidance: first, look for continuing strong winds under complete cloud cover in the period around dusk.

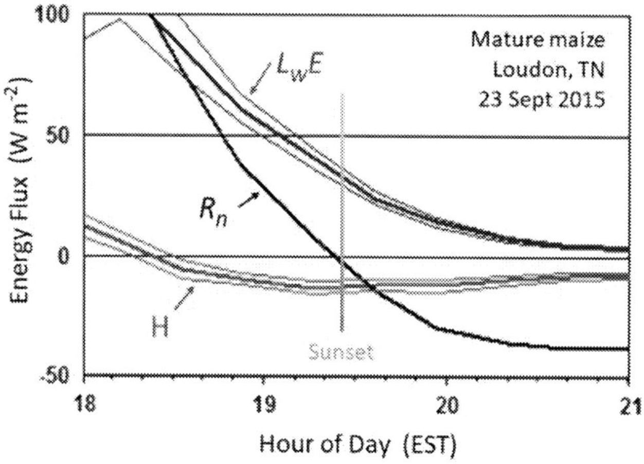

Figure 25.1. Data from one day of eddy correlation studies over a maize canopy at Loudon, Tennessee – 23 September 2017. Plots are shown of the major components of the surface energy balance; R_n, H and L_wE. The error bounds indicate the statistical uncertainty associated with the evaluation of average fluxes over a sampling period of ten minutes. EST is Eastern Standard Time, 5 hours delayed from UTC.

The question arises as to how well a period of neutrality can be identified. In this regard it is informative to consider the way in which eddy fluxes are determined and the standard errors associated with these determinations. It is recognized that the fluxes are a consequence of turbulence working at many frequencies and with a common lack of time constancy. Among the many ways in which eddy fluxes can be computed, a singularly useful methodology is to employ the digital equivalent to the analog approach used in many of the experimental campaigns of the 1960s and 1970s. In this, the mean-removal step involves the specification of a relevant mean value updated every time a measurement is made, say every second. To illustrate the statistical uncertainties associated with the evaluation of eddy fluxes at a single location, data collected during a field study at the Loudon (Tennessee) field site of the University of Tennessee will be used. Covariances shown in Figure 25.1 have been computed using the recursive filter method, using 1 Hz data with a 100 s filter. Statistical errors σ_{se} defining the error bounds shown in the diagram are derived from the n (= 900) separate measurements of the products $w'T'$ and $w'e'$ within each 15–min averaging period.

$$\overline{w'T'} = n^{-1}\Sigma w'T' \qquad (25.1)$$

and

$$\sigma_{se}(\overline{w'T'}) = [n^{-1}\Sigma(w'T')^2 - (\overline{w'T'})^2]^{0.5}/n^{0.5} \quad (25.2)$$

In general, a fifteen-minute average of the eddy fluxes measured at a single point is likely to have uncertainty associated with it such that the standard deviation associated with the determination of the average is about 10%. It would be expected that such would be the case for any location in the same area and hence that precise identification of the passage of neutrality would be impossible. From this pragmatic and instrumental perspective, it is apparent that the search for neutral conditions is likely a process of approximations, with awareness that the perfect situation is generally unattainable.

There is a considerable experimental challenge involved. The logarithmic wind profile relationship for neutral conditions requires specification of the height. Like in the development of the classical logarithmic wind profile, Monin-Obukhov similarity theory assumes that the vertical height scale of relevance is the height above the surface. As emphasized by McNaughton (2006), both developments relate primarily to the local association of gradients with eddy covariances. It is the vertical displacement of the log profile that determines the relevant zero plane displacement. In truly neutral conditions, a plot of wind speed against the logarithm of height will often be curved. The zero plane displacement (d) is then identified as that height which transforms a plot of $\bar{u}$ against $\ln(z + d)$ into a straight line. In view of the obvious susceptibility to alternative interpretations of measurements, an obvious pragmatic response routinely adopted by field experimentalists has been to select field sites such that any uncertainty that could be involved would have negligible effect on the interpretation of the experimental results. Hence, the earliest field studies were conducted over surfaces selected to avoid the need to account for a zero plane displacement. Experimental programs in Australia were conducted over well-grazed pasture (the studies at Edithvale and Kerang), over sparsely vegetated arid land (Hay) or over bare soil (Gurley). In the USA, the Kansas study and the subsequent Minnesota experiment (Haugen et al. 1971; Kaimal et al. 1976, respectively) were conducted with much the same philosophy and in Russia the site at Tsimlyansk was correspondingly favored. Having accounted for the zero plane displacement (d) and/or errors in it, the slope of the neutral wind profile and its intercepts determined the surface drag coefficient and the relevant roughness lengths.

To illustrate the various difficulties that arise, near-neutral wind profile observations derived during the Ocotillo (2013) experiment is Texas are plotted in Figure 25.2a. Average (15 min duration) wind profiles have been selected, such that $0.001 > 1/L > -0.001$. In a simple plot as in Figure 25.2a, inconsistencies are not easily detected. Figure 25.2b shows plots of wind speed against the logarithm of the height. The lines fitted to the observations are obviously curved, an indication that either the profiles are affected by a zero-plane displacement or by stability effects not yet accounted for (note that the contribution to stability of L_wE is not considered here). In Figure 25.3c, allowance is made for a zero-plane displacement of 0.7 m and the straight lines then form the basis for subsequent examination of z_0 and of the relevant friction velocity.

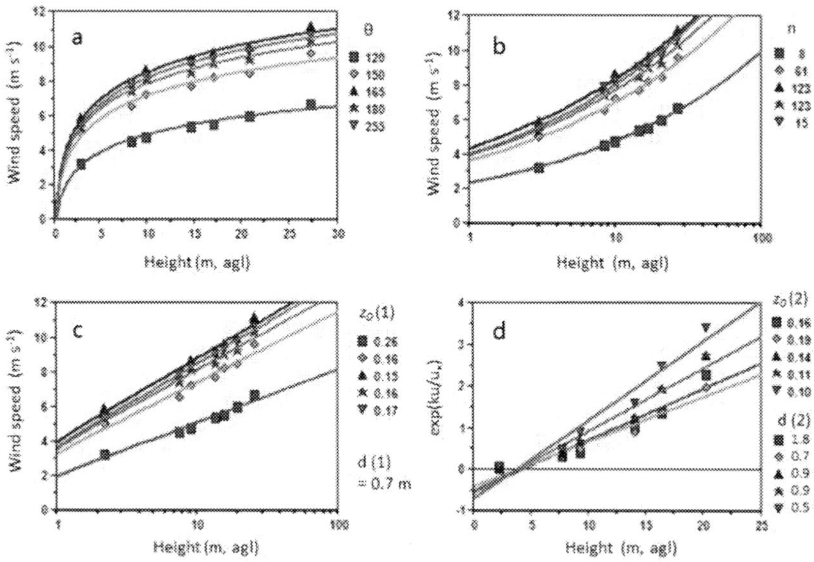

Figure 25.2. Examples of average wind profiles for Ocotillo winds from different directions. Four panels show the evolution of interrogation methods: (a) plotting wind speed against height; (b) after correction for an assumed displacement height, d; (c) plotting in the more familiar way of $\bar{u}$ versus (z – d) and (d) after allowing for the availability of $\overline{u'w'}$ measurements and revealing z_0 and d as derivable quantities. The same data are used in every plot, sorted according to direction (θ). N is the number if profiles contributing to every average.

The discussion above summarizes concerns that have pervaded micrometeorology for decades. But now that the friction velocity can be measured, it is possible to plot the quantity $k\bar{u}/u_*$ against height, z, as in Figure 25.2d. In this case, the best-fitting straight line identifies both *d* (derived from the *x*-axis intercept) and the roughness length (as the inverse of the slope of the regression line). The benefit of adopting the option illustrated in Figure 25.2d is that the quantities z_0 and *d* are both evaluated on the basis of the available observations. However, the correlation coefficients yielded by the plots in Figure 25.2c are greater than those of Figure 25.2d (typically 0.95 versus 0.90), because the uncertainties associated with quantifications of $\overline{w'u'}$ contribute in the latter case but not in the former.

Conclusion

When the atmosphere is undecided about whether it is stable or unstable, we observers tend to call it 'neutral.'

Chapter 26

Sources, Sinks and Zero Planes

"Progress is measured by the amount of complexity added to simple problems" – anon.

The twin goals of micrometeorological research are (usually) (a) to understand the processes by which heat, momentum, gases and particles are exchanged between the air and the underlying surface and (b) to formulate the association of these fluxes with the relevant gradients/profiles above the surface, i.e., to formulate effects in terms of the relevant causes. An initial step is to identify these causes. Over barren ground, the source of the sensible heat affecting the atmosphere in daytime is certainly the surface. In the case of momentum exchange, the source of momentum is the wind aloft. The sink is then at the surface. For the nighttime case, the surface and the structures on it serve as sinks for both heat and momentum. These statements represent simplifications that become poorly defensible whenever micrometeorological perfect site conditions are violated, or when vegetation and/or water surfaces interfere.

The discussion in Parts I and II of experiments conducted in and before the early 1960s related mainly to instrumentation and analytical practices predating the development of eddy correlation systems (especially sonic anemometers) for measuring the momentum flux. In the mid-1970s, measurements of the eddy flux of momentum became possible and sonic anemometers started to be used routinely. The resulting direct measurements of the momentum flux yielded a quantification of the friction velocity. As was introduced above (in Chapter 25) a plot of $\exp(k\bar{u}/u_*)$ against z will then be a straight line in neutral conditions, with slope that equates to z_0^{-1} and an intercept that identifies the displacement height d. Figure 26.1 presents examples of how this works, for two well-known forested sites – those of the Koorin experiment of 1974 (16.294S, 133.380E; 16.301S, 133.409E; see Clarke and Brook, 1979). In this study, instrumented towers were set up at two locations near Daly Waters, in the Australian Northern Territory. One site was sparsely vegetated with arid scrub, the other with a spatially uniform spattering of mainly acacia species with a height of about 10 m. Detailed site descriptions are given by McIlroy (1979). The plots show the averages obtained from all afternoon 1–hr runs for which the kinematic heat covariance was in the range $|\overline{w'T'}| < 0.001$ K m s^{-1}. There has been no other selection of which runs to include or exclude, except that no morning data are considered; experience has shown that the morning transition periods from stable to unstable are highly intermittent, whereas the evening transition from unstable to stable is typically smoother with time.

The interpretation here assumes that the wind profile extends in a logarithmic form to distances sufficiently close to the upper foliage that conventional wisdom would likely be offended. Garratt (1979) points out that the wind profiles obtained in this particular experiment failed to yield consistent values for z_0 and d when analyzed conventionally, however the observations indicated that $z_0 = 0.4 \pm 0.2$ m for site 1 and $z_0 = 0.9 \pm 0.25$ m for site 2. The present results for z_0 are 0.35 m and 0.96 m respectively. The displacement heights are about 4 m in both cases.

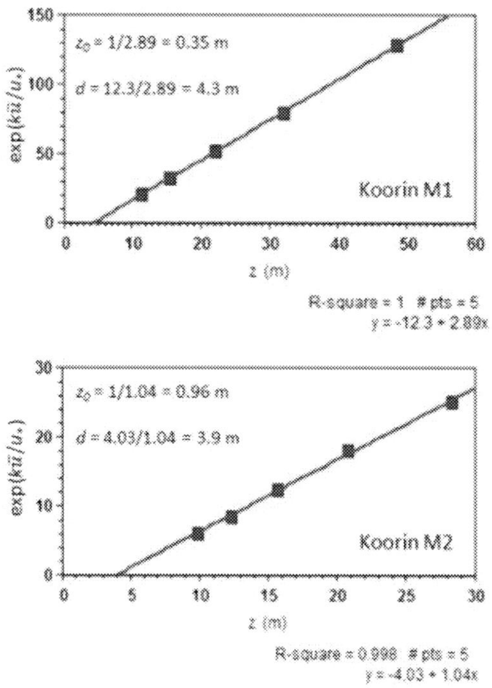

Figure 26.1. Average near-neutral wind profiles obtained during the Koorin experiment at Daly Waters, Australia, in 1974 (Clarke and Brook, 1979). Annotations show the results of linear regressions and the interpretations of them, yielding substantially different site calibrations for the two areas. Data for which $|\overline{w'T'}| < 0.001$ K m s^{-1} have been used.

Figure 26.2 shows the utility of the methodology now promoted, using (as in Chapter 25) data obtained in the Ocotillo (Texas) study. Examination of the data indicates that a logarithmic profile provides a good fit in stable conditions, provided z_0 and d are permitted to vary, but not in unstable. To select good stable-stratification datasets for examination, observations made within two hours of dawn and dusk occasions and/or for which the seven measured covariances (from 3 m to 27.5 m height) were not constant with height have been rejected. Each of the panels of Figure 26.2 presents averages constructed in groups of 50 occasions after sorting according to wind speed (measured at 10 m height) for the set of panels to the left and according to the tower-average $\overline{w'T'}$ –d join values corresponding to ± one standard error on the averages plotted. Figure 26.2a indicates that the apparent roughness length varies with wind speed, increasing as the speed increases above about 4 m s^{-1}. Figure 26.2b indicates that z_0 is approximately constant as the downward sensible heat exchange increases (as stability increases) until an extreme value is attained, about -0.05 K m s^{-1}. In practice, this high negative heat flux situation likely corresponds to the light wind extreme of Figure 26.2a, in which case the results are likely suspect (as will become apparent in the following).

The analysis indicates a variation of the displacement height necessary to maintain a logarithmic profile. The derived estimate of d depends strongly on the wind speed (in Figure 26.2c) but is apparently unaffected by the sensible heat flux (in Figure 26.2d) provided the high–flux extremes are once again disregarded. This is in accord with expectations — in stable stratification the sensible heat flux plays a role mainly as a damping influence on vertical turbulent motion and the dynamics of the lower atmosphere are almost completely at the mercy of the wind and its exchange of momentum at the surface. This characteristic is evidenced in panels e and f of Figure 26.2, where it is seen that the conventional stability property $\zeta = Z/L$ is almost completely determined by the wind speed whereas the influence of the sensible heat flux is small. The curve drawn in Figure 26.2e represents the inverse square dependence of ζ on $\bar{u}$ as required by the algebraic formulation of ζ and the near proportionality between u_* and $\bar{u}$. The shared variable syndrome strikes again.

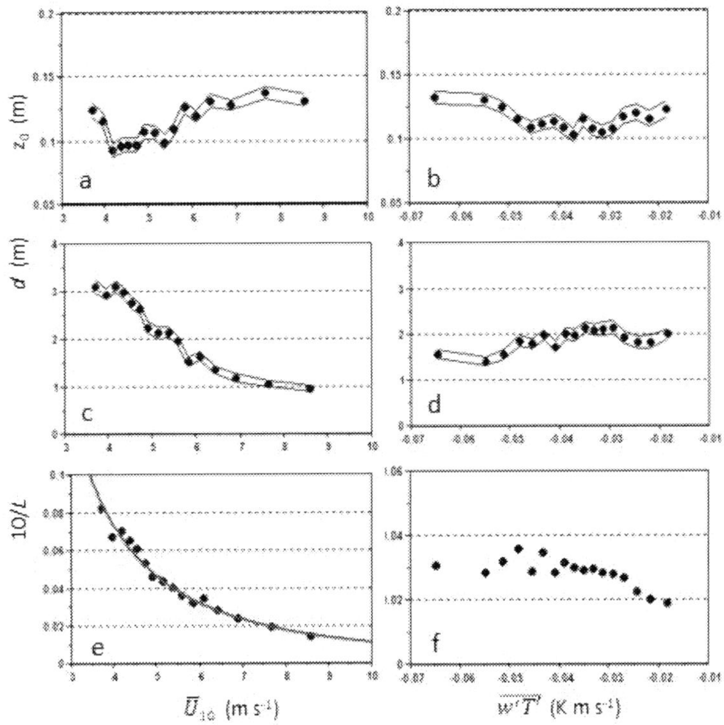

Figure 26.2. Results from fitting a log profile to stable stratification wind data from the Ocotillo (Texas) experiment. Data have been sorted according to wind speed (at 10 m height) in the case of then three panels to the left and according to $\overline{w'T'}$ for those to the right. Diagrams (a) and (b) show determinations of the roughness length (with ± one standard error bounds identified); (c) and (d) show the resulting values of the displacement height. Panels (e) and (f) show the corresponding dependences of the classical stability index, $\zeta = Z/L$ for the 10-m level.

An analysis by Zilitinkevich (2009) matches profile descriptions above the surface roughness layer with those below it, yielding an expectation that the roughness length will vary with ζ but with an emphasis on the unstable case. The current analysis shows that the logarithmic shape of the wind profile remains a characteristic of the stable SBL and indeed through the early stages of low heat fluxes into unstable (not shown here), provided the roughness length and displacement height are permitted to change. It is only after the sensible heat flux reaches some critical value (corresponding to the familiar free-convection initiation value of $\zeta \approx -0.03$ or -0.05, or some-such) that the sensible heat exchange starts to override the momentum-driven turbulent exchange process.

The results presented here are clearly at odds with classical interpretations of the stable wind profile. An obvious interpretation is that the wind profile in stable conditions is affected by local flows associated with terrain slope. As will be discussed in the following Chapter, evidence from other sources is accruing that such katabatic flows are associated with the changes in apparent d as now deduced. The Ocotillo field site was certainly susceptible. The data now available do not permit additional examination of this aspect. It must also be noted that the present dataset is limited by its lack of water vapor flux measurements and hence estimates of ζ are imperfect.

Further examination of this and other points related to the formulation of the wind profile in the stable and near-neutral regime demands more observations; the analysis here is already contradicting much of convention and is perhaps stretching credulity. The physical interpretation of Figure 26.2 (which begs for independent examination) is that the nighttime SBL wind profile could retain its basic logarithmic form when allowance is made for changes in the zero plane and the effective roughness length.

It might be expected that the precision of any such analysis will be improved if measurements are made very close to the surface. However, measurements made too close to the surface will reflect the deformations imposed on streamlines by individual surface obstacles (e.g., protruding branches). Since it is the goal to derive spatially representative values for z_0 and d, the measurements used must be obtained sufficiently far above the surface that this single-element influence can be neglected. The region within the SBL that is affected by individual surface roughness elements is conventionally referred to as the Roughness Sub-Layer (RSL). A thorough experimental approach might employ several instrumented towers, so that the role of surface element obstruction might be addressed and observations analyzed accordingly, however practicality calls for reliance (usually) on a single instrumented tower in which case a pragmatic and probably conservative 'rule of thumb' has arisen, such that micrometeorological wind speed measurements for examination of the displacement height should not be made closer than about ten times the roughness length above the displacement height. Note that it is necessary to make an assumption about the displacement height before advantageously deploying the instruments then used to determine the displacement height. Such is life.

It is clear that MOST provides a superbly practical and utilitarian answer to the obvious question about how the average relationship between fluxes and gradients changes as atmospheric stability varies through neutral stability. It might be asked whether modifying MOST would make any difference, to which the automatic response is that getting the science right is a necessary step towards generating confidence in any conclusions. The nighttime situation has been accepted as presenting the greatest challenge. The considerations presented here offer an incentive to think afresh. Or so it is hoped.

Conclusion

It is not obvious that the atmosphere agrees with textbook discussions about z_0 and d.

Chapter 27

Kinematic Isolation in Strong Stability (KISS)

"My height is not what defines me" — Napoleon

All of the legacy relationships that are now widely accepted result from theoretical considerations of simple circumstances, defining the 'perfect sites' of micrometeorology. Verification has also been concentrated on the sites where these relationships are most likely to work. Studies of situations in which the relationships are not expected to work are far fewer, in spite of the reality frequently acknowledged — that more is learned when observations disagree with expectations than when they agree.

Some relevant background: The arrival of the nuclear age in the 1940s caused urgent attention to issues of complex terrain. Nuclear research and testing of reactors and weapons caused many areas of dedicated research to be set up in the United States (and probably in other countries as well, but here the focus will be on the US experience). The fear of accidents releasing radioactive materials into the air with grave consequences downwind caused these research sites to be in carefully selected remote areas, all operated under the auspices of the Atomic Energy Commission and/or the Department of Defense. In time, many of these areas grew into the national laboratories of contemporary times, now operated as components of the US Department of Energy (e.g., in Washington, Nevada, Tennessee, Idaho and South Carolina. All such locations required meteorological attention, usually supplied through collaboration with the US Weather Bureau. A defensible and reliable methodology for predicting the dispersion of radioactive particles in the atmosphere following an accidental release was needed, for situations where meteorological predictions of surface winds were well known to be poor. Research into micrometeorology necessarily eventuated. Micrometeorological networks were set up, distributed across the reservations identified for testing but usually extending far beyond. Many of these networks are still operating.

Studies conducted in the ASCOT program had previously associated such low-level intermittency with the interaction between competing downslope regimes, all lying below the height of influence of the synoptic flow aloft (e.g., Porch, 1982; Porch et al. 1991). Key contributors to the dimensions of these competing flows included the aspect ratios of the terrain features as well as their dominant slopes, providing the winds aloft were not strong enough to overwhelm such katabatic effects (Barr and Orgill, 1989). The consequences of these various considerations do not lead to easy generalization.

The Idaho reactor testing reservation, now the Idaho National Laboratory, attracted considerable attention because of its proximity to the Rocky Mountains and the expected influence of canyons injecting gravity flows into the regional velocity regime associated with the Snake River plain. Early studies of the influence of terrain on the velocities detected by ground-level instrumentation involved the use of radar-tracked constant-level balloons (Dickson and Angell, 1968), tracer releases of various kinds and extensive numerical modelling. The balloon trajectory studies revealed frequent occurrences of helical flows of about 2 km diameter (Angell et al. 1968; Angell, 1972). These were hypothesized to be indicative of roll vortices, perhaps triggered by canyon flows. The more-energetic nocturnal canyon flows were observed to undercut the weaker downslope velocity fields affecting the Snake River plain (Clawson et al. 2007).

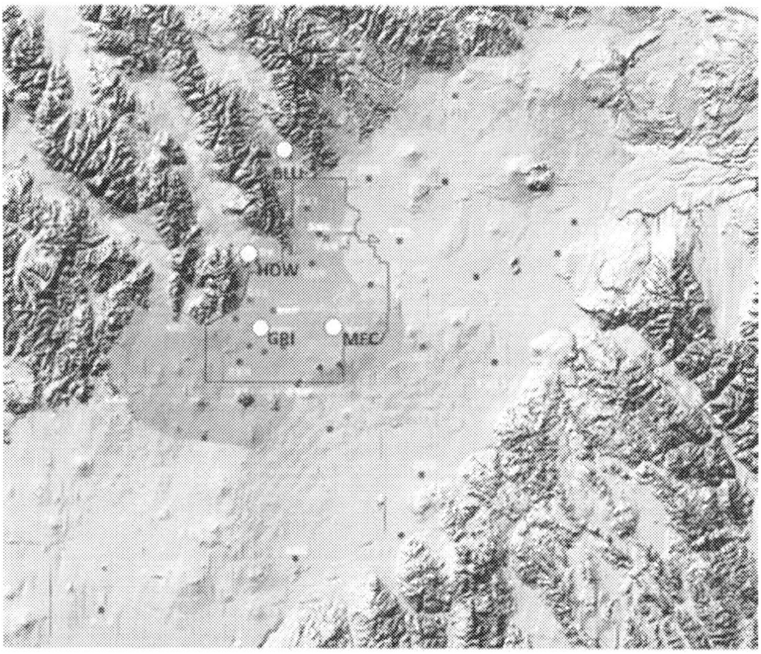

Figure 27.1. The network of 15–m towers (some taller) operated by the Air Resources Laboratory of the US National Oceanic and Atmospheric Administration in the flood plain of the Snake River, to the immediate east of the Rocky Mountains in Idaho. The network is designed to meet the requirements for initiating dispersion calculations. Four specific locations are emphasized, these being used to examine the extent of influence of canyon katabatic winds at night.

A mesoscale network operated for the Idaho National Laboratory has now been operational for best part of a century. This network, located as indicated in Figure 26.1, immediately to the east of the Rocky Mountains and in the flood plain of the Snake River, has provided copious supporting information on the influence on regional-scale wind fields of localized katabatic flows from canyons (Finn et al. 2016; Clawson et al. 2007). Figure 27.2 gives examples of observed wind roses for three of the network locations, two at the mouths of canyons (BLU and HOW in Figure 27.1) and a third near the center of the Snake River plain (MFC). Central daytime and nighttime periods are depicted. During daytime, the BLU observations (Figure 27.2a) show evidence of classical canyon upflow with some residual persistence of the downflow occurring at night. The HOW location (Figure 27.2b), being further down its canyon and expected to be more influenced by the larger-scale regional flow, shows the expected dominance of the Snake River flow field (upslope in daytime, from the south) but also with some persistence of the downflow canyon phenomenon. There is no canyon influence evident in the MFC wind rose (Figure 27.2c). At night, downflows dominate the canyon observations (Figures 27.2d and e) but evidence of their affecting the distant MFC location is absent (Figure 27.2f).

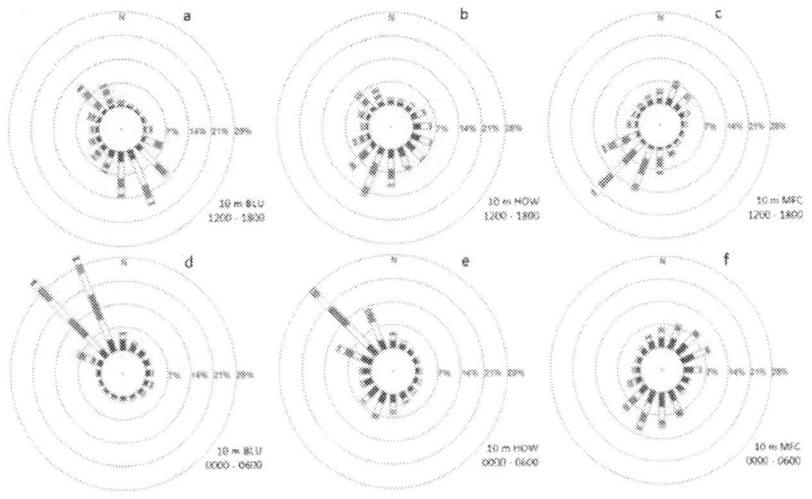

Figure 27.2. Examples of daytime (a, b, c) and nighttime (d, e, f) wind roses at 15 m above ground level for three locations as identified in Figure 27.1 — BLU and HOW are at the mouths of canyons; MFC is nearer the center of the Snake River Plain, about 80 km distant from the canyons. Velocity increments are 3.2 m s^{-1} (5 mph).

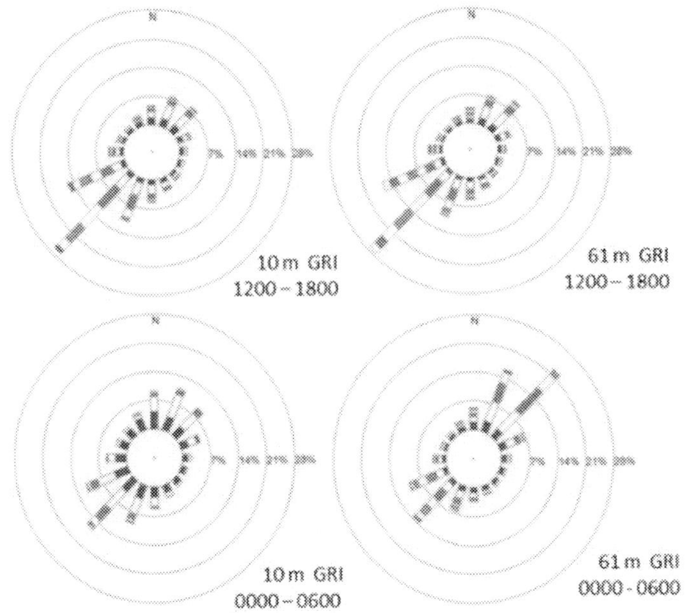

Figure 27.3. Wind roses corresponding to two heights of measurement (10 m and 61 m) at the GRI tall tower, about 30 km from the mouths of the canyons.

At location GRI in Figure 27.1, tall tower instrumentation yields evidence that the lower height of measurement (10 m) was influenced by the canyon flow, but that the 61 m level was not so influenced. Only one of the four GRI wind roses shown in Figure 27.3 indicates the influence of canyon flow persisting as far as the GRI location — the nighttime lower height (10 m) shows a greater frequency of light winds from the west (the direction of the canyons) than do the other three roses and the influences of the Snake River katabatic flow fields (upslope as well and downslope) are reduced. This provides additional evidence of the undercutting of the larger-scale flow field by the energetic canyon downflow, but the indications are strong that the effective range of disruption of the larger-scale flow field did not extend beyond a few tens of km.

All of the other nuclear installations were affected in ways that demanded individual attention. It was recognized that the range of circumstances requiring attention by research meteorologists at the various nuclear research centers (later, the array of U. S. National Laboratories) that many of the complex terrain issues being confronted would best be investigated by a coordinated, multi-laboratory program. To this end, the US DOE Atmospheric Studies of Complex Terrain (ASCOT) program was initiated. The focus was on katabatic flows occurring in canyons and affecting wind fields downstream. The field studies yielded observations supporting the Idaho conclusions that young and energetic gravity-driven flows tended to undercut the larger-scale velocity regimes. In simpler terms, at night the local terrain tended to control the local lower atmosphere.

The matter was explored in considerable detail during the Sangamon series of field experiments in central Illinois (Hicks et al. 1981), which demonstrated the characteristic nighttime decoupling of the surface layers of the atmosphere from the velocity field aloft. Low level jets were inevitably encountered by tethered kytoons, at heights in the range 200 m to 400 m above ground level. There were many notable occurrences of jet-generated turbulence intermittency, with the packages of turbulence sometimes being detected at ground level. Measurements made during these episodes were rare. On several occasions, the episodic turbulence destroyed the sensors.

Best part of a century earlier, scientific interest in katabatic winds was energized by the observations reported by Sir Douglas Mawson of exceedingly strong downslope winds observed at the lower extremities of glaciers in Antarctica (Mawson, 1915). These strong flows did not undercut the larger-scale velocity field they encountered but injected themselves into the wind field they encountered at whatever height was ordained by the topography. The physical processes involved were examined by Ball (1956), with mathematical explanations of many of the downslope flow characteristics and extending to the development of roll vortices (rotors) downstream of a glacier. A key feature of the glacier downslope winds was intermittency, with periods of lulls occurring within the otherwise continual strong katabatic flows. Looking at this behavior a century after its discovery, it seems likely that the strong downslope flows eventually exhausted the cold airmass that served as the hydraulic head, with a consequent curtailment of the downslope flow until the cold air resource was replenished.

The CASES experiment of 1999 (Poulos et al. 2002) has provided a unique set of relevant observations. The location of the CASES-99 tower cannot be described as 'micrometeorologically satisfactory' but the observed wind profiles provide an opportunity to examine the katabatic undercutting issue. In particular, three-dimensional velocity information was obtained at many heights on the tower, clearly illustrating the influence of local gravity flows while the atmosphere above their influence retained its logarithmic wind profile (see Hicks, 2022). Examination of this elevated wind profile revealed that it carried the memory of its upwind exposures, with turbulence and fluxes indicating a roughness length as expected for the area as a whole but with a computed displacement height as expected on the basis of the katabatic considerations. This Katabatic Isolation in Stable Stratification (KISS) feature is compatible with all of the previous experiences discussed above.

In particular, the CASES wind profiles show that at about 2200 LT an initial flow disruption occurred, followed by the generation of a layer of locally generated flow lasting until sunrise, with direction varying uniformly with height. The wind at a nominal height of 1 m above the surface was oscillatory, with cyclic variations in wind direction of about 50–min period. These oscillations decreased with altitude, supporting the KISS conjecture that the causative process was at the surface. At a height of between 20 and 30 m, the locally influenced flow regime merged with the larger-scale flow above it, apparently as imposed by synoptics as well as by regional-scale terrain (q.v. Hicks, 2022).

In direct contrast with the classical interpretation of wind profiles above the surface in stable conditions, the body of data considered here does not support the reliance on the height of measurement above the ground as a central variable. Rather, the relevant height measure is the height above the level of kinematic isolation of the wind field by whatever local katabatic flows might exist. It is obvious that this elevation will be site-specific and that further examinations of this (along the lines of the analysis of ASCOT observations by Barr and Orgill, 1989) would be useful. McNaughton (2009) has emphasized that a major shortcoming of classical MOST is that the height above the surface is assumed to be a central variable, whereas the development of the MOST construct is based upon considerations of gradients, not of bulk properties.

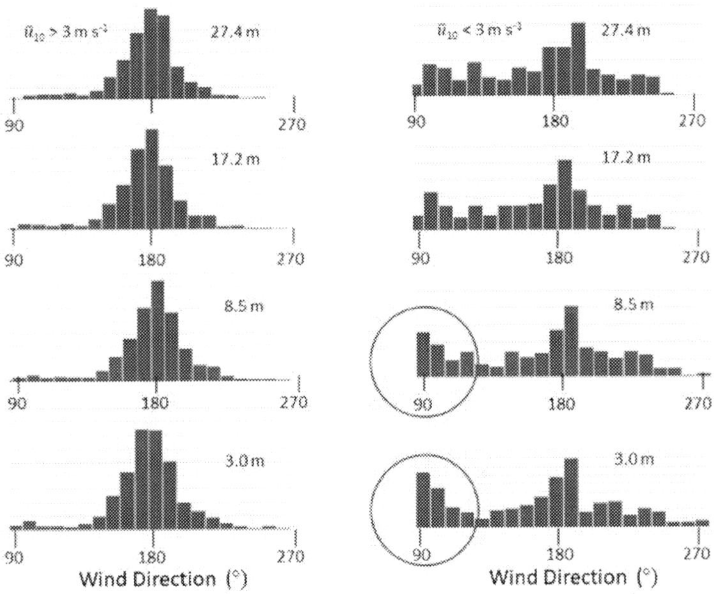

Figure 27.4. A selection of histograms showing the frequency of occurrence of winds from different directions, using Ocotillo data for the local time period 0100 to 0400 hrs. The diagrams to the left are for wind speeds greater than 3 m s^{-1}, to the right wind speeds less than 3 m s^{-1}.

The analysis of Ocotillo data presented earlier, with results presented in Figure 26.2, can now be considered with suspicion that in stable conditions the wind at the lowest height of measurement was not always well associated with the velocity regime above. The correlation analysis that led to the data of Figure 26.2a included all wind speed observations made below 28 m, lacking the higher-level observations of the CASES-99 experiment. Figure 26.7 presents an example of the way in which wind directions changed with height during the night. The histograms are based on data obtained at four of the seven tower heights during the early morning (0100 to 0400 local time). One set of histograms is confined to cases with the wind speed at 10 m height less than 3 m s^{-1}; the other is for wind speeds $\bar{u}_{10} > 3$ m s^{-1}. It is clear that in light winds a new wind field enters consideration, as indicated by the circles, flowing from the east and with influence extending upwards through 8.5 m and 10 m (not shown) but not noticeably affecting the 14.7 m velocity regime. Examination of geological maps indicates that the Ocotillo site was central in a large plain with a maximum slope downwards to the northwest. The large-scale slope was small (about 1‰) but its maximum direction was such that widespread gravity flow would be expected from the southeast, should the imposed synoptic wind field permit.

Note that a first order estimate of the depth of the katabatic flow (h) derived by assuming that the upwind surface was extensive grassland (with a friction coefficient $C_f \approx 0.12$ relative to the 10–m level) can be derived by equating the tangential gravitational component to the drag, so that

$$\alpha hg = u_*^2 \qquad (27.1)$$

where α is the slope. Hence

$$h \approx \bar{u}^2 C_f^2 / (\alpha g) \qquad (27.2)$$

from which it is estimated that the depth of the layer now considered would have been in the vicinity of 13 m. The apparent agreement with observations as presented in Figure 27.4 is probably little more than good luck, considering the crudity of the assumptions made. The development by Ball (1956) provides a better way to examine this matter but the dataset now available does not warrant such attention.

On the basis of the discussions regarding the stable wind profile and kinematic isolation, it is concluded that the conventional log-linear description of the flux-gradient relationship at night is likely a pragmatic response to a d = constant assumption when in reality d varies according to the local slope and katabatic flow. Agreement of observed wind profiles with classical log-linear expectations would then be bordering on fortuitous, unless observations were made at a site (or at sufficient height) where effects of local and regional slopes were negligible. Indeed, such sites have provided much of the historic body of data on which the log-linear expectations have been tested, with results that indicate some scatter in the constant involved that could be a consequence of otherwise disregarded terrain effects.

Conclusion

KISS is common at night.

Chapter 28

Cities and Urban Areas

"People who live in glass houses, shouldn't" — Noel E. Ridere.

Cities and urban areas present an example of complex terrain in a way that elevates its societal importance — it is in such locations that people work and live. In recognition of this reality, there have been countless intensive studies of the surface-level wind speeds involved. Tall towers are frequently installed and the data derived from their instrumentation used to examine the applicability of conventional micrometeorology. Studies of urban meteorology have yielded a substantial understanding of the peculiarities of the urban environment and have grown in importance following terrorist activities of the last twenty years. Early work was summarized in the review by Munn (1966). The extensive considerations presented by Oke (1974) were followed by many experimental programs such as the studies at Uppsala in Sweden (Högström et al. 1978) and in Chicago (DePaul and Sheih, 1985). Subsequent studies have continued to focus on detailed consideration of the many contributing processes, especially those that offer clear differentiation from 'perfect site' expectations. Among these more recent intensive studies have been Urban 2000 (Salt Lake City, Utah, Allwine et al. 2002), the 2001 Urban Boundary Layer field study (Marseille, France; Mestayer et al. 2005; Masson et al. 2005) and the BUBBLE program of 2002 (Basel, Switzerland; Rotach and Zardib, 2007). Each of these reports addressed an intensive study of the meteorology of the cities of relevance. All of them revealed important aspects of the association between external forces and the characteristics describing the surface of main interest. It remains uncertain as to how much their results could be extrapolated to other cities and/or urban areas. Much of the related research has been in the context of dispersion, as will be discussed later. Results obtained in such studies usually emphasize the complexity of the problem.

In this regard, Washington, DC, offers an enticing testbed for research, because of its constraint on building heights. With rare exceptions, building heights are limited to less than 40 m agl. Following the terrorist attacks on New York City and Washington DC on 11 September 2001 and with awareness of the results of a year-long dispersion tracer experiment in Washington (METREX; Draxler, 1985), a research network (DCNet) of micrometeorological instrumentation was installed in Washington, starting in 2002, using 10 m towers standing on the roofs of buildings. The network sensors were intentionally arranged to report on the wind fields affecting the local population and controlling exposures in the event of a release of some hazardous substance from a ground level source within the urban region. The intent was not to repeat intensive investigations of the past, but to focus instead on how routinely available observations could be used to guide urban forecasting as well as atmospheric dispersion calculations should there be a release of some hazardous material into the air.

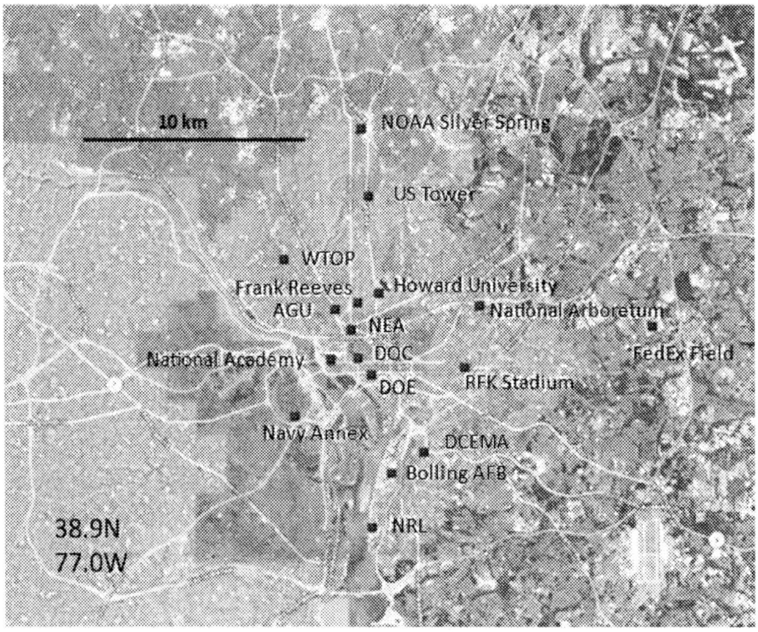

Figure 28.1. The DCNet array of roof-mounted micrometeorological towers operated as a research program in Washington, DC, and surrounding areas, commencing in 2001.

Given the considerations of representativeness presented earlier, in Chapter 23, a first question confronting the DCNet program related to the extent to which demands of first responders in the targeted area can be satisfied by a single observation. Clearly, no single set of meteorological instrumentation could hope to indicate relevant areal averages in detail. Nor can any single measurement be assumed to define the wind field at another location at the same time. However, the pragmatic issue of relevance to dispersion calculation is whether a single measurement can be interpreted as indicative of the meteorological situation at some other location within an error margin small enough to be accommodated within the consequent statistical dispersion routine without great loss of confidence. In many cities, there are measurements made of relevant quantities, but these are usually the standard meteorological variables — wind speed and direction, temperature, humidity and (perhaps) incoming solar radiation. The DCNet experiment was set up to test the utility of the expanded set of model initialization data now available from a variety of weather forecast models which provide estimates of turbulent kinetic energy and of momentum and sensible heat fluxes. To this end, DCNet deployed sonic anemometer systems at all locations.

Figure 28.1 shows the network of measurement locations as it existed in 2008. The initial focus was on studying the spatial variability of the conventional micrometeorological properties, associated with the wind and temperature regimes. The need for investigation was substantiated by some of the earliest results, which showed a distinct difference between wind fields detected at DCNet sites and the nearby airport observations (from Washington Reagan National Airport, as is evident in Figure 28.2 — frequently used in various forms to illustrate the limitations of airport observations). The routine systems of the weather forecasting community rely on the airport data and clearly there is need to correct these if the intent were to predict the winds affecting downtown Washington. These initial observations were tested further in work that will be summarized below.

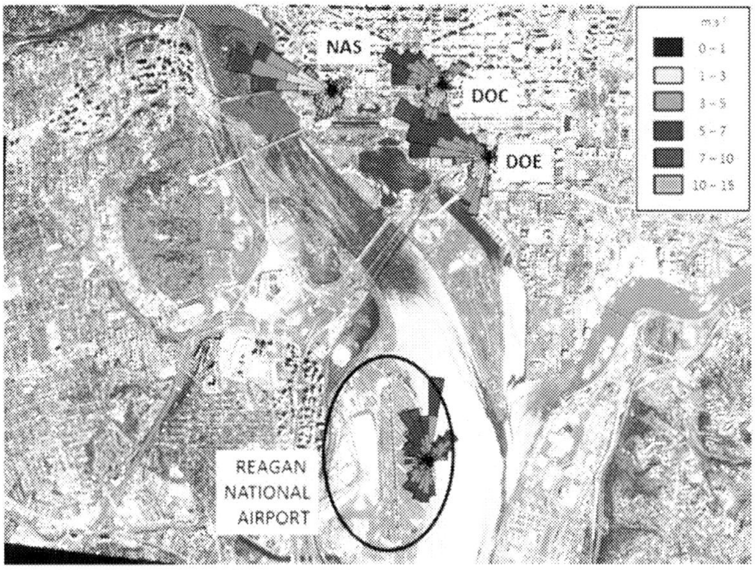

Figure 28.2. Wind roses from three DCNet sites (the National Academy of Sciences, NAS; the Department of Commerce building, DOC; and the Department of Energy, DOE) and as derived from routine observations made simultaneously at the nearby Reagan National Airport.

Figure 28.3 shows how observations made at the central business district of Washington agree. In every case, measurements were made at a height of 10 m above the local surroundings, usually above a roof. The two panels show the variation in wind speed (the upper panel) and the momentum covariance (the lower). The similarity between the two panels is striking. On the whole, it does not matter too much which of the several measurement locations within the central business district is used to represent the area. However, this conclusion should not be extended to the entire capital region (as was well illustrated by Figure 28.2).

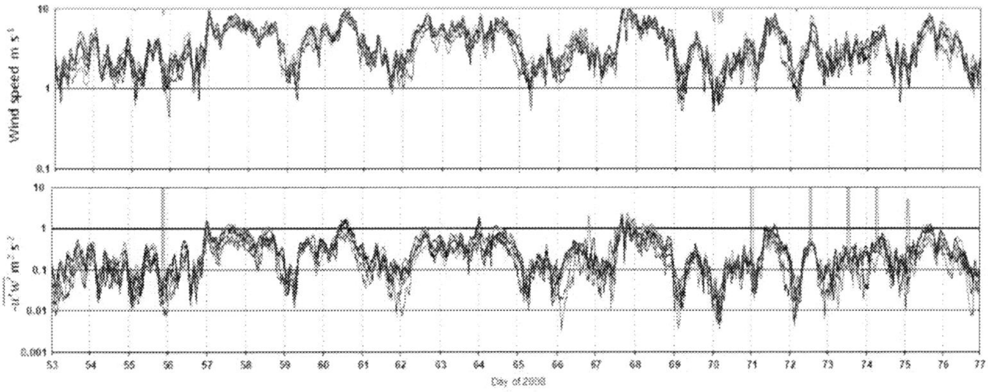

Figure 28.3. For the month of March 2008, separate measurements of wind speed and momentum covariance at seven DCNet locations within the central business district of Washington, DC (see Figure 28.1). Data have not been subjected to removal of wildly anomalous results (WARts) and the repercussions are evident in some of the records.

The METREX tracer studies (Draxler, 1985) indicated a characteristic of the Washington terrain reminiscent of that discussed earlier in the case of the CASES-99 experiment — the larger-scale wind field tends to overflow the region of influence of local obstacles. Wind profiles measured by SODAR above a building in central downtown Washington (the Department of Energy, DOE, building identified in Figure 28.1) reveal the wind field decoupling in more detail. Analyses by Vogel and Pendergrass (private communication) are summarized in Figure 28.4. Average wind profiles are plotted, for different wind direction sectors to show that the behavior is independent of the characteristics of the upwind surface. Before averaging, profiles have been normalized using the SODAR observations for a height of $z_h = 4h_m = 120$ m, where $h_m = 30$ m was the height of the building. This normalization height has been selected because it introduces the most order into the profiles then resulting. Independent of the wind direction, the normalized profiles reveal a point of inflection at about 55 m, 25 m above the building. Further analysis reveals that above this height the average wind profiles approximate the neutral expectation with d ≈ 25 m. The similarity with the CASES-99 observations already presented is striking, but likely somewhat fortuitous. On the whole, the most that can be said is that the urban roughness area appears to cause the oncoming wind field to be elevated above a locally controlled velocity regime in two distinctly different ways. Firstly, the Potomac River valley imposes a katabatic flow regime below the larger-scale velocity field; secondly, the urban area of increased surface obstruction and roughness imposes its presence on the larger wind field.

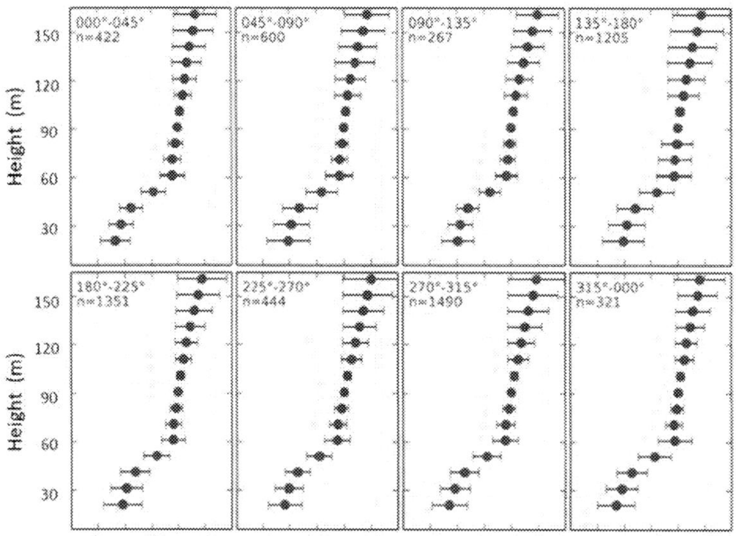

Figure 28.4. Average normalized wind profiles observed using remote probing above the Forrestal building of the Department of Energy (DOE in Figures 28.1 and 28.2). This figure is based upon unpublished results derived by Chris Vogel and Will Pendergrass of NOAA/ATDD, to whom thanks are due.

The installation of a wind profiling system on the roof of the H. C. Hoover building in downtown Washington (identified as DOC in Figures 28.1and 28.2) has enabled more information to be gathered about the wind profile above the city, extending upwards to more than 2 km. Figure 28.5 summarizes observations made during May 2021. Each profile drawn represents the average of observations made at the same time of day — hourly from 1200 to

1600 local time and hourly from 0000 to 0400. Fine lines show example profiles obtained at times between these day and night extremes. The daytime profile appears to be linear, up to some indeterminant height apparently slightly above 2 km. Nighttime profiles are also linear, but only up to about 300 m. Although such data above 2 km are comparatively rare, at heights above 2 km the daytime average profiles tend to asymptote to the prevailing free-stream PBL wind speed of about 10 m s^{-1}. The same limit is evident for heights above about 300 m for the nighttime cases.

The near-surface detail derived from the SODAR observations made from the DOE location is not evident in the DOC profiles of Figure 28.5. Whether this results from a difference in the sites and their surroundings or from the greater height resolution available using the SODAR system remains to be seen.

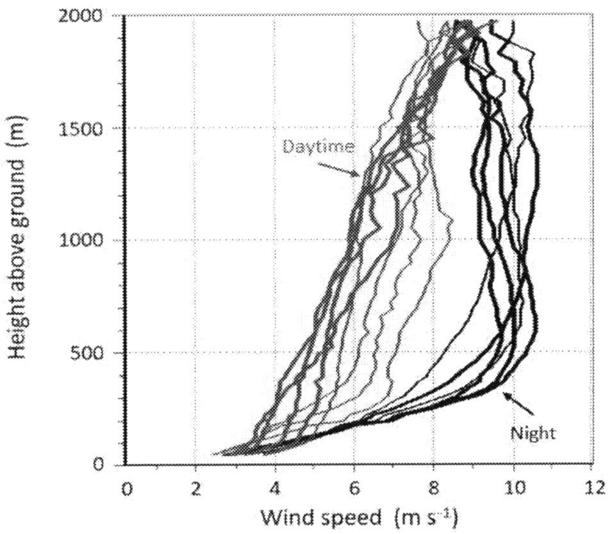

Figure 28.5. Wind profiles derived in May 2021 using LIDAR above the DOC location of Figure 28.1. Heavy lines represent results obtained at night (0000 to 1400 local time) and in daytime (1200 to 1600). Fine lines are examples of profiles at intermediate times.

In Chapter 27, discussion of kinematic isolation in strong stability (KISS) introduced the benefits of computing surface displacement and roughness characteristics from plots of $\exp(k\bar{u}/u_*)$ versus height. The same procedure can be used to examine the low-level velocity data extracted from the profiles illustrated in Figure 28.5. The discussion of Chapter 27 leads to the conclusion that stable wind profiles can be considered as logarithmic but raised in the vertical by some local features, thus affecting the displacement height, *d*. Adopting this same understanding to the lidar wind profiles derived at the DOC location, the results illustrated in Figure 28.6a are obtained. Three candidate profiles are shown, each representing a 20–min wind profile obtained at midnight. Two are in moderate winds (plotted against the left-hand axis) and one is for light winds (using the right-hand axis). The regression analysis results indicate averages of $z_0 = 0.44$ m and $d = 8.0$ m, with $R^2 = 0.82$.

Figure 28.6b illustrates the conventional technique for examining neutral wind profiles. In this case, near-neutral has been defined on the basis of low sensible heat fluxes and wind speed exceeding 4 m s^{-1}. The solid line in Figure 28.6b is drawn through the average wind speeds that are plotted, assuming that the relevant displacement height was 25 m — about 80% of the height of the rooftop on which the lidar system was mounted and about 60% of the height at which sonic anemometry was installed. This is in line with conventional expectations, resulting in an estimation of the friction velocity (0.66 m s^{-1}) for the assembled dataset, with a roughness length of 0.75 m. However, inspection of the data indicates that a curve would be a better fit and hence that a different value of *d* could be an improvement. The points and the dashed line drawn through them correspond to the results derived from Figure 28.6a. There has been no manipulation of the data other than to change *d* from 25 m to 8 m in the presentation. The data no longer display curvature, so verifying the relevance of the regression method now promoted. Moreover, note that the regression analysis of Figure 28.6a was of data collected at midnight, whereas the data of Figure 28.6b relate to near neutral. Once again, the evidence points to a stable wind profile that is no more than the neutral profile elevated to a new displacement height (q.v. Chapters 15, 16 and 25).

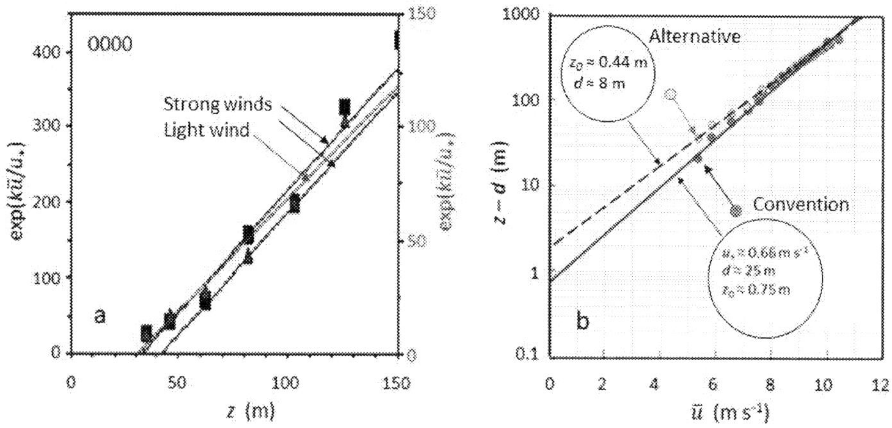

Figure 28.6. Two interpretations of observations made by a lidar wind profiler atop the Department of Commerce (DOC; H. C. Hoover) building in Washington, DC, in May 2021. To the left is a representation (a) of the regression method preferred here (see Chapters 25, 26 and 27); to the right (b) is seen a more conventional interpretation of near-neutral observations and the way this would be modified by the alternative approach now promoted.

Within the community of DCNet stations, the DOC case is an extreme. It appears that the building climate control can be a dominant factor at night. Maintaining a tolerable working environment within this building requires considerable heating in winter and cooling in summer. Figure 28.7 illustrates the consequences of air conditioning for several locations by showing how the nighttime values of $\overline{w'T'}$ vary with the outside air temperature. Data from six locations have been selected for illustration. The DOC example represents one of the oldest and largest CBD structures, largely poorly insulated stone exterior. AGU and NEA are more modern buildings, with less apparent air conditioning consequences. The CSPN data were obtained at a greater height (84 m) on a convenient tower and represent a larger and more residential area.

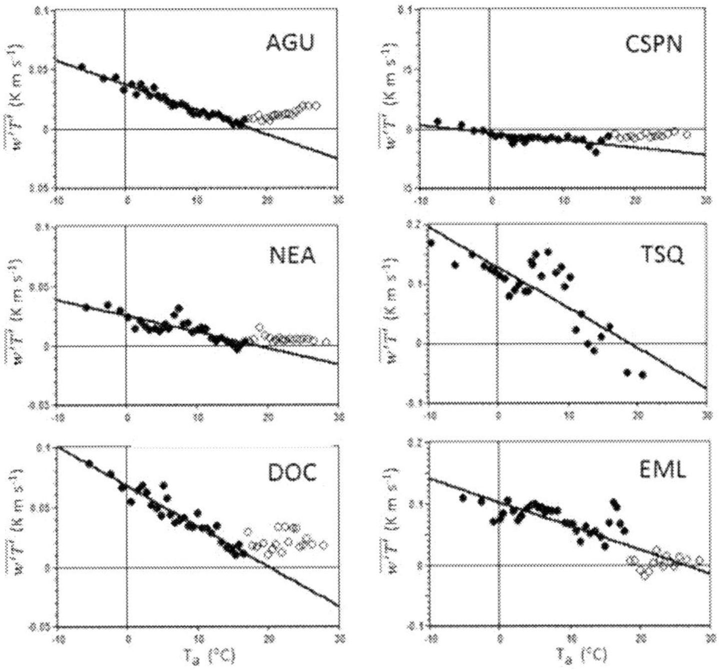

Figure 28.7. Relationships between measured kinematic heat fluxes, $\overline{w'T'}$ and prevailing air temperature, exemplifying the consequences of working space heating in winter. Note that TSQ and EML were two sites in New York City, data from which are plotted with different vertical scales. Details of DCNet deployments are given in Appendix 3.

Also represented in Figure 28.7 are results from two locations in New York City, operated identically with stations of the DCNet array. One of these locations (TSQ, instrumentation at 125 m agl) was above a building near Times Square. The other (EML, at 36 m agl) was in lower Manhattan, near Houston subway station. The surrounding areas are different, with Times Square characterized by irregular tall buildings and with the other location being in an area of comparatively homogeneous smaller structures much like DOC in Washington. The TSQ data provide the best evidence of spatial nighttime heat modification so far encountered. The TSQ observations were obtained at a height of 125 m; they represent a larger and more heavily structured and populated area than EML, somewhat like the CSPN data from the Washington area. For the CSPN data, the spatial heat imposition due to winter air conditioning appears to result in a slope of about 0.0005 m s^{-1} (the slope of the regression line drawn in Figure 28.7). For the Times Square observations, the regression has a slope of about 0.007 m s^{-1}. This appears to relate well to first-order expectations: the buildings of the Times Square area of New York indeed occupy ten to twenty times the volume of those typifying the DC urban area (estimated per unit surface area). In all cases, the observations indicate little demand for additional heating when the air temperature exceeds about 20°C.

Conclusion

Studies of urban wind field appear to be generating more questions than answers.

Chapter 29

Agricultural Checkerboards

"Imaginary obstacles are insurmountable. Real ones aren't" — Barbara Sher.

In reality, micrometeorology is little more than a tool for describing the association of the surface with the air above it. The discussion of fetch (Chapter 23) emphasizes the way in which the various fluxes respond quickly to changes in the surface. The effects of these changes on the air are seen most quickly nearest the surface. Such effects are delayed as the height above the surface increases and there is a smoothing that results. In general, it is fair to say that fluxes respond quickly to changes in the surface, whereas wind speed, temperature and humidity respond more slowly. As was seen already, this difference causes insecurity about the specification of such conventional considerations as fetch to height ratios.

There are two main reasons for addressing the issue of how variations in the texture of the surface affect the atmosphere. First, it is in the SBL (which is defined as that region of the atmosphere where changes in the surface become dominant considerations) that people are most exposed. Specifically, it is at a height of several meters above the ground that exposure to high winds and extreme temperatures become most worrying. However, for applications at a larger scale it is the average drag at the surface that affects the synoptic wind structure, especially its profile above and through the PBL (the envelope containing the radiation-induced diurnal cycle). It is in this latter category that much of weather forecasting research has been focused, with an additional incentive related to the safety of air travel. Consideration of a checkerboard pattern of varying surface texture is a first step towards coupling the two interests, with an obvious practical application related to agricultural areas growing different crops.

As a first step towards generalization of the homogeneous-site legacy relationships, the case of adjacent fields of different surface vegetation was selected for consideration. Consider the situation of a wind speed u_1 influencing an upwind surface. The friction velocity over this surface (u_{*1}) is determined by its roughness length (z_{01}), the displacement height being small enough to disregard in the present *gedunken* exercise. The basic micrometeorological characteristics of flow across the adjacent field of roughness length z_{02} will be bounded within two extremes. First, the flow could adjust so that $u_{*2} = u_{*1}$. Second, the surface stress would compensate to leave the wind speed unaffected: $\bar{u}_2 = \bar{u}_1$. These two constraints represent bounds on a distribution of actual situations, expected to have a log-normal distribution such that the most likely result would be that $\bar{u}.u_*$ = a constant specific to the particular circumstances (a function of the height of measurement and the prevailing stability).

The matter was examined in field studies conducted as part of the Sangamon experiments (Hicks et al. 1981), conducted in the 1970s by staff of Argonne National Laboratory (under the leadership of Paul Frenzen) over cropland in Sangamon County, central Illinois. In one of the many Sangamon experiments (Hicks and Wesely, 1977), eddy flux systems were installed at the border between adjacent fields of maize and soybeans (when z_0 (soybean) was 5 cm and z_0 (maize) was 17 cm). Identical instruments were used for both quantifications, each being set up at a height of 6 m above the ground at the edge of the field so that the upwind fetch was more than 330 m for each measurement. Data from these instruments are illustrated in Figure 29.1. The results verified the average expectation, that

$$\overline{u1}\cdot u_{*1} = \overline{u2}\cdot u_{*2} \tag{29.1}$$

leading to

$$\tau_1/\tau_2 = \overline{u2}^2/\overline{u1}^2 \tag{29.2}$$

or

$$\tau_1/\tau_2 = \ln(Z_2/z_{02})/\ln(Z_1/z_{01}). \tag{29.3}$$

where (as before) the capitalized heights Z_1 and Z_2 represent the height of measurement above the relevant zero plane. The expected slope of the linear plot shown in Figure 29.1 is about 1.18, whereas the slope of the regression line plotted is 1.17 ± 0.03.

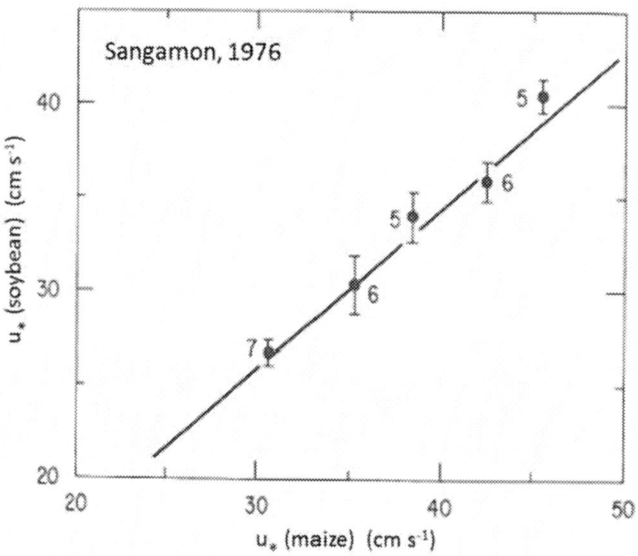

Figure 29.1. Results of a comparison between simultaneous measurements of the momentum flux over two adjacent fields, one of maize and the other of soybeans. This was a component of the Sangamon experiment of 1976, conducted by Argonne National Laboratory at a field site about 30 km south of Springfield, Illinois (q.v. Hicks and Wesely, 1977).

In concept, this simple relationship could be of considerable utility in examining (for example) how best to describe grid-cell averages for numerical models when only a single surface station is available. Indeed, this goal was the reason for the field study described above. If the fine-scale surface structure of a target area is well described in terms of the variation of friction coefficient, yielding (say) n values of the friction coefficient, C_{fi}, $i = 1$ to n, then a first-order quantification of an areal average when only a single average wind speed observation is available $\bar{u}_0$ at a location where C_f is known to be C_{f0} can be derived as

$$\bar{U} = (\sum \bar{u}_0 \cdot (C_{f0}/C_{fi})^{0.5})/n. \quad (29.4)$$

A model prediction for some specified height above the grid-cell average zero plane would need to be evaluated using an appropriately computed data set, calculated using such tower observations as might be available and extending these to a relevant grid cell average by first allowing for spatial variations in the zero plane and then by application of Equation 29.4. In actual application, an initial step would necessarily be to take the distribution of surface roughness into account with full allowance for variations in the displacement height and translating these into variations of C_f relative to the selected (10 m?) height. Clearly, the process can be as complex as the application warrants, but an informative first test (before comparing derived results) would be to compare model assumptions describing a grid-cell surface with observations or with the extrapolation of whatever observations might be available. A familiar warning is appropriate. The result expressed by Equation 29.4 is based on limited data and has subsequently been found to be a poor description of situations in which surface roughnesses vary greatly. Additional (and independent) study would seem appropriate.

The issue of extending SBL micrometeorology to variegated surfaces has been addressed in many presentations, such as Mason (1988) and Philip (1996). The wind direction relative to the orientation and edges of otherwise uniform areas has been identified as an issue of importance (as indeed was the conclusion drawn on the basis of the field studies discussed above and by several early studies over row crops). However, a major analytical result is that the deviations of covariances with height show more influence of surface inhomogeneity than wind speed, temperature and humidity. The height at which flux variability diminished to a level of little importance was defined as the blending height, of the order of tens of meters above ground level.

In this regard, note is taken of the conclusion expressed in a review of model forecast verification procedures (Casati et al. 2008) that there is need "to extend spatial verification methods to sparse observation networks." An observer would note that many numerical simulations have SBL formulations that defy logic, such as the complete omission of consideration of a surface vegetation canopy, omission of any considerations of a displacement height of any kind and assumptions that there are no lags involved in the relationships between above-surface and below-surface temperatures. While such disregard might result in an outwardly pleasing comparison against observations made in early afternoons (when the uncertainty consequences of the above considerations are swamped by other errors inherent in the simulations), the result will obviously be of great relevance when canopies are heating or cooling most rapidly. It is expected, therefore, that claims of good model performance will tend to be based on midday observations. The nighttime case remains to be studied.

Conclusion

If the ground is fairly flat, surface complexity seems a tractable issue.

Chapter 30

Onwards Toward Complexity

"The importance of details diminishes with distance from the observer" — (?)

First, it is appropriate to define what is meant by the term 'complex terrain.' In practice, the term is used to refer to areas over which standard micrometeorological formulations are not expected to work. There are processes involved that are external to the familiar micrometeorological assumptions of flat, uniform, endless terrain. The matter of a variegated surface was considered in the previous Chapter. Most often, meteorology considers 'complex terrain' to be a landscape where hills, rivers, lakes and mountains impose the need to consider processes arising from their presence, both individually and in concert. Air-surface exchange can be controlled by local penetration of canopies on slopes, for example. The concept of an aerodynamic resistance to transfer is then a mite alien. The relevance of a scale length for turbulence, controlled by vertical processes, is likewise difficult to support. In unstable conditions, extending the understanding of momentum exchange relative to a sloping surface (in the near vicinity of the surface, tangential to it) to the case of turbulent dominance by buoyant eddies (responding to the influence of gravity) imposes difficulties that currently confound field experimentalists and numerical modelers alike. However, it is clear that modern measurement capabilities are potentially capable of addressing the complex terrain situation in several potentially rewarding ways.

The various aspects of complex terrain meteorology carry a range of societal ramifications. People need information related to (a) their individual exposures, for which it is the local surface boundary layer that must be understood and at a larger scale (b) the synoptic-scale weather affecting them. At an intermediate level, regional changes in the atmosphere introduced by the presence of local terrain play an important role. None of these considerations can be considered independently of the others, a fundamental reason for the inability to generalize the landscape-scale attributes of complex terrain as it affects the atmosphere. As an obvious complex interaction affecting valleys, consider that flow experienced at a valley floor will likely be determined by pressure differences and gravity, on which a diurnal variation will be imposed by the way in which parts of the terrain receive and apportion solar radiation. Additional allowance must be made for the wind field high above, from which momentum will be extracted as modified by the orientation of surface features with the flow vector aloft. This matter was addressed earlier.

The complex terrain issue is a matter of different scales. Consider an area in which the topography can be described in terms of a mean slope (∇z_c) and a spatial standard deviation σ_{zc} of height z_c above sea level. A relevant horizontal length scale L_c can be defined in many ways, but for the moment consider L_c to be the dominant length scale associated with the topography at average elevation. The ratio

$$A_c = \sigma_{zc}/L_c \qquad (30.1)$$

could serve as an index of terrain complexity, perhaps presented as an isopleth description of a selected area (say 100 km x 100 km). The ASCOT findings summarized by Barr and Orgill (1989) offer an opportunity to extend this mapping of surface complexity to describe flow fields above and below the elevation demarcation defined as z_c (or some other level deemed appropriate on the basis of field observations). However, the matter of flow separation is not yet incorporated. A possible methodology to account for this would be to consider the finer-grid contents of the larger area once again and to identify those smaller-scale areas where the local slope differs from the surroundings by more than one part in seven (this representing a classical 'rule of thumb' for predicting when flow separation might be observed). Questions of form drag and viscosity arise; also, of thermal winds and such factors as low-level jet formation and intermittency. Flat earth micrometeorology would certainly work best for small values of A_c. For large values, something better would be needed. All such considerations would be influenced by convection, which works vertically regardless of local slope.

If the terrain is complex, the concept of a vertical aerodynamic resistance to transfer would likely be inappropriate near the surface. Granted, there would be a handover at some height above the terrain and there must be a diffusive exchange of some sort in the near vicinity of the roughness elements constituting what the atmosphere sees as 'the surface,' but for the present let us think in terms of surface boundary layer slope-generated wind penetration.

The repercussions of complex terrain extend far downwind. Earlier in the discussions presented here, there was attention given to the matter of intermittency as a sometimes dominating factor in both very unstable and very stable regimes. There are many causative processes, details of which remain to be described with the detail necessary to warrant confident extrapolation beyond circumstances of observation (q.v. Sun et al. 2012; Mahrt et al. 2020, for example). However, whatever the controlling mechanisms might be, the frequency of occurrence of such events will increase as terrain complexity increases and the variety of topographically induced local wind regimes expands. Moreover, the influence of terrain complexity must be expected to extend downwind of the terrain itself. To demonstrate this, Figure 30.1 presents average diurnal cycles of the probability of detecting TKE exceedances at the GRI tower in Idaho (see Figure 24.1). 'Exceedances' are defined as occasions for which a 5–min determination of TKE exceeds the 25–min average in which it is nested by more than two standard deviations. The results are presented as average diurnal cycles for sequential three-month periods. The intent is to seek evidence of turbulence exceedances downwind of the Rocky Mountains, the foothills of which constitute the western edge of the site in question. In this context, daytime data are explored as well as nighttime. The causative mechanisms suspected to influence the occurrence of intermittency in an area directly influenced by severe terrain complexity (such as the Idaho site) include (a) the generation of gravity waves, their breaking when the wind shear becomes sufficient and the transport of the resulting packets of turbulence to the lower part of the atmosphere where they can influence the transfer characteristics of the surface boundary layer, and (b) the shear generated when canyon downslope winds impose themselves on a flow field dominated by other factors (such as the characteristic mesoscale flows caused by the larger-scale topography of the area, as characterize the present case, see Finn et al. 2016; Frisch, 1985; Nappo, 2002; El-Madany et al. 2014).

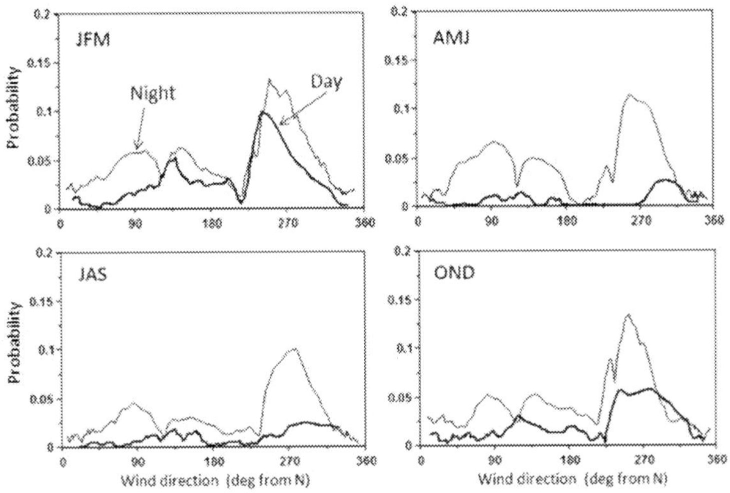

Figure 30.1. The probability of encountering an exceedance in turbulent kinetic energy at a site closely downwind of the Rocky Mountains, on the reservations of Idaho National Laboratory (see Clawson et al. 2007). Daytime data are for 1000 to 1600 hrs local standard time; nighttime for 2200 to 0400 hrs.

A striking feature of Figure 30.1 is the recurring high probability associated with wind from the west, with upwind mountains. That this maximum is more evident for the nighttime than for the daytime data is as expected, since the generation of gravity waves and their breaking is primarily a nocturnal feature (q.v. Nappo, 2002) and it is also then that the downslope canyon winds maximize. However, some exceedances are also found in daytime and when winds are from other directions. It is noted that the easterly-component exceedance maximum is a nighttime feature, with minimal occurrence in daytime. There is no obvious way to distinguish among the possible causative mechanisms. Evidence presented elsewhere supports the expectation that low-level jets are generated in the area of the INL site, with a direction dictated by the slope of the valley floor but modified by flows from the canyons feeding into it.

The influence of upwind topography was studied in many similar field experiments. For example, Davidson (1963) reported the intermittent occurrence of extreme turbulence at distances of many kilometers downwind of a ridge, maximizing at a height about halfway between the downwind valley floor and the upwind ridgetop but extending throughout the vertical regime. These occurrences were reported for daytime periods of strong winds but also for nighttime.

While conventional micrometeorology warns against measurements in complex surroundings, reality suggests tolerance. For example, consider observations made at a forested site at the eastern boundary of the Appalachian range in western Virginia (Lee et al. 2015). The site in question was about 100 m east of the famous Blue Ridge Parkway, in an area ('the Pinnacles') surrounded by mature mixed deciduous trees, at an altitude of 1017 m asl. A 15–m tower carried a sonic anemometer and supporting fast-response sensors to yield state and flux determinations for momentum, sensible and latent heat fluxes, and carbon dioxide. Measurements of net radiation (R_n) permit the nighttime data to be sorted and grouped according to R_n. Figure 30.2 shows a selection of results, for the nocturnal case and for the warmest periods of the year. The examples show convincing correlations among measured fluxes when net radiation drops below about –30 W m^{-2}, in accord with the results derived earlier for data from less challenging circumstances. The causes of the associations between fluxes are not revealed by the Blue Ridge dataset, but the pragmatic conclusion is that the complex terrain of this particular location is conducive to the behavior patterns detected in less demanding situations, for the warmest months.

The behavior seen in Figure 30.2 is strikingly similar to that observed over 'better' micrometeorological sites, as discussed earlier (see Chapter 18). Further discussion of such matters must be left until more data are available. For the present, available data indicate that the presence of upwind complex terrain can increase the likelihood of encountering turbulence intermittency. Moreover, the evidence now available suggests that describing the coupling between the air aloft and the surface in situations affected by terrain complexity might not be as difficult as is often imagined (c.f. Liang et al. 2014). Over a large enough grid cell, the consequences of terrain complexity will be evident as a reduction in the overall resistivity involved in the coupling. It is proposed in the discussion above that an appropriate indicator of this complexity is A_c, as in Equation 30.1. Perhaps it is only as the size of the area being considered diminishes that valley flows start to cause concern.

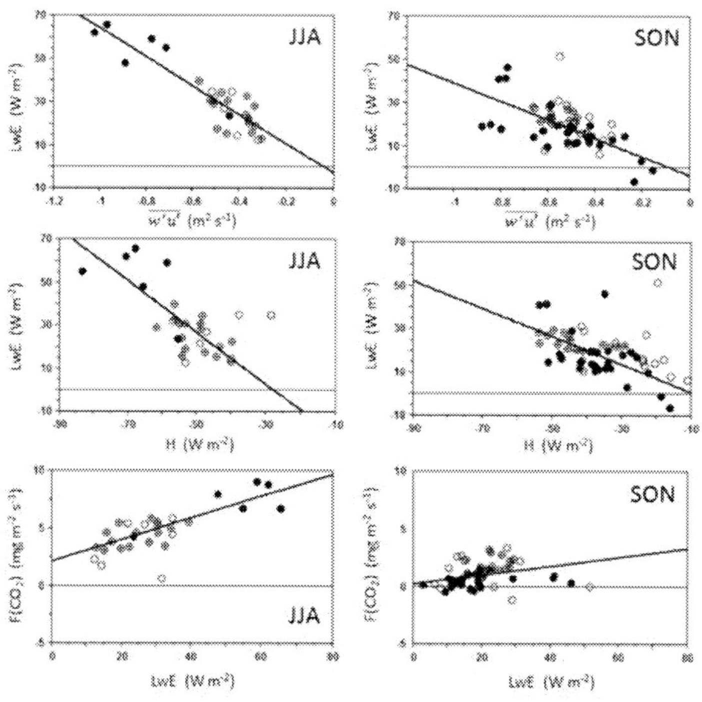

Figure 30.2. Comparisons between eddy fluxes in stable conditions. For each three-month period, data have been grouped according to net radiation (open circles, $-10 > R_n > -30$ W m^{-2}; Circles and crosses, $-30 > R_n > -50$ W m^{-2}; solid circles, $R_n < -50$ W m^{-2}) and averages of groups of 50 measurements plotted. Data are from a site at 1017 m elevation on the eastern ridge of the Appalachians in Virginia (q.v. Lee et al. 2015). Regression lines are derived using all data for $R_n < -30$ W m^{-2}.

Conclusion

Adequate understanding is not the same as detailed knowledge.

Part IV: The Role of Vegetation

Chapter 31

More on Zero Planes and Roughness

"Assumptions are made, and most assumptions are wrong" — Albert Einstein.

It is accepted that the momentum zero plane, like the roughness length, varies with the kind of canopy and maybe with additional factors such as terrain complexity. However, over water and barren land, it is fair to assume that $d = 0$, and over grassland a good approximation is that d is of the order 75% of the average height of the canopy. This is not likely to be in error by more than 50%. Note, however, that the examples given above (Chapter 26) of forest and maize canopies show exceptions to this expectation. In the case of the Koorin study, the foliage was about 10 m high but was very sparse (as is seen in the aerial photographs presented in the Koorin report: Clarke and Brook, 1979). As became evident in the analysis leading to Figure 24.1, the zero-plane displacement appropriate for both of the Koorin sites (as derived using the analysis favored here) was about 4 m. The corresponding ratio $d/h \approx 0.4$ differs greatly from the common 'rule of thumb,' that $d/h \approx 0.75$ to 0.8. Clearly the spatial density of the surface roughness elements is a major consideration in the real world, as has been demonstrated in many wind tunnel simulations and numerical models (e.g., Verhoef et al. 1997; Belcher et al. 2003).

Field studies have shown that the organization of surface roughness elements is a similar issue. Studies over agricultural row crops have shown that the surface exchange characteristics vary with wind direction (Hicks, 1973; Kustas et al. 1989; Verhoef et al. 1997). As expected, the roughness properties (z_0 and d) maximize when the wind is orthogonal to ordered rows. Studies over vineyards are among the best (and the most enjoyable) to demonstrate this phenomenon.

The assumption often made is that the zero plane, once determined for a specific location, is common for all fluxes, representing the effective level of action of the drag force (Thom, 1971) and the effective level of origin of other fluxes (particularly heat, water vapor, CO_2, etc.) in daytime. However, consider the cases of radon and water vapor exchange. In the first case, the source is certainly the soil. If a vegetation canopy is present, in daytime the source of evaporating water vapor is probably open stomata at a height determined by the distribution of plant biomass and the penetration and intensity of incoming radiation. Whether a refinement of the conventionally assumed constancy of d is warranted is certainly debatable. However, an obvious ploy is to continue to accept the familiar constraints about using data obtained too close to the surface. In the case of a field of maize with a crop height of $h = 2.0$ m, d (as revealed by measurements made well above the canopy) is likely to be about 1.5 m above the ground surface. With an assumed error margin of ± 50% on the decrement ($h - d$; 0.5 m here) and with the roughness length approximated as 10% of the crop height ($z_0 \approx 0.2$ m) it follows that the flux/gradient relationships *per se* should not be employed below about 3.75 m above the ground surface. If, however, the surface of interest is a forest with trees 16 m high, then the same procedure would indicate that safe application of the legacy relationships should be constrained to data obtained above about 31 m agl. In general, it is best to explore details of data obtained at heights below twice the canopy height with caution, expecting to find site-specific departures from legacy predictions. An obvious exception is the use of the well-known Bowen ratio energy balance methodology, in which it is not assumed that the diffusivities indicated by gradient measurements of temperature and humidity relate closely to assumed fluxes, but only that the turbulent diffusivities are the same. A later Chapter will explore this further.

From the perspective of the turbulent exchange affecting the atmosphere well above whatever canopy might be present, it might be that the effective levels of origin of quantities being transferred upwards by eddies will appear to be the same since it is the same eddies contributing to the transfer. However, in the case of momentum the surface constitutes a sink, not a source; the transfer of momentum is downwards. Likewise, in stable stratification the eddy flux of heat is downwards and it appears safe to assume that the same eddies then contribute to both $\overline{w'T'}$ and $\overline{w'u'}$

will be some restorative influence of the temperature changes associated with the vertical fluctuations.) However, in unstable conditions, the eddies transferring heat (upwards) are different from those associated with momentum exchange (downwards). Otherwise, the eddy diffusivities would be the same. Given these viewpoints, the assumption that the level of the effective sink for momentum is the same as the effective source of the sensible heat flux seems appropriate for the stable case but arbitrary, although convenient, for the unstable.

Assume, for the present, that the zero plane for momentum exchange is constant and is as determined by examination of wind profiles in neutral conditions. Extension of this understanding to heat fluxes and temperatures gradients immediately introduces concerns about the consequences of solar angle and radiation intensity following sunrise and the thermal structure of the surface. The uppermost foliage of surface vegetation will be the first influenced by the rising sun, but the consequences will be small because there is then little radiative energy involved. As the sun rises, solar radiation increases and the depth of penetration of the radiation into the canopy also increases. If we imagine, following the same line of thinking as in the case of momentum, that there is no single level at which the sensible heat flux originates but that the source is distributed over the volume of foliar heating, then it is clear that the effective reference level for describing profiles above the surface will change, but perhaps insignificantly. Quantifying the consequences presents a challenge, since any model of this behavior must include consideration of such factors as leaf area distributions, leaf orientation, etc. as have been considered at length by researchers in the forest meteorology community (e.g., Oliphant et al. 2004).

An additional complexity is that plants respond quickly to an input of radiation. Following dawn, leaves at the top of canopies would be expected to start transpiring sooner than leaves deeper in the canopy. From the perspective of the SBL, in air above the vegetation, the buoyant driving force combines the sensible and latent heat fluxes. Hence, regardless of whether different layers of a canopy contribute differently to the heating of the air and its water vapor content, the turbulent regime above will likely display a commonality of heat and humidity diffusivities due to the role of both in generating a common driving force – buoyancy. There remains, however, no persuasive argument to support the assumption that the effective sink for momentum should coincide with the effective source of specific heat. The matter is mostly one of convention and convenience.

Figure 31.1 shows three foliage temperature profiles obtained during a study of the mixed deciduous canopy of the Oak Ridge (TN) Walker Branch watershed. The first profile shows that in the early morning the tops of the trees were then warmer than the bulk of the canopy beneath. The third profile shows that the treetops were then (about an hour after noon) cooler than the ground, but that the rest of the foliage was essentially the same temperature as the ground. The leaf area index of the canopy was sufficiently small that the ground surface was directly heated by solar radiation in the midday period. These data were obtained using sideways-looking hand-held infrared thermometers operated by students climbing a scaffolding tower. The results show that in this case the height at which the temperature of foliage maximizes changed through the morning hours. If one relates the effective zero plane to the origin level of sensible heat, then it would follow that d_H decreases with time after sunrise.

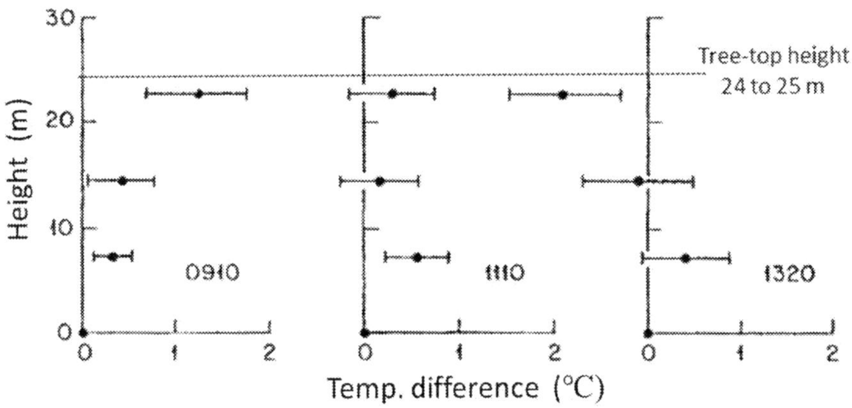

Figure 31.1. Foliage temperature differences relative to the ground through a mixed deciduous canopy at the Walker Branch watershed in Oak Ridge, Tennessee. All temperatures were measured using hand-held infrared thermometers, looking horizontally in successive ascents on ladders. Measurements were made on three days in May 1983, centered around the average times shown. Error bars show ± one standard deviation.

Clearly, the results shown in Figure 31.1 cannot be considered more than a single-site demonstration that the distribution of foliage temperature could vary through the day. However similar behavior was detected in studies of pine plantations, such as are reported by Hicks et al. (1975) and Luvall and Holbo (1989). Figure 31.2 extends consideration to a field of mature maize. In the simplest possible concept, the extent of penetration of incident solar radiation will determine the level within the canopy at which the foliage temperature maximizes. Note that this is not necessarily indicative of sub-canopy air temperatures or of the level at which convection generates, but it remains indicative of possible changes with time of day. The concept carries with it an expectation that the density/transparency of the canopy is a key factor.

Extensive studies of the origins of convection have resulted in a different view of the sub-canopy air temperature regime. Experimental and theoretical studies (initiated mainly by the need to maintain near-zero temperatures in liquid helium, in support of studies of fundamental particle physics) have indicated that convection follows molecular-scale consolidation within a fluid that is being heated (Charlson and Sani, 1971; Feigenbaum, 1979; Gollub and Benson, 1980). Convective cells do not arise from immediate contact with individual heated surfaces. Analogy with the case of bubbles generating in a heated pan of water is inappropriate, since the convection is then driven by phase changes occurring at favored locations at the bottom of the pan. Hence, the view now adopted is that the interpretation by Thom (1971) relating the zero plane for momentum exchange to the effective level of action of the drag on roughness elements of the overall canopy remains defensible for conditions insignificantly different from neutral and perhaps extending into stable conditions but the unstable case requires a different view. For the unstable case (assumed here to be for $\zeta < -0.03$ or thereabouts, in accord with the discussions already held) it is now concluded that the zero plane for heat transfer will be within the sub-canopy airspace. Since all eddy exchanges are then dominated by convection, it seems appropriate to extend this to the cases of water vapor exchange and any trace quantity of interest. For momentum, however, the matter remains obscure because there is then a need to consider the role of convection as it might, or might not, influence the exchange of momentum.

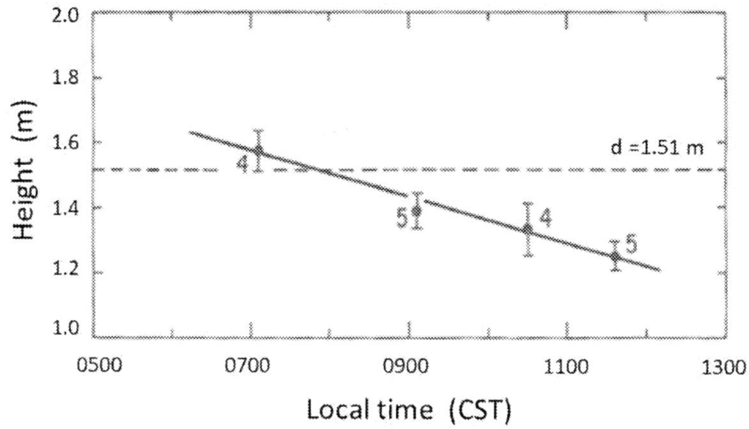

Figure 31.2. The variation with time of day of the height of maximum foliage temperature of a field of maize, measured as a part of the Sangamon series of studies in central Illinois in 1976 (Hicks and Wesely, 1977).

If it is assumed that free convection dominates exchange during most of the daytime then measurements of temperature differences between heights above the vegetation canopy can reveal the height of effective origin of the buoyant eddies. Figure 31.3 shows results derived in this way, using observations obtained at the pine plantation on the reservation of the Savannah River Laboratory (discussed earlier, in Chapter 23). The plot shows the variation with time of day of the height at which free convection should have originated if free convection reigned, given the temperature data at 36 and 61 m. The smooth line has been drawn by eye and is therefore susceptible to the criticism that it portrays the preferences of this author — that there is evidence of a variation (admittedly small) of the effective level of origin of freely convective exchange. As a point of reference, note that the displacement height of this forest canopy was independently determined to be $d \approx 27$ m (Weber et al. 2012).

There is no obvious analogy appropriate for the case of the distribution of the momentum sink, but there are several ways in which resulting uncertainties can be addressed. From the viewpoint of examining and /or utilizing the profiles of wind and temperature near the surface, (1) the variation in zero plane displacements could be investigated and above-canopy heights of measurement corrected accordingly, (2) the zero plane could be considered a constant and flux-gradient relations adjusted accordingly for application elsewhere, or (3) the measurement of gradients could be confined to heights such that any related uncertainties could then be ignored (q.v. Hicks et al. 1979; Raupach, 1979).

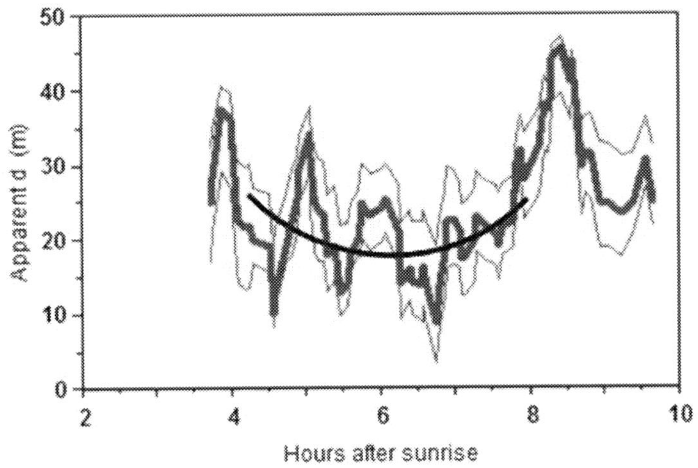

Figure 31.3. The change with time of the height of apparent origin of the sensible heat flux observed over a South Carolina pine forest (q.v. Figure 23.5). The heights were computed on the assumption that free convection was the process affecting temperature measurements at 36 m and 61 m above ground level. The canopy height was 32 m. The finer lines indicate the ± one standard error bounds around the averages that are plotted. The smooth line is drawn by eye.

A further alternative relates to the extended analyses of the problem following the presentation by Garratt (1980) of near-surface modifications to the familiar flux-gradient relationships, found to be necessary within the roughness sub-layer (RSL). Subsequent research verified that describing flow and fluxes within this sublayer required consideration of more than appears in the conventional formalism, as well illustrated by Harman and Finnigan (2007) who introduce a new property with forest-canopy RSL relevance — the vorticity thickness. However, the practical and pragmatic response has usually been aligned with the third of the alternatives listed above. The studies now presented contribute little to the discussions of the basic issues of near-surface flow above a forest canopy, but in the cases of the data now considered it appears a simple expedient to allow the zero plane for sensible heat and latent heat applications to vary as the canopy heats in daytime and to adjust the height scale accordingly.

Hicks and Eash (2021) examined the relevance to air-surface exchange of the infrared temperature (T_0) as observed from above a canopy in the formulation of heat exchange between a vegetated surface and the air, on the one hand and between the surface and the soil on the other. A regression of available data reveals that the temperature most associated with both of these heat exchanges, into the air and into the ground, is not T_0 but a temperature more indicative of air temperature within the maize canopy that was then under examination. If this finding is substantiated, then the common supposition that foliage temperatures determine convection may be wrong. The information now available indicates that it is the well-mixed attributes of the sub-canopy airspace that play the major part.

Conclusion

That z_0 and d are constant 'calibrations' of a vegetated surface might be text-book gospel but nature often seems to disagree.

Chapter 32

Roughness Lengths and kB^1

Introducing z_{0H} and questioning it.

The preceding discussion indicates that there are good reasons to dispute assumptions concerning the time constancy of zero plane displacements. The matter is reminiscent of the discussion of CASES-99 observations (see Chapter 27) which demonstrated that the wind profile can be displaced by local but stronger katabatic flow. In practice, the simulations most often used in simulations assume that d and z_0 remain constant with time (and are the same for heat and momentum cases) but vary from place to place. However, a first step in refining this assumption has been well documented. As prescribed by Garratt and Hicks (1973) the proposed methodology describes the offset of z_{0H} (the roughness length associate with the kinematic heat flux) from z_0 (the conventional roughness length for momentum exchange) over the surface of relevance. The algebra is straightforward. Consider the familiar flux-gradient relationships

$$\partial\bar{u}/\partial z = [u_*/(kZ)] \cdot \phi_m \qquad (32.1)$$

and

$$\partial\bar{\theta}/\partial z = [\overline{w'T'}/(ku_*Z)] \cdot \phi_H \qquad (32.2)$$

which define the dimensionless gradients ϕ_m and ϕ_H. As before, $Z = z - d$ and it is therefore assumed that d is the same for both momentum and sensible heat considerations. Integration with respect to z yields

$$\bar{u}_z = (u_*/k) \cdot [\ln(Z/z_0) - \psi_m] \qquad (32.3)$$

and

$$\bar{\theta}_0 - \bar{\theta}_z = [\overline{w'T'}\ \ ku_*)] \cdot [\ln(Z/z_{0H}) - \psi_H]. \qquad (32.4)$$

Here, the ψ-functions are as introduced earlier (Chapter 8) — functions of stability that can be approximated in several ways. Equation 32.4 can be re-written as

$$\bar{\theta}_0 - \bar{\theta}_z = [\overline{w'T'}\ \ ku_*)] \cdot [\ln(Z/z_0) + \ln(z_0/z_{0H}) - \psi_H], \qquad (32.5)$$

in which the term $\ln(z_0/z_{0H})$, alternatively referred to as kB^{-1}, represents the difference between the roughness lengths appropriate for momentum and heat eddy fluxes. The experimental challenge is to measure temperature differences and eddy covariances sufficiently well to enable kB^{-1} values to be derived with acceptable confidence. Central in this methodology is the measurement of the surface temperature, $\bar{\theta}_0$. In most experimental studies from which estimates of kB^{-1} have been derived the measurements of surface temperature have been by use of infrared thermometry. Since the initial publication of a plot showing how kB^{-1} varies according to the kind of vegetative cover (Garratt and Hicks, 1973), results of many studies have converged on a value of about two for kB^{-1} (q.v. Garratt, 1992).

Figure 32.1 shows how a selection of model predictions agree with results from field programs (in this case, the original Garratt and Hicks data from 1978) and from two experiments conducted recently, over fallow land and over pasture. Plotting against the roughness Reynolds number ($Re_* \equiv u_* z_0/\upsilon$) follows from experience in laboratory studies, especially of pipe-flow, and appears to result in a conceptually acceptable ordering of surface texture, from flat, barren surfaces to crops and forests. However, the combination of (a) an eddy-covariance statistic (u_*), (b) the result of extrapolating the wind profile downwards (z_0) and (c) the molecular-scale diffusivity (υ) in the construction of Re_* appears to challenge credibility regardless of its apparent widespread acceptance. Moreover, the graphical presentation as in Figure 32.1 invites consideration of the familiar shared variable syndrome, with both u_* and z_0 appearing in both axes. To explore the matter further, a dataset was constructed using random numbers constrained to lie in the ranges observed in daytime studies over short grass. The heavy black curve is drawn through the (smoothed) results derived using this artificial dataset. Further discussion is unnecessary. The quasi-horizontal lines to the lower right side of the diagram are drawn by eye, to show two alternative interpretations of the behavior as the surface complexity increases. Regardless of which alternative might be preferred, the data appear to indicate that z_{0H} approaches z_0 as the surface becomes rougher and as the wind speed increases.

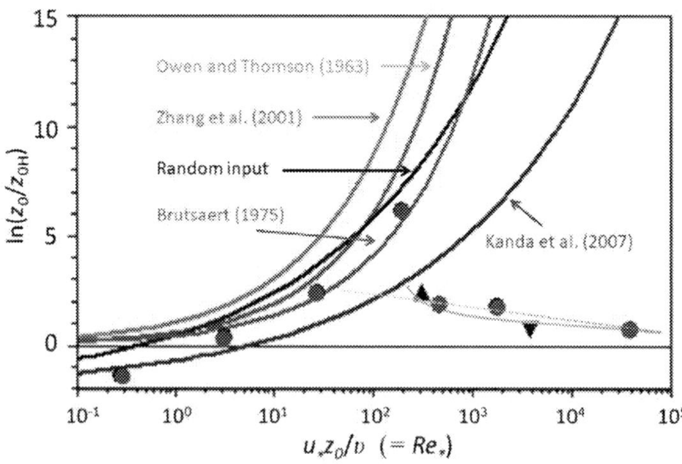

Figure 32.1. The difference in roughness lengths for heat and momentum, as represented by $\ln(z_0/z_{0H}) = kB^{-1}$, with surface texture, the latter quantified in terms of the roughness Reynolds number. Four examples of published model predictions are illustrated, as well as the result obtained when random numbers are used to replace the original dataset. The solid circles relate to data as originally presented by Garratt and Hicks (1973). The upward triangle relates to a study over fallow land in Zimbabwe (O'Dell et al. 2015) and the inverted triangle to grazed pasture in Alabama (Hicks et al. 2018). An earlier version of this diagram was published by Hicks et al. (2018).

Two recent results shown in Figure 32.1 (the triangles) present new information, because neither case involved a significant vegetation canopy and the relevant displacement heights (for both momentum and heat) were undoubtedly small in comparison to the heights of measurement. The data are for daytime, when it is clear (from analysis presented earlier) that the heat exchange is the controlling property. The difference between z_0 and z_{0H} (approximately an order of magnitude) has long been conceptually difficult to accept. In this context, note that all of the results presented in Figure 32.1 make use of infrared surface temperature observations and assume that the temperatures so measured are those that determine z_{0H} and d_H.

The indications of inequality of z_0 and z_{0H} derive from analyses of field data that include quantification of the surface temperature that serves similarly to the zero-wind speed extrapolation to z_0 in the case of the wind profile. There are conceptual difficulties involved, many of which will become apparent as this discussion proceeds. For example, the potential role of molecular-scale diffusivity as an eventual contributing property is conveniently overlooked, in spite of the convincing evidence of pipe-flow studies that the matter cannot be ignored. Allowance should be made for differences in molecular diffusivity by injecting a reliance on the Schmidt number, via $Sc^{-2/3}$ or Sc^{-1}, depending on how the literature is interpreted ($Sc \equiv \upsilon/\kappa$, where υ is air viscosity and κ is the thermal diffusivity of air). Central among the issues of concern is that downward-looking infrared temperatures are used as measures of the temperature serving the same conceptual role as the zero-wind speed extrapolation of the SBL wind profile. The temperature profile measured above the surface (or above a vegetated canopy) is extrapolated downwards to where it is found to equate to T_0 as measured by infrared thermometry. In practice, this requires the acceptance of the stability-correction terms represented by ψ_m and ψ_H in Equations 32.3 and 32.4. These terms involve the downward extrapolation of the empirical and familiar ϕ-functions to below their acknowledged height of applicability, leaving a severe distaste in the mouth of the user.

Concerns also arise about the infrared measurements of temperature. It has been commonly assumed that the surface temperature indicated by a downward-looking detector represents the appropriate reference level for extrapolation of the temperature profile. This was specifically assumed in early work by Garratt and Hicks (1973) and has been accepted in most subsequent examinations of the kB^{-1} matter. A better temperature could well be the relevant effective interfacial temperature (T_e) between the air and the soil. To explore this aspect further, data obtained in the 2015 Ohio agricultural study of a maize crop have been used. As reported by Hicks and Eash (2022), measurements were made of soil temperatures ($\bar{T}_s$) within a few cm of the surface, as well as of air temperature above the canopy ($\bar{T}_a$) and downward-looking infrared surface temperature ($\bar{T}_0$). Figure 32.2a shows the average diurnal variations of ($\bar{T}_0 - \bar{T}_a$) and ($\bar{T}_0 - \bar{T}_s$) for the period from 30 June not 13 July, 2015. Consider the relationship that would correspond to a linear relationship (with no offset) between ($\bar{T}_0 - \delta\bar{T}_0 - \bar{T}_a$) and ($\bar{T}_0 - \delta\bar{T}_0 - \bar{T}_s$), where $\delta\bar{T}_0$ is a correction to $\bar{T}_0$ in order to derive this proportionality. Then a regression of ($\bar{T}_0 - \bar{T}_a$) against ($\bar{T}_0 - \bar{T}_s$), as in Figure 32.2b, would yield a result of the form

$$(\bar{T}_0 - \bar{T}_a) = a + b(\bar{T}_0 - \bar{T}_s) \quad (32.1)$$

from which it is seen that

$$\delta\bar{T}_0 = a/(1 - b) \quad (32.2)$$

Figure 32.1b plots these two temperature differences against each other, yielding a regression best fit as illustrated. Figure 32.2c shows the results derived by applying Equation 32.2, indicating that T_e was less than T_0 by an amount approaching 5°C during the peak of the daylight hours. An earlier analysis (Hicks and Eash (2022) made use of an approximation different from Equation 32.2, leading to a lower value for $(T_0 - T_e)$.

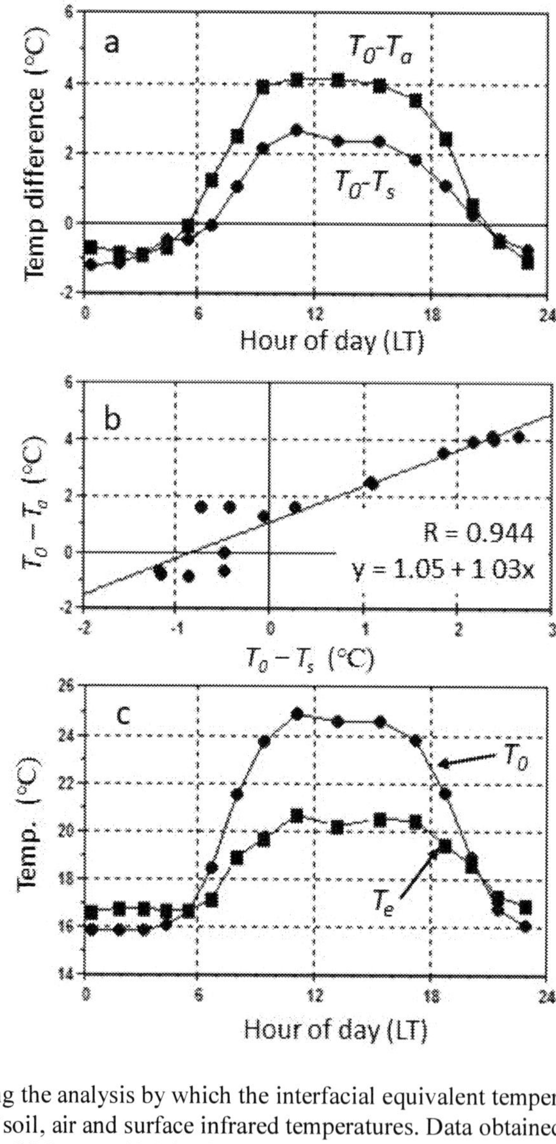

Figure 32.2. Illustrating the analysis by which the interfacial equivalent temperature T_e can be derived from measurements of soil, air and surface infrared temperatures. Data obtained in the Ohio (2015) field study are used, for 30 June to 13 July, 2015.

For the Ohio 2015 experiment, the sequence of diurnal cycles of T_0 and of T_e derived as above is illustrated in Figure 32.3. Note that the diagrams share the same coordinate system, and that Figure 32.2c has been included. The sequence is continuous, except for a period of missing data influenced by rainfall from 28 July to 11 August. Three of the two–week periods illustrated provide examples of how apparent anomalies can arise, as a result of crop management (the first of the sequence shown) and second during and after heavy rainfall periods (mainly in August). Otherwise, it appears that T_0 exceeds T_e in the convectively active hours, by about 3°C while the crop was growing and increasing to about 5 °C after reaching maturity (in July).

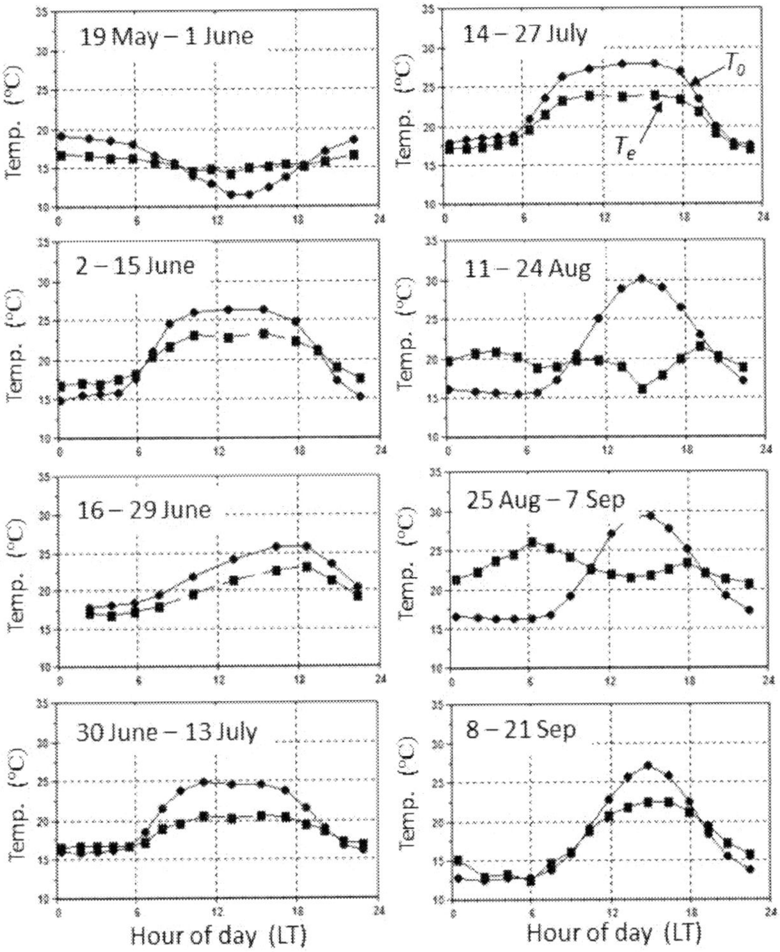

Figure 32.3. Comparisons of diurnal cycles of T_0 and T_e for the sequence of two-week averages during the maize growing season, Ohio in 2015. Heavy rainfall affected observations during much of August. The maize grew to a height of 2.9 m before harvesting on 22 September.

It is expected that such factors as solar elevation, radiation intensity, wind direction and wind speed will contribute, as well as the many technical and deployment issues related to the infrared measurement (q.v. Verhoef et al. 2007). The present data are not adequate to explore the influence of such factors. Nevertheless, these observations could help explain why interpretations of surface-to-air exchange observations yield an apparent order of magnitude between roughness lengths z_0 and z_H. Examination of the nighttime observations is less rewarding, but a general conclusion that $(\bar{T}_0-\bar{T}_e) = O(1\ °C)$ then seems appropriate.

Conclusions of this kind are not new. Many others have examined the relationship between infrared surface temperature measurements and the thermal regimes associated with the initiation of convection. The matter was the subject of considerable attention in the 1990s, when infrared measurements techniques were evolving with potential satellite applications. The topic is intimately associated with the depiction of air-surface exchange of heat, as depicted in numerical simulations and as estimated from satellite observations. Matsushima (2005) provides an examination of the subject that includes errors in the assessment of the effective surface temperature as an explanation for the roughness length inequality that imposes the result that $kB^{-1} \approx 2$ as is currently assumed in many applications.

Conclusion

The concept of a relevant 'surface temperature' needs refining.

Chapter 33

Canopy Breathing — Sweeps and Ejections

The cutting edge — Occam's razor

Exploring the way in which forests interact with the atmosphere has been the target of many experimental programs, which in general have failed to yield an understanding that is universally applicable. The structure of a forest canopy, its location and the prevailing meteorology all have central roles to play. If the intent is to make use of forest characteristics to describe the air-surface exchange of a forested landscape, then there are several basic properties of obvious relevance.

1. How extensive is the forest and how tall are the trees?
2. How (dynamically) transparent is the canopy?
3. How obstructed is the subcanopy airspace?
4. Are the trees fully growing, or adversely affected by drought or some other stressor?

Any one of these considerations could rule out a generality developed in other circumstances. In practice, forest micrometeorology tends to be focused on *my* forest, rather than *yours*.

New technologies have resulted in a general awareness of the sporadic coupling between sub-canopy air and the atmosphere above it, a feature that was accepted in early experimental programs that relied on real-time observation of micrometeorological features. Before the arrival of sonic anemometers, knowledge of the sub-canopy wind field was the result of observations using cup anemometers. The sub-canopy average wind profile was rendered, accordingly, as a continuous curve increasing from zero at the ground and asymptoting to the free stream above at some height within the upper canopy (Cionco, 1972; Fritschen, 1985). However, cup anemometers report the average of the instantaneous wind speed and, in simple concept, cannot be considered indicative of the flow of air through a sub-canopy airspace. Visual observations indicate that the assumption of a coherent and well-ordered speed profile has little relevance to the sub-canopy velocity regime. To convince a critic of this, zero-lift balloons were released within the sub-canopy airspace during the 1972 Mt. Gambia experiment (Hicks et al. 1975) and were observed to repeatedly drift away from the location of release and then return to it, with little coherence in the drift direction. At the same time, the wind observed above the canopy was about 3 m s^{-1} from a constant direction.

Observations indicating that forest subcanopy wind fields can be dominated by turbulence with minimal mean flow arose in the late phases of World War II (as already discussed in Chapter 1), when a field study led by Frank Pasquill in Queensland's northern forests showed that nerve gases potentially released in jungle warfare would not spread far. In subsequent years, the matter of sub-canopy wind and temperature behavior has been the subject of considerable investigation, with growing awareness that the situation is subject to variety of considerations such as canopy structure, local topography and prevailing meteorology. The availability of sonic anemometry has been crucial in these studies.

A behavior characteristic of the development of eddies within forest canopies became evident in the studies reported by Shaw et al. (1989) — temperatures within the trunk space of the forest they studied rose in thermal 'ramps' until suddenly reverting to background levels. This serves as confirmation of canopy thermal behavior as expected on the basis of the low-temperature convection initiation (as noted previously, e.g Feigenbaum, 1979). In essence, when the subcanopy air mass is heated, through contact with surfaces already heated by incoming radiation, convective cells will generate. These embryonic cells will not penetrate a dense canopy above them unless (a) they are sufficient buoyant, (b) they encounter a location where the canopy is thin enough to permit easy penetration, or (c) both. The sub-canopy airmass temperature will slowly increase until a trigger point is reached, at which the cell in question will have buoyancy sufficient to penetrate the canopy and be replaced by a downwelling sweep of air from above. A transfer of momentum follows. In this concept, momentum exchange and the heat flux are results of exchanges influenced by a common phenomenon but involving considerably different eddies, a matter that pervades almost all daytime micrometeorology. At other times, wind gusts create dynamic pressure differentials that cause air to move through the canopy (given sufficient dynamic transparency of the canopy), constituting a momentum–driven mechanism for eddy exchange of all atmospheric quantities.

Some visual consequences of the convective cells that result were anecdotally reported by observers participating in the BOREAS campaign in northern Canada (Sellers et al. 1995; Ron Dobosy, private communication). Columns of pollen-carrying air were seen to arise intermittently from the expansive spruce forest then under scrutiny.

The concept of turbulence sweeps and ejections has dominated much recent attention to exchange between forests and the atmosphere. As described by Shaw (1985), occurrences of turbulence bursts arising from daytime forests are compensated by slower downward sweeps of faster-moving air from above. Building upon the understanding generated in earlier studies, regarding the initiation of convection, a simple picture of the overall air/canopy exchange phenomenon arises.

Examination of detailed sub-canopy observations made in the Amazon rain forest (Chowdhuri et al. 2022) support the present view of intermittent exchange and its association with the ejection–sweep cycle. Understanding of the complex ejection–sweep phenomenon continues to grow with detailed descriptions arising from many intensive field studies (e.g., as reported by Katul et al. 1997). The details of the behavior are such that generalization appears difficult, especially when the area of study is such that that edge effects become central considerations (see Irvine et al. 1997, for example) or when terrain complexity issues arise. On the whole, edge effects influencing sub-canopy flow fields can be expected to extend to little more than 20 times the height of the canopy (based on a review of available information) and the height of the forest-dominated atmospheric environment appears to extend beyond it to as far as two to three times the height of the trees, but rarely further. But it must be noted that the number of numerical simulations of forest edge effects far exceeds the number of observational datasets.

Conclusion

If a result seems counter-intuitive, rethink the issue — even fifty years later.

Chapter 34

Once Again — the 'Big Leaf'

"If it was good enough for Adam and Eve . . ." — Noel E. Ridere

Whereas much discussion of air-surface exchange concentrates on the case of momentum, much pragmatic interest relates to the scalar fluxes, particularly of water vapor, sensible heat and carbon dioxide. In these cases, details of plant physiology are often determining factors. There is no need to review the details of leaf structure, stomata and the related fundamentals of the physiology of plant life. The topic is well summarized in many readily available sources, starting with common textbooks and encyclopedic websites. The present focus is on the way in which plant life modifies the exchange of various attributes of the atmospheric environment, with obvious relevance to the fluxes of heat, water vapor, carbon dioxide and momentum. Later, the matter of trace gas exchange and particle exchange will be considered — atmospheric deposition and emission.

Studies of the reasons that plants (especially crops) respond adversely to weather and climate have been commonplace for centuries, usually driven by the need to understand the requirement for water by growing crops (Meidner, 1987). Earlier, Chapter 5 reviewed (in brief) the early studies of stomatal and cuticular factors that influence the exchange of water vapor and heat on a leaf-area basis. Experimental investigations of the characteristics of stomata abounded, with extensive consideration of how stomata respond to various controlling agents, especially photosynthetically active radiation (PAR), water stress and high temperatures. These early experimental investigations started with determination of stomatal resistance per unit area of plant leaf surface (r_s), using porometer instrumentation whose development (e.g., Penman, 1942) contributed to a rapid increase in understanding of differences among plant species. Stomata are not the only components of the conceptual resistance model describing exchange between the air and a leaf. Cuticular resistance r_{cut} formulates the exchange associated with the skin of foliage, the plant cuticle. For the case of water vapor exchange, the role of the cuticle is minor and hence experimental effort has concentrated on describing r_s as it depends on such basic quantities as PAR, water stress and plant speciation. For some plant species, stomata populate both sides of a leaf, whereas most often only a single side. Accordingly, values of LAI could be reported as either one-sided or two-sided, depending on the plant species. The framework in which such matters are considered is usually such that distances relate to the surface of an individual leaf, with the exchange in question being normal to the leaf surface. The case of conifers is left for contemplation.

At night, stomata are usually closed and any water vapor flux associated with a leaf results either from incomplete stomatal closure or cuticular exchange, or perhaps both. In direct contrast, heat exchange involves contact between the skin of leaves and the air; it is then that an additive laminar boundary layer resistance, r_b, arises as a central concern. This additional quantity derives from studies of the wind drag on flate-plates which result in the identification of a laminar boundary layer in which molecular exchange properties are of concern. Heat transfer from the skin of a leaf will be influenced by this laminar boundary layer effect. In the lack of convincing experimental evidence, whether this laminar boundary-layer resistance applies equally as a modifier of stomatal exchange remains conjectural.

The view of a leaf as a flat surface (as is assumed whenever accepted approximations for such laminar processes are extrapolated to foliage) raises issues related to the size and temperature of individual leaves as well as their morphology, flexibility and orientation relative to any velocity streamline that could be identified. Figure 34.1 shows an example of the complexity that nature imposes. Confronting the issues that arise is necessarily in the hands of those with the largest computers.

Figure 34.1. A photograph of hirsute foliage surviving in the author's garden.

The discussion so far relates to the view from below, regarding individual components of the plant canopy. In the meteorological context, it is not the values of r_s, r_{cut} or r_c that are relevant, these being determined per unit area of the leaves, but the integrated consequence on the air above of all of the contributing foliage. In the meteorological context, it is the view from above that is of main interest. A 'big leaf' model of water vapor fluxes has evolved, in which the integrated effect of individual leaf stomatal resistances (R_s) is used to quantify the resulting canopy-level attribute of opening and closing stomata. Transfer through the air beneath some level of observation imposes the need for additional resistance to be added, primarily the aerodynamic resistance R_a which describes the exchange between some height of measurement (typically 10 m) and the zero plane, *d*. A resistance R_b is included to represent the overall extrapolation from the level *d* to the canopy-wide mix of vegetation surfaces. (In analogy with the terminology used to describe flate-plate exchanges in laboratory studies, R_b is referred to as the quasi-laminar layer resistance, a terminology coined by Paul Frenzen.

The question then arises as to how to derive estimates of the canopy-level quantity R_s from the basic leaf-level measurements of r_s. There have been many detailed examinations of this matter (e.g., Norman, 1979; Jarvis and McNaughton, 1986) that succeed in illustrating the complexity of the overall problem. However, it is relevant to observe that a central contributing factor is the leaf area index, LAI. In concept, two extreme situations can be considered. In the case of a grass-like surface (or a sufficiently dense canopy that turbulence does not penetrate it) it might well be adequate to assume that the relevant relationship should be such that the upper limit on R_s would be

$$R_{s1} = r_s. \tag{34.1}$$

But if the vegetation is distributed thinly through the available subcanopy airspace, then a better approximation might well be

$$R_{s2} = r_s/LAI. \tag{34.2}$$

When there is no way to decide which extreme might be preferred, a best estimate is likely to be

$$R_s \approx r_s/LAI^{0.5}. \tag{34.3}$$

For a uniformly vegetated canopy, Noilhan and Planton (1989) addressed this issue and proposed that

$$R_s = r_s e^{-LAI} \tag{34.4}$$

and this has been used to characterize an extensive surface with uniform plant coverage (maize, in the experimental program of relevance, see Irmak and Mutiibwa, 2009). In the case of a mixed-species canopy, the reader is left to determine which approximation is most appropriate, with awareness that Equation 34.4 is unlikely to extend to the case of forests or other surfaces with large LAI.

In the case of the sensible heat flux, the assumption that the origin of the flux is the surface of heated leaves is conventional (Garratt, 1992). It follows that R_{cut} can be approximated similarly to R_s, so that

$$R_{cut} \approx r_{cut}/LAI^{0.5}. \tag{34.5}$$

Just as r_s and r_{cut} work in parallel when considering the heat balance of a leaf, so a canopy-level integrated surface resistance R_c can be calculated as

$$R_c = (R_s^{-1} + R_{cut}^{-1})^{-1}. \tag{34.6}$$

In agricultural research, the quasi-laminar resistance R_b is usually subsumed into the aerodynamic resistance term R_a, so that the simplified version of the big leaf model then involves only two resistances, one associated with transfer through the air and the other with the role of the surface itself. The Penman-Monteith approach (see Chapter 4) is based on this separation.

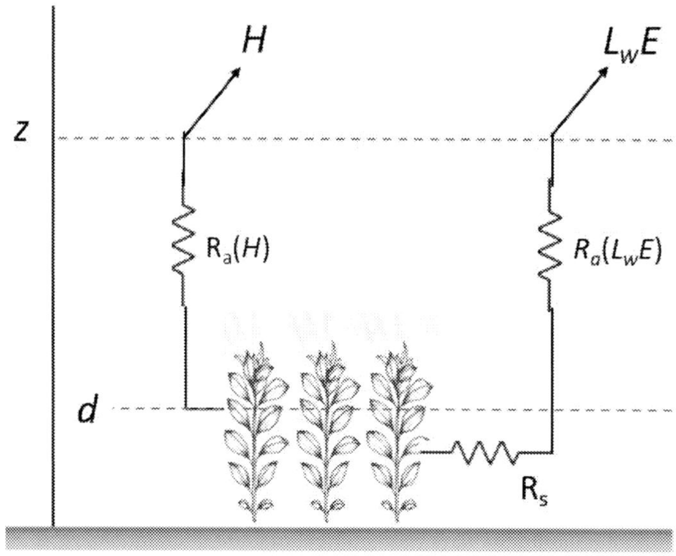

Figure 34.2. The essence of the big leaf model, with the in-air transfer of both heat and water vapor being represent by an aerodynamic resistance R_a and stomatal resistance by R_S (following Alves et al. 1998).

There have been many embellishments of the big leaf model, with the obvious opportunity to 'refine' it by addressing sub-canopy exchange in several layers. It is immediately obvious, however, that any such layered approach extending into the canopy imposes the need to account for changes in PAR, LAI, etc. through the sub-canopy air space. Such matters are necessarily specific to a plant species and to the location and time of day as well as time of year and prevailing weather conditions. The complexities involved quickly become intimidating, but nevertheless continue to attract those who instinctively associate complexity with progress. In the meantime, it is quite acceptable to consider the big leaf model as a conceptual and parsimonious construct suitable for guiding considerations of the roles of different influences and to help understand the characteristics of areas in which it has been tested and calibrated. However, as stated by Raupach and Finnegan (1988), 'Single-layer models of evaporation from plant canopies are incorrect but useful, whereas multilayer models are correct but useless.'

Conclusion

The big leaf approach has distinct advantages, once its limitations are overlooked.

Chapter 35

Heat Energy Apportionment

"The secret of getting ahead is getting started" — *Mark Twain.*

An application of frequent experimental interest concerns the use of micrometeorological methods to study fluxes from a test plot, such as a managed area of some crop surrounded by other vegetation. The Bowen Ratio Energy Balance (BREB) methodology is often used (q.v. Tanner, 1960; Dugas et al. 1991). This makes use of measurements of temperature and humidity differences (δ's) between two heights above the canopy, z_1 and z_2, to determine the Bowen ratio,

$$\beta \equiv H/L_wE = (c_p\delta\bar{\theta})/(L_w\delta\bar{q}) \\ = \gamma(\bar{\theta}_{z2} - \bar{\theta}_{z1})/(\bar{q}_{z2} - \bar{q}_{z1}) \tag{35.1}$$

Note that the temperature difference between two levels near the surface involves the potential temperature $\bar{\theta}$ rather than the absolute temperature $\bar{T}_a$. The difference is usually not important but the temperature differences should best be corrected to account for the dry adiabatic temperature gradient, a process that involves the slope of the dry adiabat: $\Gamma' = g/c_p$ (= 9.77 K km^{-1}). The most common application of Equation 35.1 relates to positive values of H and L_wE and with z_1 smaller than z_2. In accord with meteorological convention the temperature difference is then negative, as will be the specific humidity difference in midday situations. Here, as elsewhere, c_p is the specific heat of air at constant pressure, L_w is the latent heat of evaporation of water and γ is the psychrometric constant: c_p/L_w. Using measurements of the net radiation, R_n and the ground heat flux, G, estimates of the latent and sensible heat fluxes are then derived as

$$H = (R_n - G)/(1 + \beta^{-1}) \tag{35.2}$$

and

$$L_wE = (R_n - G)/(1 + \beta). \tag{35.3}$$

The attraction of this methodology is that it makes no assumption about the quantification of eddy diffusivities. All that it assumes is that these are the same, for $\overline{w'T'}$ and $\overline{w'q'}$. In passing, note that this matter is of particular relevance in the case of experiments making use of the aerodynamic method for evaluating H and L_wE, in which diffusivities appropriate for deriving H and L_wE from measurements of $\delta\bar{\theta}$ and $\delta\bar{e}$ (or $\delta\bar{q}$) are calculated using accepted flux-gradient relationships. This latter approach lacks the insensitivity to height of measurement that BREB offers and hence is susceptible to fetch limitations (Malek, 1993). (Note that the water vapor terminology used here involves either e or q, depending upon whether the agricultural or the meteorological side of the author's brain is working).

Clearly, there are problems that arise with the standard BREB methodology. The canopy storage term is missing and the measurement of G is demanding. Such difficulties are often overlooked, because in daytime conditions when R_n is greatest, the associated errors are likely to be small. However, the question remains as to how well the results derived might be affected by areas beyond the focal area of the study.

In many cases, measurements of the surface temperature ($\bar{T}_0$) are also available, derived from downward-looking infrared thermometers. It is the availability of such surface temperature data that enables a statistical examination of the heat storage term, S. In rethinking the crop heat balance problem, a first step is to address the common issue that Equations 35.2 and 35.3 break down when $\beta = -1$, as is frequently the case in situations near dawn and dusk. In practice the transfer of physical relevance is that of the total heat content of the air, the flux of specific enthalpy, η

$$F_\eta = H + L_w E \tag{35.4}$$

and the vertical difference against which this should be interpreted is

$$\delta\bar{\eta} = c_p \cdot \delta\bar{\theta} + L_w \cdot \delta\bar{q} \tag{35.5}$$

The diffusivity revealed from consideration of Equations 35.4 and 35.5 is that which expresses the transfer of the total heat content of the air. It is now assumed that the diffusivity derived by consideration of specific enthalpy is shared by both the sensible and latent heat flux but is better quantified than by consideration of either of them alone.

Inherent in both the BREB method and the Augmented Bowen Ratio Analysis (ABRA) now being introduced is the assumption that the eddy diffusivities of sensible heat and water vapor are the same. This assumption is compatible with the examination of effective surface temperature ($\bar{T}_e$, q.v. Chapter 32), but is at odds with the Penman-Monteith approach which assumes dependence on separate cuticular and stomatal resistances.

The form of the surface energy balance used here includes two terms that are commonly ignored: S (the thermal storage rate in the biomass above the ground) and P (the energy use in photosynthesis):

$$R_n - G = H + L_w E + S + P. \tag{35.6}$$

Sinclair et al. (1975) included consideration of the photosynthetic term in their examination of possible errors in BREB calculations but concluded that its contribution is small. They also emphasized the susceptibility of the computation to errors when H is small. Ohmura (1982) proposed criteria for rejecting data but did not consider the possibility the observations could be right but their interpretation lacking.

Data obtained at a site in Africa have been addressed elsewhere (Hicks et al. 2015; O'Dell et al. 2015) and will be used here. Four BREB systems were established in June 2013 on adjoining 0.64 ha plots at the International Maize and Wheat Improvement Center (CIMMYT) at Mt. Pleasant (near Harare), Zimbabwe. The site in question was at an altitude of 1460 m asl, on the Mafungabusa Plateau to the east of the Kalahari Desert. The availability of infrared surface temperature measurements enabled the standard BREB procedure to be augmented so that the heat storage term could be quantified. Data obtained include frequent nights of nocturnal intermittency (Hicks et al. 2015a) but such do not affect the present analysis of primarily daytime observations. In the following, data obtained in 2016 for two of the four plots will be considered: plots 1 and 2 as shown in Figure 35.1. Plot 1 was left fallow and was treated with weedkiller. Plot 2 carried a crop of pigeonpea (*Cajanus cajan* (L.) Millsp.) that grew to a height of about 1.5 m before harvest. In other years the plantings differed; the present was the only pigeonpea crop during the multi-year Zimbabwe study. Most often, the main focus was on maize as considered by O'Dell et al. (2015) and Hicks et al. (2015a).

Figure 35.1. The BREB site of the University of Tennessee in Zimbabwe. The site is a part of the International Maize and Wheat Improvement Center (CIMMYT) near Harare, Zimbabwe. In March, 2016, plot #1 was fallow and #2 was growing pigeonpea. The satellite image relates to a different time.

Consider a statistical analysis of a long sequence of observations. Since the intent is to describe the average diurnal cycle of the fluxes during a specific period, observations made during the period are first ordered according to time of day. Sequential groups of 100 experimental runs are then constructed, with statistical analysis applied to each grouping. In the following example, 10–min data samples ('runs') have been used and two

of these data have been constructed. The ABRA method now described applies multiple regression analysis to a selection of key variables derived from the observations within each data grouping. Each ensemble contains 100 measurements of the central properties. Consider the relationships involving the following three quantities:

$$X_1 = R_n - G \tag{35.7}$$

$$X_2 = dT_0/dt \tag{35.8}$$

$$X_3 = d\bar{\eta}/dz \tag{35.9}$$

X_1 is the basis for the computations relating fluxes to gradients. The quantity R_n is measured, using standard instrumentation. G is conventionally based on measurements within the soil, averaged over the measurement period of relevance (in the present case, 10 min). It should not be confused with the quantity S, which relates to the rate of heating or cooling of the above-ground biomass and possibly of some part of the air mass within the vegetated canopy. No matter how such complexities might arise, it is clear that $(R_n - G)$ is indeed a first-order and credible measure of $(H + L_wE + S)$ for each 10–min period. In line with Sinclair et al. (1975), the contribution of P is presently ignored.

X_2 expresses the rate of change with time of canopy temperature — indicative of the storage term, S. Downward-looking infrared thermometers were deployed centrally above both Plots #1 and #2. The characteristics of the plots of land now considered are such that for Plot #1 the temperature is of the uppermost soil surface. For Plot #2, the temperature $\bar{T}_0$ is assumed to be that of the vegetation (pigeon pea). It is possible that the relevant term could involve factors other than are quantified by measurements of changes in infrared surface temperature, such as the heating of the sub-canopy air mass.

X_3 is the appropriate measure of the gradient defining the transfer between the levels of measurement, z_1 and z_2. In the present example, the differences are measured over a height interval of 1.5 m, at a minimum of 1 m above the tops of the vegetation. (As mentioned previously, proximity to the surface is not a major constraining consideration, because there is no need to rely on any specific formulation of the flux-gradient relationship. All that is needed is an assumption that the diffusivities of heat and water are the same. In practice, proximity to the surface is a benefit, because it reduces fetch requirements.)

It must be emphasized that the relevant surface temperature property that enters the present procedure is its change from run to run. The examination of $\bar{T}_0$ measurement as might determine the effective interfacial temperature $\bar{T}_e$ (as in Chapter 32 above) does not greatly influence the results now derived but could account for the low partial correlation coefficients derived when expectations were otherwise.

The linear multiple regression analysis used here yields estimates of the coefficients in the equation

$$X_1 = a + b_{12.3}X_2 + b_{13.2}X_3, \tag{35.10}$$

along with quantification of the relevant correlation coefficients R_{12}, R_{23} and R_{32} and the corresponding partial correlation coefficients $R_{12.3}$, $R_{23.1}$ and $R_{32.2}$. The correlation $R_{12.3}$ expresses the relationship between X_1 and X_2 if variable X_3 is held constant, i.e., if $d\bar{\eta}/dz$ is constant and hence if $(H + L_wE)$ is constant. The corresponding slope, $b_{12.3}$ quantifies the dependence of S on dT_0/dt. The specification of the storage term is then quantified as $b_{12.3}\cdot\langle dT_0/dt\rangle$. (The angle brackets denote the ensemble average.) Likewise, $R_{13.2}$ determines a slope $b_{13.2}$ which relates $\langle H + L_wE\rangle$ to $\langle d\bar{\eta}/dz\rangle$ when S is unchanging. The analysis also yields an assessment of the ability of Equation 35.10 to represent $(R_n - G)$, by quantifying the proportion of the variance in X_1 explained by the combined influence of X_2 and X_3:

$$R_{1.23}^2 = R_{12.3}^2(1 - R_{13}^2) + R_{13}^2 \qquad (35.11)$$

as was used earlier in Equation 16.1.

Figure 35.2a (for the month of March, 2016) shows the total proportion of the variance in $X_1 = (H + L_wE + S + P)$ that can be accounted for by the consideration of the two variables X_2 and X_3. This is typically about 0.4, for the pigeonpea crop; for the fallow surface the proportion is considerably lower. This is explicable in terms of the contributions of the surrounding terrain, as was previously considered in the case of the Ohio data considered in Chapter 32. There is a reality of any test program like those considered here: the size of the test plots is a central consideration — the larger, the better.

Figure 35.2 (b) shows the quantifications of the storage terms then obtained for the two surfaces. In the case of the fallow (and barren) surface, there is no substantial storage term resulting from this analysis. In the case of the vegetated surface, the storage appears to follow expectations. In the morning, heat continues to be accumulated in the biomass (and in the soil and sub-canopy air space) until soon after solar noon, after which time the stored heat is released as the previously heated media cool.

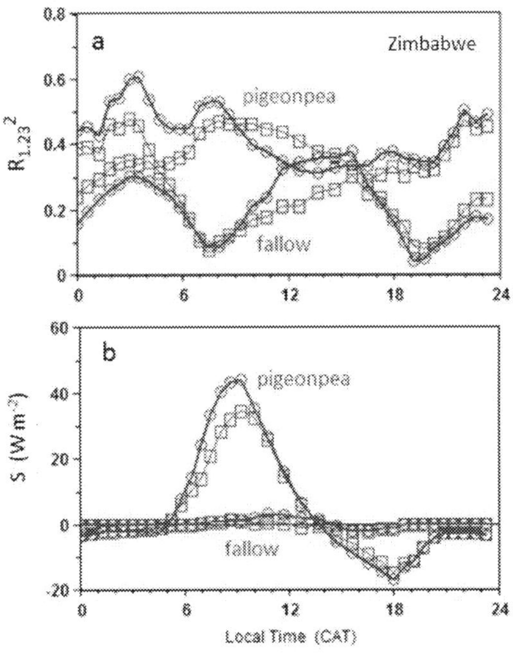

Figure 35.2. Results of an initial partial correlation examination of Zimbabwe BREB data, following the ABRA protocols now advocated. Data are for March, 2016. In diagram (a), the lines with circles show the extent to which the variances in $X_1 = (H + L_WE + S + P)$ are accounted for by the variables $X_2 = dT_0/dt$ and $X_3 = d\bar{\eta}/dz$. The lines with squares result when a modified form of X_3 is used: $X_3' = \bar{u}\cdot d\bar{\eta}/dz$. In diagram (b), the corresponding quantifications of the storage term S are depicted.

Included in the quantification of the term S using the ABRA methodology must be a contribution by photosynthesis, P. In practice, P will increase through the early morning sunlit hours and continue at a relatively constant rate (see Meyers and Hollinger, 2004) provided PAR continues adequately and the leaves do not get too hot. Hence, P and $d\bar{T}_0/dt$ will be positively correlated during the first part of the morning and will be negatively correlated during the hours approaching dusk. The determinations of S will reflect this, so that the high value indicated in Figure 35.2b in the early morning represents the sum of thermal heat storage and the equivalent heat consumption by photosynthesis (CO_2 uptake by photosynthesis requires an energy consumption of about 11.2 W m^{-2} for every mg of CO_2 m^{-2} s^{-1} exchanged, q.v. Nobel, 1999).

Consideration of the basic relationships associated with classical air-surface exchange would suggest that the wind speed should be included as a contributing variable. Accounting for an effect of wind speed can be tested by replacing the term $X_3 = \mathrm{d}\bar{\eta}/\mathrm{d}z$ by $X_3' = \bar{u}\cdot\mathrm{d}\bar{\eta}/\mathrm{d}z$ in the partial correlation analysis as already explained. Results obtained by separate analyses of the Zimbabwe data are illustrated in Figure 35.2. There are differences apparent, but in general these differences appear to amount to less than about 10 W m^{-2} in S. This could be viewed as an indication of the error associated with the evaluation of the storage term using the ABRA technique.

Figure 35.3a shows how the storage 'calibration' ($b_{12.3}$) varies with time of day, for the pigeonpea crop and the fallow. As before, data are for March, 2016. Also shown, as the dashed line, is the result of a correlation (not partial) of X_1 on X_2. Comparison between the two curves indicates that the peak in the total correlation seen during the daytime disappears when the contribution of X_3 is accounted for. On the whole, the storage term coefficient shows no substantial indication of changes with time of day for the vegetated surface. However, the change with time is convincing for the fallow case and a similarity with the daytime pigeonpea result is intriguing. The matter is confounded (or perhaps explained) by consideration of the association of air and soil temperatures, to follow.

At this stage, there are two different ways to arrive at the desired product — a best estimate of the average diurnal cycle of fluxes during the month of March, 2016. An obvious method would be to use the ensemble-mean quantifications of $b_{13.2}$ to derive a set of L_wE evaluations with 10–min time resolution and then to average them. This would impose the risk of generating a statistical distribution of L_wE estimates dominated by extremes, due to errors or other inconsistencies in the core measurements of $\delta\bar{T}$ and $\delta\bar{q}$. A preferred approach is to apply the methodology already described to the 100–point ensemble averages rather than individual 10–min data points. In this regard, note that the quantifications of $(R_n - G)$ and $\mathrm{d}T_0/\mathrm{d}t$ are largely immune to errors associated with upwind conditions. It is the effects of limited fetch that cause much of the aleatoric variability in $\delta\bar{T}$ and $\delta\bar{q}$ and it is the dominance of uncertainty of this kind that hinders the determination of levels of statistical certainty associated with any of the final products. The challenge is therefore to generate robust evaluations of $\delta\bar{T}$ and $\delta\bar{q}$ from which to derive the desired flux quantifications. The results illustrated in Figure 35.3b are the products of this second approach, in which 100–point averages of the basic quantities have been used as a smoothed dataset, after correction for the storage term S. Hence, L_wE has been computed as

$$<L_wE> = <R_n - G - S>/(1 + (c_p/L_w)\cdot(<\delta\bar{\theta}>/<\delta\bar{q}>) \qquad (35.12)$$

where the angle brackets indicate ensemble averages. Figure 35.3b also shows the determinations of the latent heat flux produced by the conventional BREB analysis methods (Perez et al. 1999), including the data selection criteria recommended by Ohmura (1982). Differences are generally small, but the consequences of allowing for heat storage are clearly evident. Of passing interest is the appearance of a slight lag in the evaporation from the crop, peaking about an hour after the fallow data indicate. The scatter after dusk presents issues that will not be addressed immediately. It was during this period that intermittency became most apparent; its repercussions have yet to be assimilated into the procedures now utilized.

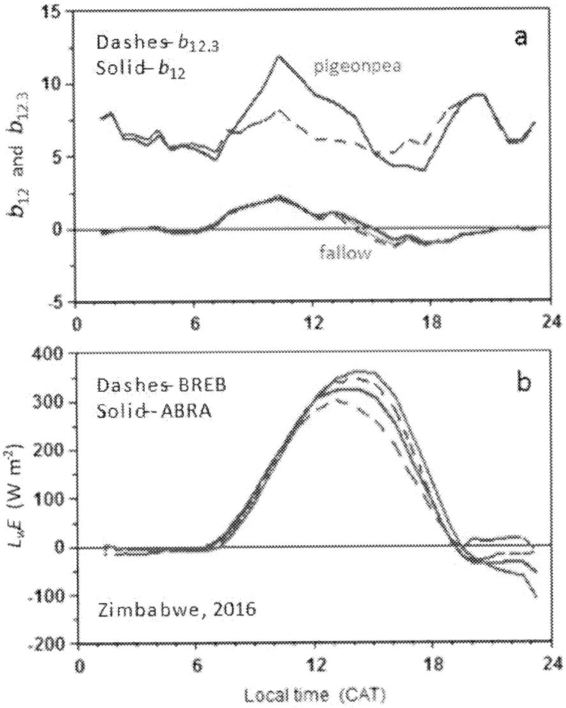

Figure 35.3. Month-long average diurnal cycles of (a) the "calibration" coefficients for deriving the heat storage term S from measured changes of surface temperature with time and (b) the latent heat loss from the surface.

Using observations made at the same location in 2015, when area #1 was left fallow and area #2 was planted with maize (*zea mays L*), the storage term derived for the maize area was similar to that depicted in Figure 35.2b for pigeonpea, peaking at about 40 W m^{-2} in the mid-morning. A study at a site in Ohio (see O'Dell et al. 2018; Hicks et al. 2020,) resulted in a storage cycle much the same as was for Zimbabwe, peaking at about 50 W m^{-2} for the mature maize case.

A subsequent experiment conducted at a larger field site (see Figure 35.4) in Loudon, Tennessee, in 2019 confirmed many of the Zimbabwe and Ohio results. Two independent sets of observations were obtained, using identical instrumentation deployed at different locations in the field of maize, identified as the north field (N) and the south (S) in Figure 35.4. Each set of observations contained duplicated measurements of soil temperature at 1.5 cm and 4.5 cm depth, from which a mean soil temperature (T_s) representing 3 cm depth has been computed (with four separate measurements contributing). Measurements of air temperature (T_a) were based on measurements by two sensors, one referring to a height of about 1 m above the top of the maize and the other 1.6 m higher. Averages constructed over sequential 10–min periods have been used.

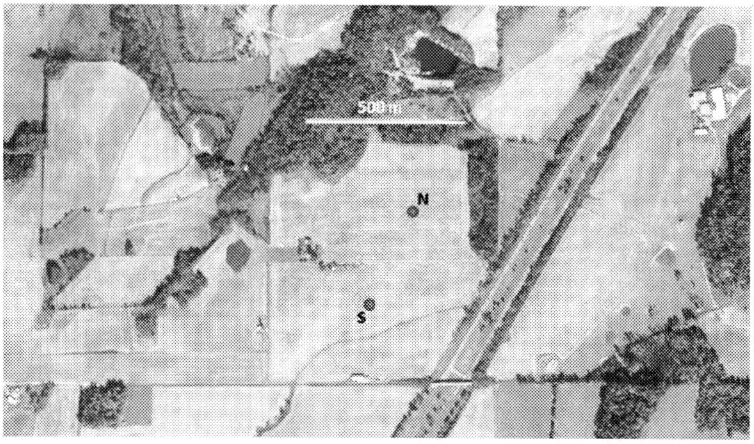

Figure 35.4. A satellite view (from Google Earth) of the Loudon (Tennessee; 37.67 °N; 84.45 °W) field site used in the expanded field study of a maize crop in 2019. Locations of North and South BREB and EC instrumentation are identified.

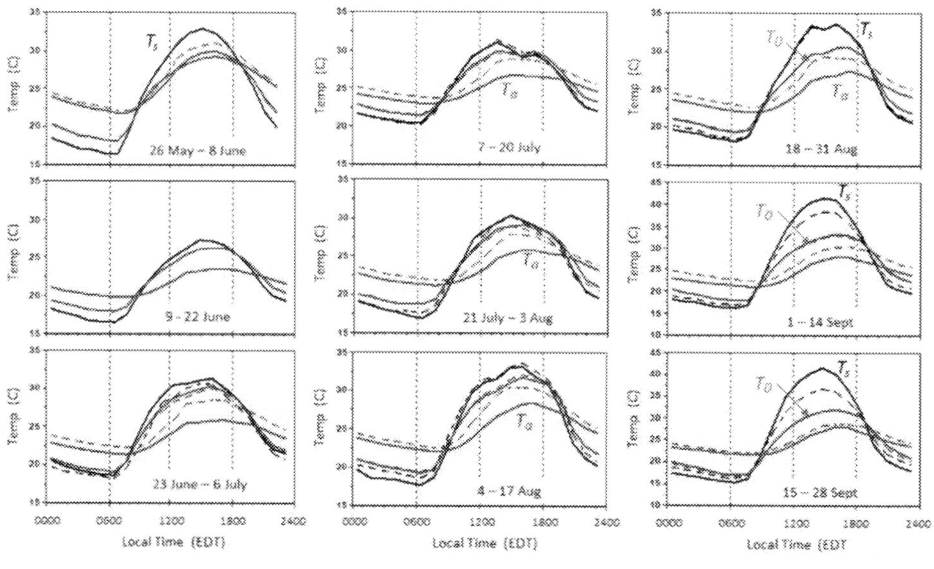

Figure 35.5. Average diurnal cycles computed over sequential two-week periods. Solid lines represent data from location N in Figure 35.4, broken lines for location S. Measurements of infrared surface temperature (T_O), air temperature (T_a) and soil temperature (T_S) are shown. Local time (EDT) is 4 h behind UTC.

Figure 35.5 shows how average air temperatures and soil temperatures varied with time of day for a sequence of two-week periods starting with 26 May 2019 (about 8 weeks after seeding, on 28 March) and extending to after harvest (on 22 August for the north field, 4 September for the south). Before 23 June, instrumentation at site S was not operational.) The solid lines represent observations made at installation N in Figure 35.4; the dashed lines are for measurements at site S. The differences between observations at the two locations are to be expected, since measurements were certainly influenced by details of the deployments, such as the location relative to crop rows and the amount of bare soil influencing the infrared temperature measurement. Crop emergence was noted on 11 – 13 April. The crop reached maturity (minimal further growth) in mid-July. As the crop grew the air temperature heights of measurement were adjusted to maintain a constant height above the crop, but with no requirement that the precise height of measurement was the same for both installations. This last consideration is a contributing cause of the large differences in measurements of T_a at the two locations. This difference decreased as the crop grew and essentially disappeared when the maize was senescent (after mid-August). A possible explanation is that a different variety of maize was planted in the southern field. There was no discernable difference in the two measurements of T_0 plotted in Figure 35.5. Soil temperatures differed considerably after harvest, presumably a result of mechanical disruption of some of the subsurface sensors during the harvest.

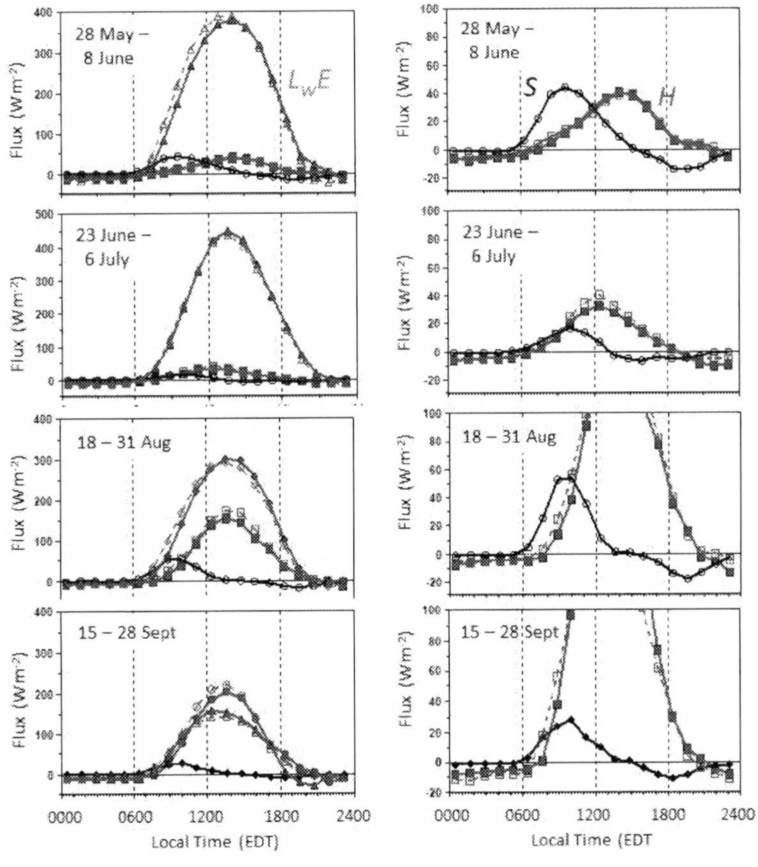

Figure 35.6. Average diurnal cycles of three components of the surface heat energy balance: the latent heat of evaporation L_WE, the sensible heat flux H and the heat storage term S, for four of the two-week periods examined during the Loudon 2019 field study over maize. Solid lines indicate data derived using the ABRA methodology, which enables quantification of S, broken lines represent results derived using standard BREB protocols, which assume S = 0. Data are from the northern field of maize as identified in Figure 35.4.

Figure 35.6 permits a comparison between the results of the standard BREB methodology (which imposes data-rejection criteria and assumes $S = 0$) with the ABRA approach as outlined here. On the whole the differences are small, necessarily of magnitude less than the magnitude of *S*. However, for L_wE, ABRA results lag behind BREB. Also, the maxima of L_wE for BREB exceed those for ABRA. For *H*, ABRA also lags behind BREB, but not consistently. Perhaps most surprising (and not relevant for further discussion here) is that the ABRA results near dawn and dusk indicate that evaporation commences well before sensible heat exchange after sunrise and extends well beyond it after sunset.

Conclusion

The amount of heat not considered in standard BREB analyses is associated with the canopy but exceeds that accounted for by changes in biomass temperature. The extra component is related to photosynthesis.

Chapter 36

The Origin of Convection

"The warm-up is critical" — Usain Bolt.

The observations summarized so far result in some insecurities about contemporary understanding of how plant canopies interact with the atmosphere. In particular, the heat storage term in the energy balance equations seems (on the basis of many separate studies) to be more than can be accounted for by changes in the heat of the biomass. The reason for this has been explored in Chapters 32 and 35, confirming that the apparent heat imbalance (S) is partially due to the omission of photosynthetically active radiation in the logical construct underpinning the ABRA methodology.

In conventional concept, the molecular-scale mechanisms that control transfer between the surface itself and the air in contact with it correspond to the expectations of classical laminar boundary layer assumptions. Geiger (1950) initiates related discussion with consideration of exchange between the air and a flate-plate. An extensive series of studies in Germany in the 1930s resulted in a clear recognition of what is now recognized as turbulent intermittency in unstable stratification, controlled by the strength of the surface heating. The studies involved many aspects of the overall problem, extending from dispersion over a heated smooth plate to examination of the lowest layers of air in the Gobi Desert. The matter of the initiation of convection was studied extensively a century ago and the studies continue. Here, it is the extension to a vegetated canopy that is of interest.

In daytime, exposed surfaces (e.g., the leaves of plants) are heated by solar radiation. Their heat is transferred to the air in contact with them by diffusion, at a molecular scale. Convention decrees that thermal gradients evolve and, following the expectations of Fickian diffusion, fluxes result. As a free convection embellishment, Priestley 1959 advocated that heated packets of air combine until their collective buoyancy is sufficient to overcome the restraining drag force, so that convection initiates sporadically, without extensive Fickian involvement. Many times in the discussion presented here it has been demonstrated that the free convective description of the sensible heat flux is a good representation of it, almost regardless of how much the surface is heated (provided that it is indeed heated). Building on the developments regarding the role of chaos (Feigenbaum, 1979; 1980) and a browsing of relevant progress in materials science, molecular dynamics and nucleation physics (e.g., Bodenschatz et al. 1991; ter Horst et al. 2011), the following is presented as a modification of the usual view of the sub-canopy thermal environment.

Once heated by solar radiation the foliage of a vegetated surface will start heating the air in contact with it. At a very fine scale, packets of heated air will evolve at local hotspots, originating as determined by plant physiology or perhaps no more than randomly. Once these packets of heated air break loose from their point of origin, they will diffuse through the available airspace under the influence of whatever atmospheric flow might exist. Clusters will combine (coagulate, nucleate), possibly gaining small heat increases as the process continues (due to the entropy enhancement as in the case of gas-to-particle conversion through nucleation). Eventually, the clusters will grow sufficiently that the buoyancy resulting from the density difference from their surroundings (due to their higher temperature) will be such that convection can be initiated (as described by Priestley, 1959; Feigenbaum, 1979).

This conjecture is no more than an extrapolation of results found in other circumstances, involving other disciplines. However, its consequences are tentatively testable. In Chapter 32 above, the concept of the effective surface temperature (T_e) was introduced, with results obtained during the 2015 Ohio experiment (O'Dell et al. 2018) showing that T_e was less then T_0 in daytime (unless external circumstances intervened, such as rainfall).. Other experiments have yielded similar results. In concept, T_e could well determine the starting point for convection, or the origin for applying Priestley's free convection formulation.

Observations made in 2019 in the study of a maize crop addressed earlier, located in a eastern Tennessee (Loudon, TN; see Figure 35.4) provide an opportunity for exploration. Data were archived the same way as in the Ohio experiment conducted in 2015. Analysis of the Loudon data follows the procedures described in Chapter 32. As before, ensembles have been constructed to contain 100 members, this time after ordering according to the heat energy available — (R_n – G). As is seen in Figure 35.4 two sites were monitored. One of these (the northern site, N in Figure 35.4) was planted several weeks before the other site (S) and harvest was correspondingly earlier. Results are shown in Figure 36.1, in a series of panels permitting comparison between the north and south results. Because of the different planting and harvesting schedules and the occurrence of periods of missing data, not all sampling periods yield results from both locations. The site was operated remotely and instrumentation malfunctions were not always quickly detected.

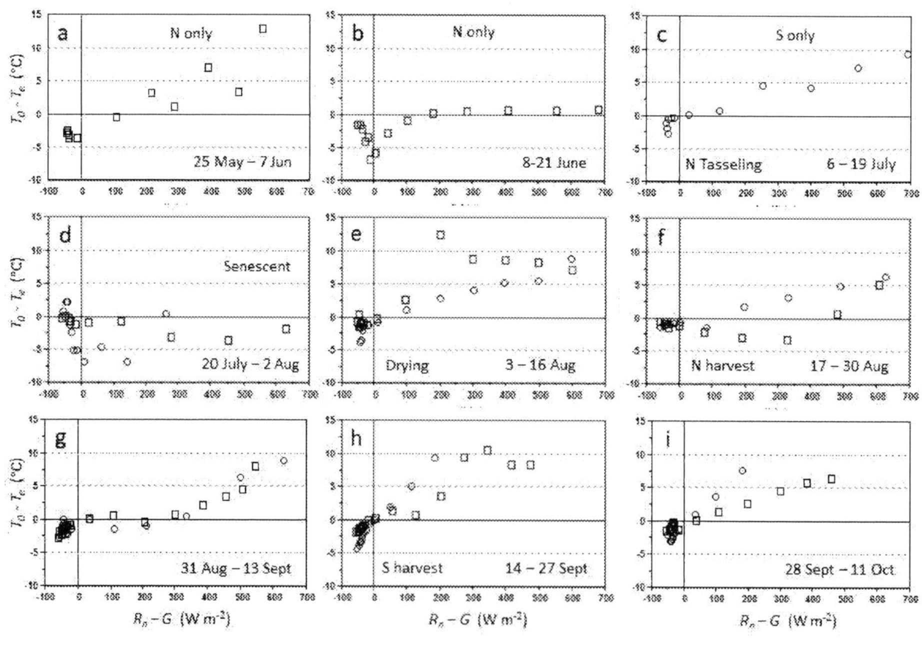

Figure 36.1. Plots of the amount (two–week averages) by which the equilibrium interfacial temperature (T_e) was less than the measured infrared temperature (T_0) for the 2019 maize crop study conducted at Loudon, Tennessee. Circles indicate results from the southern measurement site (see Figure 35.4), squares from the northern.

A striking feature of the panels of Figure 36.1 is the near-zero departure indicated in Figure 36.1b. At the time illustrated, the crop was emerging and the surface was largely bare soil, tilled. One tentative interpretation of this result is that when no crop is present there is no reason to expect T_e to differ from T_0. The nighttime results are largely uninformative, generally indicating that the IR temperatures lower than the values of T_e derived from the analysis presented here. Observations made at other locations indicate that this particular result might not prove to be an endemic feature of the surface boundary layer. On the whole, the available observations indicate differences in the interpretation of IR temperatures that could well result from simple but consistent errors of measurement or sensor deployment.

Most recent attention to air-surface exchange characteristics of this kind have concentrated on forests, where many relevant studies have been conducted under the auspices of several international programs such as Fluxnet (Baldocchi et al. 2001). Research has mostly been with a model of air-surface exchange in mind, differing somewhat from the visualization now advocated. The widely accepted approach is to address the matter of convective initiation in a sub-canopy airspace in terms of temperature gradients and their interpretation — the gradients are set up and convection follows. At the opposite extreme the current hypothesis is that chaotic motions result in the generation of buoyant elements from which convection develops — gradients still exist but their coupling with fluxes is then indeterminant. Further examination of this matter will require experimental attention that is presently invited.

Conclusion

In daytime, the air temperature affecting convection appears to differ from IR surface temperature.

Chapter 37

Introducing the Soil

"We are too much accustomed to attribute to a single cause that which is the product of several"— Marcus Aurelius.

Over land, sunrise is followed by a slow rise in surface temperature as the incoming solar radiation increases, modified according to soil thermal characteristics and vegetation. The discussion in Chapter 35 of heat energy apportionment and the ABRA methodology depended on quantification of dT_0/dt, but usually simultaneity is assumed between fluctuations in air temperature, soil temperature and infrared surface temperature. To minimize the effects of thermal lags, soil temperatures are conventionally made near to the surface. In the 2013–2014 Zimbabwe experiment (as described earlier), soil temperatures were measured at 4 and 16 cm depth. Figure 37.1 shows how air temperature, surface temperature and subsurface soil temperature changed for two different occasions, each a two–day period. The first of these periods was of negligible cloud cover, the second with continuing cloud coverage. In both cases, the subsurface temperatures lag the surface temperatures considerably, by an amount that increases with depth. This demonstrates that the two-media (atmosphere and soil) system cannot be described well without taking time changes into account. In general, shallow soil temperatures will display a diurnal cycle that will diminish in amplitude and increasing lag behind surface temperature as depth increases, a feature that is barely evident in the plots of Figures 37.1a and 37.1c.

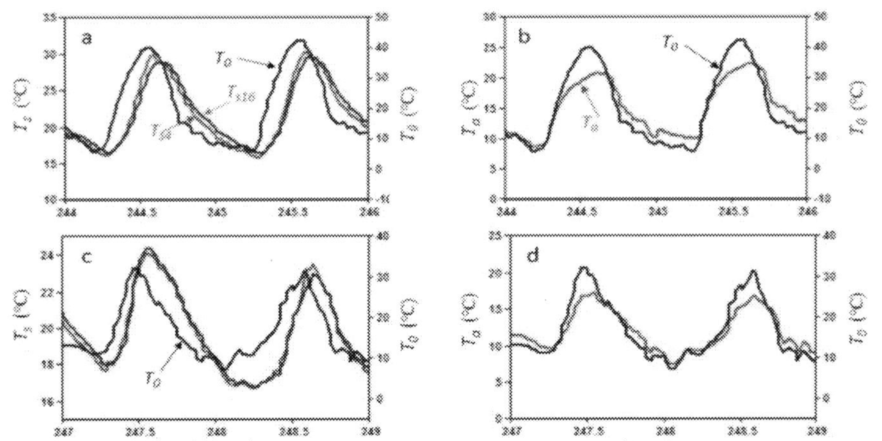

Figure 37.1. Examples of how soil temperature (T_S) and air temperature (T_a) lag behind infrared surface temperature (T_0). Plots a and b relate to cloudless skies, c and d to a period with considerable cloud cover. To the left, soil temperatures (T_S) are plotted for 4 cm and 16 cm depths. To the right, air temperatures (T_a) are plotted. Data are from the Zimbabwe field study.

Figure 37.2 presents an uninterrupted time sequence of T_0, T_a and T_s derived during a 2017 field study at a maize field site in Tennessee (at 35.696° N; 84.387° W; a different location from the 2019 site already considered). This time, the soil sensors were located nearer the surface, at depths of 1.5 and 4.5 cm. The two subsurface temperatures differed by more than was evident in Figure 37.1, as expected on the basis of simple diffusion. Figure 37.2 shows that the infrared temperature of the surface follows closely behind the variation in net radiation. However, there is an obvious delay in the atmospheric response and an even greater delay in the response observed a short distance below the surface. In practice, the temperature profile above the surface must be expected to be much the same as that below the surface (with directions opposed), with appropriate normalization being by incorporating the relevant thermal diffusivities (Priestley, 1959).

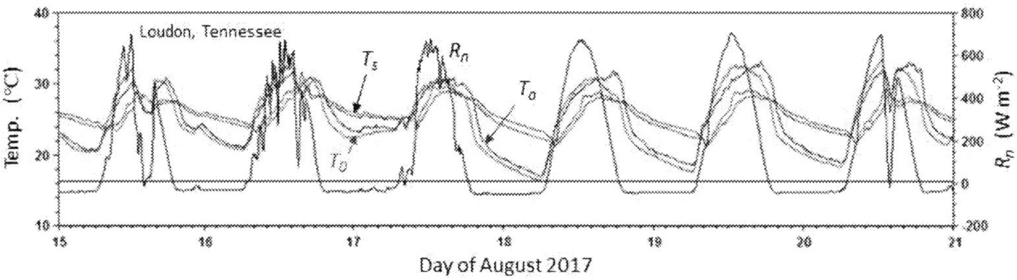

Figure 37.2. A six-day sequence of measurements made during August 2017 at a field site of the University of Tennessee. Plotted are net radiation (R_n), infrared surface temperature (T_0), air temperature (T_a) measured at 1.5 m above the maize canopy and soil temperatures (T_s) at depths of 1.5 cm and 4.5 cm.

In general, incoming radiation imposes a diurnal cycle on the surface temperature of the soil, of amplitude decreasing with depth until eventually the diurnal cycle becomes undetectable. Remaining changes are then due factors of longer time scales, such as seasonal or longer term. (In this regard, it has long been advocated that measurement of soil temperature profiles at remote locations would be an excellent basis for detecting and studying climate change.) The rate at which changes occur in near-surface soils is determined by the thermal conductivity of the soil and the temperature difference between the surface and some depth below the surface where the diurnal cycle is not detected. The analogy with the planetary boundary layer of the atmosphere is obvious and the physical considerations are indeed similar. However, in the subsurface case there is no equivalent to the convective transfer that dominates the daytime PBL and in the subsurface the role of water is critical. The moisture content of the soil is frequently the major determinant of the thermal conductivity of relevance and is clearly difficult to predict and challenging to measure. One might anticipate that the prediction of soil moisture content will be a greater challenge than the prediction of precipitation and such has indeed proved to be the case. It is for this reason that efforts to incorporate soil moisture into the modeling procedures intended for weather forecasting have been notably unproductive.

The lag between soil and surface temperatures has been the subject of research for decades, usually basing analysis on the change in incident net radiation rather than its derivative surface temperature (e.g., Lettau, 1951; Priestley, 1959; Geiger, 1965). Such early work has led to an understanding that is well explained in standard textbooks (e.g., Stull, 1988; Garratt, 1992).

Simulations that couple the atmospheric medium with vegetation and the soil beneath have become common.

Many of the features affecting air-surface exchange have consequences strikingly evident in the time sequences shown in Figure 37.2. The first three days of the diagram were affected by cloud cover and rainfall, the effects of which are obvious as the downward spikes in the net radiation record. The rainfall certainly wetted the soil; the soil warmed more slowly after sunrise and cooled more slowly after sunset on the days following rainfall. The final three days of the record shown in Figure 37.2 were of largely clear skies, but with periods of scattered cumulus which slightly affected the net radiation. During this period, the nighttime surface and near-surface air cooled to a much lower temperature than when clouds were present and the net radiation loss from the surface exceeded that on other nights.

The lag between the occurrence of daytime net radiation maxima and surface temperature maxima is best seen in the cloudless sky data: about one hour. However, air temperatures continue to rise after the time of peak surface temperature, maximizing in mid-afternoon (as is common knowledge). The surface in question was a mature maize crop. The analysis in Chapter 36 already revealed that the central consideration regarding the overall surface exchange of heat might not so much be the infrared surface temperature as a sub-canopy temperature typically about 5°C cooler in daytime.

A central concern of models linking the atmosphere and the subsurface is that the moisture of the soil introduces complexities that have proved difficult to overcome. However, it seems that the data resources are sometimes capable of addressing the issue in a way that invites further attention. If the variations evident in sub-surface temperatures are a consequence of the diurnal cycle of solar radiation, then the lag associated with the peak subsurface temperature can be expressed as

$$\Delta t = (\Delta z/2)\cdot(t_{24}/(\pi\kappa))^{-1/2} \qquad (37.1)$$

where Δz is the distance between the two levels of soil temperature measurement (12 cm in the case illustrated in Figure 37.1, 3 cm for Figure 37.2) and κ is the thermal diffusivity of the soil (see Geiger, 1965; Stull, 1988). The appropriate time scale is determined by the diurnal cycle: t_{24} = 24 hrs. If Δt was 30 minutes as was the case in the Zimbabwe study (Figure 37.1), then κ would have been about 30 mm^2 s^{-1}.

Inspection of Figure 37.2 shows that for the 2017 Loudon campaign the lag was about 30 minutes on the 16 and 17 of August (when there was rainfall), increasing to about one hour by 19 August. This increase in the magnitude of the lag indicates that the thermal diffusivity of the soil changed from about 2 mm^2 s^{-1} for the rainfall-wetted soil to about 1.5 mm^2 s^{-1} as the soil subsequently dried. Published values for the thermal diffusivity of soil are in this range (Abu-Hamdeh, 2003). If routine observations of air temperature and insolation were accompanied by soil temperature measurements, daily estimates of soil thermal conductivity could be derived and soil moisture content subsequently approximated. The matter is central to land-surface parameterization schemes as proposed by Noilhan and Planton (1989), for example. The potential benefits to both agriculture and weather forecasting cannot be assessed without more extensive examination than is presently appropriate.

Two conceptual obstacle confounding many attempts to simulate the coupled air/surface/subsurface exchange of heat are the large step-function differences in density and thermal conductivity. Lettau (1951) and subsequently Priestley (1959) pointed out that the effects of the dissimilarity can be moderated by adopting new distance scales above and below the surface. Instead of z_a and z_s (respectively), make use of the products $z_a\sqrt{\kappa_a}$ and $z_s\sqrt{\kappa_s}$ where κ_a and κ_s are the thermal diffusivities of air and soil. The conductive capacities of air and soil (defined by Priestley) offer a streamlined way to smooth through an otherwise intimidating boundary.

Conclusion

The diurnal cycle of solar radiation affects both the lower atmosphere and the upper lithosphere. Considering one without the other seems incomplete.

Chapter 38

An Eclipse Case Study

"It is worth exploring what happens when the light goes out" — Noel E. Ridere.

In the early afternoon of 20 August, 2017, the Loudon site was affected by a complete solar eclipse, the effects of which are clearly evident in Figure 37.2. Some details of the eclipse were presented earlier, in Chapter 16. The observations permit a closer examination of the way plant physiological processes affect the surface energy balance.

Figure 38.1 shows how net radiation varied through the period affected by the eclipse, together with air (3 m high), surface (IR) and soil (3 cm depth) temperatures. It is immediately obvious that the surface temperature followed the trend in incoming energy with a lag of about 5 min. Air temperature lagged by 20 min and soil temperature by about 40 min. The last of these estimates is poorly determined, but it is clear that the soil at 3 cm depth (an average of measurements at 1.5 and 4.5 cm) responded more slowly than the air at 3 m height.

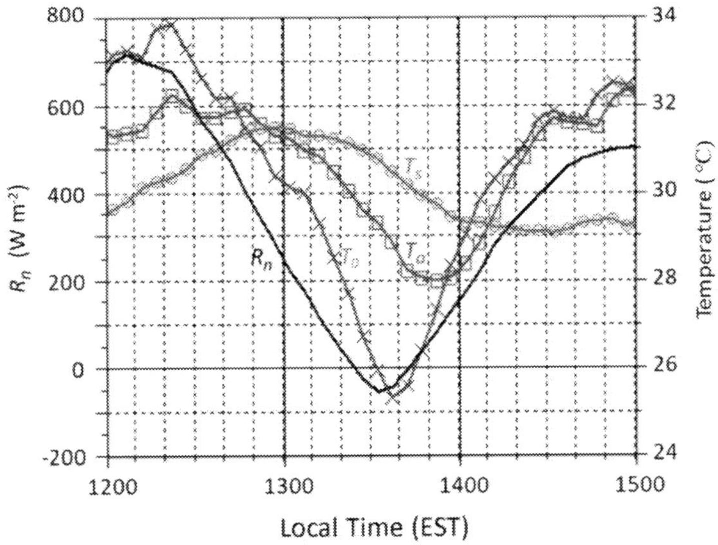

Figure 38.1. The variation of net radiation (R_n), soil temperature (T_s, 3 cm depth, magenta circles), air temperature (T_a, 3 m height, red squares) and infrared surface temperature (T_0, green crosses) during the passage of the complete solar eclipse of 20 August 2017, as observed at the Loudon, TN, site. The site was vegetated — mature maize about 2 m high.

For the period through the peak of the eclipse, Figure 38.2 plots the time sequence of several variables of relevance. In every case, the values now plotted are one-minute averages, plotted every thirty seconds. Soon after local noon, Figure 38.2a shows that incoming radiation (R) started to decrease approximately linearly but interrupted a few times by the passage of fair-weather cumulus clouds that also appear to have influenced the wind speed. Note that the wind speed increased shortly after the passage of the total eclipse.

Figures 38.2b and 38.2c show the changes in short-term covariances during the period of the eclipse: $(\widetilde{w'u'})$, $(\widetilde{w'T'})$, $(\widetilde{w'e'})$ and $(\widetilde{w'C'})$ — associated with the fluxes of momentum, sensible heat, water vapor and CO_2 respectively. The data plotted in Figure 38.2b,c are one-minute covariances, plotted every 30 seconds. The covariances plotted are not indicative of the total eddy frequency spectrum, as is emphasized by the use of a wavy overbars. The intent is to look at the time evolution, using a reliable indicator of the relevant physics, not to examine details of the energy balance. The values plotted are raw determinations of the covariances derived from the recorded instrumental outputs and (except for the incoming solar radiation, *R*) without adjustment of units or representation in more familiar terms (e.g., as W m^{-2}). In Figure 38.2b, it is seen that the sensible heat covariance $(\widetilde{w'T'})$ dropped to near zero when the incoming radiation (as seen in Figure 38.2a) dropped to below about 310 W m^{-2}. The $\widetilde{w'T'}$ covariance stayed near zero until the radiation recovered to ≈ 200 W m^{-2}. As the eclipse approached, the drop in $\widetilde{w'u'}$ (in black) occurred even though the wind speed did not yet drop sharply. It is concluded that the local atmosphere was affected by both the sudden elimination of the convective mixing that was previously contributing to the transfer of the various quantities, including momentum, and by advection from upwind areas unaffected by the eclipse. (In classical micrometeorological terms, the association with the onset of stability is obvious, but classical micrometeorology is scarcely relevant given the prevailing rapid changes with time.) Once the convective mechanism was re-introduced, the momentum flux also recovered, although substantially reduced because of the decrease in the wind speed.

Plant physiological studies indicate that the guard cells around stomata will cause the stomata to open slowly when energized by radiation, but to close less slowly when the radiation intensity drops or when a threat is detected (e.g., exposure to some dangerous chemical). The time interval from 1310 to about 1346 of Figure 38.2c is of special interest in this regard, since it appears that transpiration continued with reinforced vigor after *R* dropped below about 300 W m^{-2}, at which time the $\widetilde{w'T'}$ covariance had already approached zero. CO_2 was still being taken up while water vapor was being emitted during this period. The burst of exchange (upwards) of water vapor and (downwards) of CO_2 seen during this period appears to have been unassociated with a change in the radiation input, but clearly associated with the negative swing of the sensible heat flux. An immediate interpretation (that remains to be examined in detail) of these observations is that the energy requirement of photosynthesis was accommodated by extracting heat from the air and the soil, so utilizing the diminishing PAR to the fullest extent possible.

The inset in Figure 38.2c shows the variation of the water vapor and carbon dioxide fluxes as the central period of the eclipse progressed. The two covariances switched from being negatively correlated before about 1342, to strongly and consistently positively correlated until they reverted to negative at 1400. Evaporation continued at a low level through the eclipse. It is evident that the maize stomata were closed at about 36 min, at which time the covariances evident in the Figure 38.2 were probably associated with the soil or the biology resident in it. This is most convincingly evident in the $\widetilde{w'C'}$ trace, which went positive and so indicated that the plants were no longer taking CO_2 from the air but that CO_2 was being emitted, presumably from the soil.

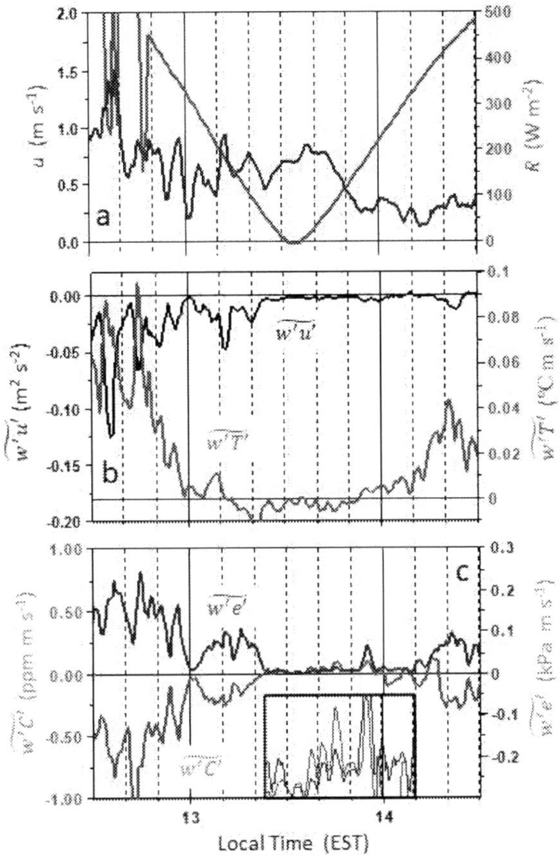

Figure 38.2. Results obtained during the solar eclipse of 20 August, 2017, (a) showing the variation in incoming radiation and the wind speed, (b) comparing the short-term covariance $\widetilde{w'T'}$ with $\widetilde{w'u'}$ and (c) showing the corresponding $\widetilde{w'e'}$ and $\widetilde{w'C'}$ results (water vapor and CO_2 exchanges respectively). Solar radiation is plotted in units of W m^{-2}. All other plots express the variables in basic SI units, except for e (kPa) and CO_2 (ppmv). The site carried a mature maize crop.

These observations are in agreement with plant physiological expectations. In general, stomata close quickly and open slowly. However, the present indications are that stomata closed as expected but re-opened some minutes later, in association with an increase in the momentum flux (comparing Figures 38.1b and 38.1c) and a reversion of the sensible heat covariance (Figure 38.2b). The implicit suggestion that heat from the soil served as an agent to help facilitate photosynthesis is conjectural. This matter is not amenable to detailed investigation using the present dataset because of the time constant involved in the measurement of the ground heat flux by flux plates.

Conclusion

Plant physiological characteristics can influence the surface heat energy budget in complicated ways.

Part V: Chemical Sources, Sinks, and Dispersion

Chapter 39

Dry and Wet Deposition

"It is always dangerous to reason from insufficient data" — Sherlock Holmes.

First, it is necessary to define what is meant by 'dry deposition' and to differentiate it from 'wet.' Removal of some trace constituent of the atmosphere is classically by a combination of three mechanisms: in-air chemical reactions, scavenging associated with the precipitation process ('wet' deposition) and the sequestration of pollutants as they contact surfaces and are retained by them ('dry' deposition). Irving (1990) assembled a multi-volume 'State of the Science' review of these and related matters, as they refer to the acid rain concerns of the 1970s and 1980s. If understanding of details is desired, then the present recommendation is to refer to the acid rain documentation and to regard the material presented here as little more than a summary of what is a complicated issue.

Radioactive fallout studies of the 1950s through the 1960s demonstrated, with confidence, that over a period of months the wet deposition of these materials far exceeded dry, with only minor departures from this generalization. A factor of ten was widely acknowledged to be a reliable estimator of the ratio of annual wet deposition of radionuclides like ^{90}Sr to dry deposition. Once scientific interest of these matters diverged into matters of industrial pollution, like sulfur (initially present as SO_2) in the context of acid rain, the radioactive fallout ratio was accepted by many as being relevant, whereas others anticipated a ratio closer to unity. Without discussing this matter further, at the risk of opening wounds that have long been salved if not healed, the central consideration can be expressed as follow.

- The radionuclides generated by nuclear weapons tests and measured in fallout studies (e.g., ^{90}Sr, ^{137}Cs) were usually of far distant origin, generally carried by winds at upper tropospheric or stratospheric levels. Rainclouds have easier access to them than the surface.
- In contrast, industrially generated pollutants (e.g., SO_2) are injected into the lower atmosphere and are largely confined within the PBL where convective mixing (in daytime, at least) ensures continuing contact with the surface.

These two simple considerations lead to the expectation that the ratio of dry deposition to wet will depend on the location of measurement, its proximity to the causative source of pollution, the height of injection into the atmosphere and the prevailing rainfall regime. Regardless of all other factors, it would be hard to explain a finding that wet deposition of sulfur species was greater than dry in the Australian desert downwind of Broken Hill (a major mining center and source of considerable sulfurous pollution).

There is an intermediate categorization of deposition that cannot be easily associated with the simple terms 'wet' and 'dry.' Mountain ranges, in particular, can experience deposition associated with passing fog and/or cloud droplets intercepted by foliage. The chemical deposition that results is clearly the consequence of the physics and chemistry of clouds and cloud droplets, but the mechanisms by which deposition occurs are closely aligned with dry deposition processes. The matter is of importance in the context of high-altitude forests, where exposure to clouds and fogs is frequent and interceptive scavenging by forests (especially by conifers) is anticipated. A dedicated Mountain Cloud Chemistry Program (Mohnen, 1989; Li and Aneja, 1992) demonstrated that the deposition of sulfur compounds by cloud droplet interception along the Appalachian range of the eastern USA was about the same as by conventional wet deposition. However, similar studies of deposition to the Serra do Mar range in Brazil (Vautz et al. 2003) showed wet cloud droplet deposition about an order of magnitude less than standard wet deposition. The difference appears to lie in the difference in elevation. The Appalachian range extends to greater heights than the Serra do Mar. Fowler et al. (1989) and others (e.g., Saxena and Lin, 1990) emphasize that a central consideration is the altitude.

The fundamentals of wet deposition processes are well known, but in essence they break into two distinct categories depending on the process by which raining clouds access air pollutants. In the case of heavy rain from a convective system as typifies the warm-weather US midwest, a convective cell serves as a chemical reactor which allows trace chemicals entering via the inflow mechanism to be associated with droplets which then coalesce and fall as rain. In this case, the concentration of chemicals in the precipitation collected at ground level would be expected to remain fairly constant as the precipitation event continued. On the opposite extreme, rain from a stratified system at higher levels would tend to fall through polluted air below and slowly extract soluble trace chemicals from it. A continuing rainfall would then slowly diminish the amount of pollution in the atmosphere below cloud level, in which case the concentration in rain at the surface should decrease as the rainfall persists. In the convective extreme, the concentration of a pollutant in rain should be remain relatively constant. At the opposite extreme, it would decrease as the amount of rain that fell. On the average, it is then likely to that the amount deposited in rainfall should vary midway between these extremes, i.e., on the average most likely as the square root of the amount of rain that fell during a particular period if the circumstances favored both mechanisms approximately equally. This anticipation is borne out by many field experiments, but not by all. The matter is not directly relevant to the present focus on micrometeorology and the matters that relate to dry deposition, however it is of interest to note that once again a pragmatic view of reality can be derived by considering opposite extremes and assuming that the real world lies somewhere between.

Figure 39.1 provides an example of early data that confirmed the expectations summarized above and provided a starting point for work in years to follow. The dashed-and-dotted line is a best fit, with a power law exponent of 0.33, not the value 0.5 if the two kinds of precipitation scavenging occurred with equal probability and if a dry deposition component could be ignored. The dashed line represents what might be explained if dry deposition accounted for 5 pCi to each sample collected through the period of the study. When this is taken into account, the curved line results with a high-rainfall characteristic more closely approximating the 0.5 exponent. Further discussion of this is not warranted, save to point out that ramifications of this kind of behavior have been of concern to the deposition community for a long time, with most emphasis being on describing gross quantifications at specific locations, as in Figure 39.1. No more will be said about this, here, for to do so would be to gild a severely wilted lily.

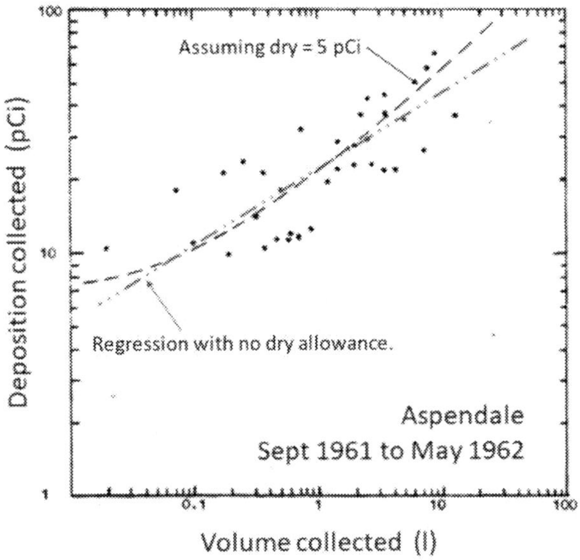

Figure 39.1. From the distant past — weekly concentrations of radionuclides in rain against the amount of rain collected (in a standardized bucket) during the period of weapons testing moratorium from September 1961 to May 1962. Measurements were made at Aspendale, Victoria, by the present author.

Conclusion

Associations of dry with wet deposition are likely either location-specific or fortuitous.

Chapter 40

Deposition Collection

"Pans, plates, pots and other kitchenware" — Noel E Ridere

In the days of radioactive fallout, when the focus was on particles that originated (usually) from an eruption or fireball, many exceedingly simple methods were deployed to determine the rate at which the resulting particles deposited to the ground. Initial interest was mainly in the hazard to closely downwind inhabitants, mainly associated with exposure to the large 'hot' particles that soon 'fell out' and carried most of the hazardous radioactivity. Simple flate-plates were often deployed to collect the near-field fallout. When it was recognized that particles that landed on a flate-plate could be blown off by a wind gust, the plates were made sticky so that a deposited particle would stay deposited. Buckets were also widely used, on the implicit assumption that the deposition of radioactivity would be dominated by the fall of large particles and scavenging by the precipitation process. (The use of as simple bucket yielded the data of Figure 38.1). These collection vessels remained in widespread favor until present times; they collect whatever falls into them. Eventually, these collection systems were updated to distinguish between dry and wet deposition of particles. The wet/dry collector evolved, using two buckets, one covered when it was raining and the other covered when it was not raining — the HASL wet/dry collector (named for the US Health and Safety Laboratory in New York; see Bogen et al. 1980). Later developments omitted the 'dry deposition' collector entirely, thus immunizing operators from simple questions such as "Do you measure the deposition to the inside of the container or to the outside, and why?" Such comments were often disregarded, because the radioactive fallout experience had proved that wet deposition usually exceeded dry by an order of magnitude unless measurements were made in the near downwind region of a nuclear test. It took many years for it to become accepted that this finding did not extend to air pollutants distributed within the PBL.

The use of the wet/dry bucket collector was not the only extension of the radioactive fallout experience. Several surrogate surface methods developed to measure particle deposition (trace gases disregarded). Discussion of these requires awareness of the three modes of the atmospheric particle size distribution — the nucleation, accumulation and primary size ranges (see any of countless references, e.g., Friedlander, 1977). The nucleation size mode contains those particles generated by reactions among trace gases in the atmosphere (typically less than 0.2 μm diameter). The accumulation size range is that containing particles formed by aggregating smaller entities but not strongly influenced by gravitational settling (0.2 to 3 μm or thereabouts). Particles injected into the air by mechanical disruption of solids and whose deposition is dominated by gravity are simply referred to as 'large particles' (generally > several μm diameter). These size distinctions vary with circumstances and with the composition and shape of the particles. Most attention centers on particles in the accumulation size range, because these remain airborne for longer periods: they do not deposit as well as the larger (gravitational) "bricks" or the smaller and younger particles that are susceptible to Brownian motion and gas-like exchange. As interest in the deposition of industrial pollutants grew, the focus on large 'hot' particles waned. However, reliance on the familiar wet/dry collectors continued until the inherent conceptual difficulties proved too difficult to dismiss.

A convenient collection device often used in early studies was the familiar petri dish. It is not clear how the results obtained could be extrapolated to a natural landscape — rarely is the terrain highly populated with petri dishes nor are they considered to be at risk of damage by pollution deposition. The same criticism relates to all deposition monitoring devices, regardless of their details. If the goal is to measure the rate at which particles are depositing from the atmosphere, it must be assured that the method of measurement does not interfere with the flux quantification that is desired. There is no doubt that a surface or vessel deployed horizontally will collect the large particles that fall on them. But these large particles are also moved with the wind and impaction upon the sides of any collecting vessel could be an effective contribution to the total deposition of the particles in question. Nevertheless, deposition surfaces became popular in the 1970s, when the distinction between the processes contributing to turbulent exchange and to gravity remained unappreciated in many communities. The expectation that artificial surfaces might provide a collection methodology for monitoring the deposition of accumulation-size aerosols remains difficult to explain.

Vegetated surfaces received special attention. Programs arose in which leaves were washed and chemical deposition by particles deduced from analysis of the leachate. At a meeting in the early 1980s, leaf-washing specialists were asked whether they washed the entire leaves or simply the upper surfaces. At the follow-up conference two years later, all of the experts had addressed this question. All had washed particles from the upper and lower leaf surfaces, separately. Some had computed the effective deposition as the sum of these two, others as the difference. No consensus was reached.

There are many separate physical processes that influence the deposition of small particles, including a range of 'phoretic' effects which in many cases tend to ensure that the smaller the particle the more likely it is to be removed from the atmosphere (gravity excepted). In the context of human biology, it is the accumulation-size particles that escape the protection afforded by the inhalation pathway of the human body and which can enter and damage lungs. Likewise, it is them that mostly contribute to the deposition of worrisome particles such as those carrying sulfates of industrial origin or tetraethyl lead resulting from use of treated gasoline.

Canopy throughfall studies evolved in the 1980s, with a focus on forest ecology and initial emphasis on the measurement of deposition of sulfur species (Lovett and Lindberg, 1984) but subsequently expanding to address such ecologically relevant chemicals as mercury (Choi et al. 2008). Measurement protocols of the kind already considered provide information over time scales long in comparison to those that would help describe the contributing processes. It is in this shorter-time context that micrometeorological methods rose to the fore.

Modern research related to dry deposition falls into two convenient categories: risk assessment and process examination. In the first, activities are focused on quantifying the dry deposition input of hazardous substances to areas where such information is needed (ecosystems, in particular). In the lack of measurement capabilities suitable for routine measurement of the fluxes of airborne chemicals and particles to the surface on a routine basis, numerical models have become a focus of much research. These models provide estimates of dry deposition rates to surfaces of different kinds and require measurement programs using sophisticated methodologies to both utilize and evaluate them. The second category relates to the elucidation of the contributing processes, for which the same complex experimental capabilities are employed in dedicated and intensive field studies. Progress has been hampered by the rarity and expense of measurement methods that are accurate, robust and suitable for field use.

Conclusion

The apportionment between wet and dry deposition necessarily depends on location and the weather.

Chapter 41

Laboratory and Pipe Studies

"Lab studies come first. Relevance can be questioned later" — Noel E. Ridere

In early work it was assumed, and verified, that the rate of deposition of some chemical present in air in trace quantities to an exposed test surface is determined by the concentration of the chemical of interest in the air. For common pollutants like sulfur dioxide (SO_2) the proportionality that results between the flux f_c and the airborne concentration $\bar{C}$, defines the familiar deposition velocity, v_d:

$$f_c = v_d \cdot \bar{C}. \qquad (40.1)$$

Straightforward extrapolation into the natural environment led to the widespread acceptance of the familiar deposition velocity V_d (as used in air pollution considerations) as a chemical-specific indicator of the rate at which air concentrations are diminished as polluted air passes over a particular area. Once again, there is need to distinguish between the cases of deposition to a specific surface or obstacle and that from the atmosphere to the assembly of such surfaces that constitute the 'surface' that the atmosphere sees. In the case of exchange with some particular component of the surface (e.g., a specific obstacle), it is local considerations of the surface and the air affecting it that are crucial. The linkage between this view from below and the atmospheric view from above has been addressed already (Chapter 32). Following the same convention as was discussed earlier in the context of plant physiology and the big leaf concept, the use of uppercase symbols to refer to the quantities associated with determining fluxes from the atmosphere has become accepted. Lowercase symbols refer to the parallel issue of fluxes to the surfaces of individual components exposed to the passing air. As was addressed earlier in this document (in Part IV), a major challenge has been to relate the lowercase quantities to their uppercase counterparts. It is the uppercase quantities that are of use in the case of numerical simulations of atmospheric transport and deposition of industrial air pollutants, for example.

Laboratory studies generated much of what was (and remains) known about the pollution uptake characteristics of different materials. In the lack of contradictory field evidence, v_d and V_d were often seen as cardinal characteristics indicative of a specific trace gas, determined on the basis of laboratory studies or some conceptual extrapolation of their results. This over-simplification was applied without impunity in simulations that today would be unacceptable.

Understanding of the way in which trace quantities in air deposit to smooth surfaces arose largely as a consequence of studies responding to industrial requirements, like the transfer of fluids in pipes and the containment of chemicals in vats. In such contexts, the methods of fluid dynamics have taken root. These reduce the complexity of the overall problem by considering the contributions of the many relevant variables when they are combined into a smaller number of dimensionless quantities. Pipe-flow and laboratory studies of chemical engineering led to acceptance of two central dimensionless quantities, the Reynolds number ($Re \equiv Ud/\nu$, to indicate when flow was turbulent) and the Prandtl number ($Pr \equiv \nu/\kappa$). As an additional generality, the Schmidt number ($Sc \equiv \nu/D$) allows for the molecular-scale transfer of a trace gas to be considered in the same way. Here, U is the flow speed, d the relevant distance scale such as the diameter of a pipe or the distance from a leading edge, ν is kinematic viscosity, κ is thermal diffusivity and D is the molecular (or Brownian) diffusivity of the trace substance being considered. The resulting formulations involving combinations of these dimensionless quantities were intended to describe how transfer across the surface 'contact layer' changes as different trace gases (or very small particles) are considered, with their diffusivities referenced to what is assumed for the momentum flux (involving Pr and Sc, see Owen and Thompson, 1963; Chamberlain, 1966; 1968).

In every case the transfer across the laminar layer in contact with the flat surface under consideration is assumed to be controlled proportionately by the molecular-scale diffusive quantities (q.v. Prandtl, 1932; Geiger, 1950; and many others). If the transfer rate to a flat surface in some particular situation is known, then the deposition velocity of a trace gas with diffusivity D' should differ from that appropriate for some reference gas with diffusivity D according to the ratio D'/D.

Now consider the situation with flow through a pipe or conduit. Repeated experimental tests have verified that in this case the corresponding correction factor is not the ratio D'/D, but $(D'/D)^{2/3}$ (a result commonly known as the Chilton-Colburn analogy (Chilton and Colburn, 1934). The physical difference between the two situations is obvious. Whereas the relevant deposition velocity in the flate-plate situation is defined in terms of the rate of exchange at the surface and the free air concentration at a distance away from the surface, in the case of pipe-flow the deposition velocity of relevance is determined by the exchange rate with the wall and the relevant property (concentration) in the stream of air flowing through the pipe. It is intuitively apparent that the rate of deposition of trace gas in a container will depend on the amount of surface area per unit volume, with the Chilton-Colburn analogy being an appealing consequence. A more robust examination of the issue of deposition to the walls of pipes is presented by Friedlander (1977) who addresses the matter with relevance to particles. The result substantiates the Chilton-Colburn analogy, but at the same time serves to demonstrate that the result is particular to confined flow as in conduits or pipes. In practice, allowing for the effect of changing diffusivities is usually by introducing the Schmidt number (or the Lewis number, $Le \equiv \kappa/D$, depending in what is known), raised to the power (either -1 or $-2/3$)
researcher's view of the surface in question. If a natural surface can by likened to a closed volume, then the 2/3 power would seem worth considering. Otherwise, the unit power would likely be more appropriate.

Some history is now relevant. In the 1970s and 1980s, when descriptions of the processes contributing to air-surface chemical exchange were starting to be understood, the only quantified and accepted information available was that derived from pipe and wind tunnel studies. The engineering literature was adamant that the Chilton-Colburn analogy worked and it was accepted (e.g., by Wesely and Hicks, 1977) as a first-order description of how air-surface exchange of one chemical differed from that of another over a natural landscape, without awareness (at that time) of the fundamental difference between flate-plate and pipe-flow exposures. In retrospect, the growth of the discipline was not well served by the acceptance of the pipe-flow guidance. There is now little doubt that the appropriate correction factor derived from laboratory studies and appropriate for outdoor application would be better expressed by the use of Sc^n (or Le^n) where $-1 < n < -2/3$ (probably), with a precise value depending on the characteristics of the surface in question. The matter remains uncertain, however. Field observations (as discussed below) have led to insecurity about the value of n — that the very rare data now available indicate that n is sometimes near zero.

Interest in air pollution and in the residence time of various air pollutants (particulate as well as gaseous) generated an increased interest in the pipe-flow understanding and its extrapolation to the real outdoor environment. The theoretical treatment of deposition to the walls of pipes provided by Friedlander and Johnstone (1957) generated a spate of studies addressing the deposition of particles from fluid streams — a matter of considerable industrial relevance. Results from many early studies conducted in pipes and wind tunnels are presented in Figure 41.1. The well in the curve is a familiar feature. Theoretical considerations proposed by Friedlander and Johnstone (1957) substantiated the empirical pipe-flow deposition results: at the low end of the size range, the dependence on Schmidt number is a dominant consideration; at the high end, gravity starts to take control. The minimum deposition velocity lies between. Particles of intermediate size (the accumulation size range) are too small to fall rapidly through the air yet are sufficiently large that inertia limits their response to the rapid changes in air motion that transfer other atmospheric quantities to exposed surfaces.

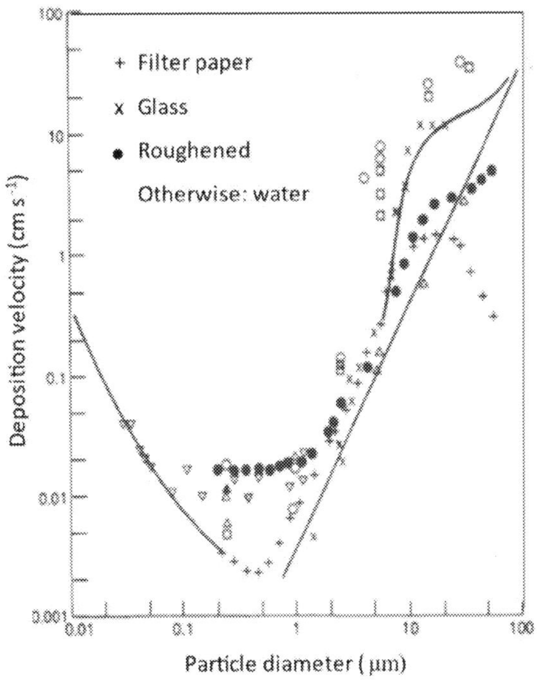

Figure 41.1. Results of early laboratory studies of particle deposition to relatively smooth surfaces. Data for glass are from Liu and Agarwal (1974); filter paper and roughened surfaces from Clough (1973). Results for water surfaces are from Möller and Schumann (1970) and Sehmel et al. (1973).

A first experimental test of this view of deposition of small particles to a natural landscape (Wesely et al. 1977) showed that the pipe-flow predictions do not extrapolate well. Faith in the universal relevance of the pipe-flow results was such that the real-world data (measured using eddy covariance) were perceived by some as evidence that micrometeorological methods were inappropriate. Subsequent experiments abound (q.v. Saylor et al. 2019), confirming that the small particles appear to be largely unaffected by the molecular diffusivity constraints so clearly revealed by the laboratory experiments. There have been many models proposed, all employing some version of the molecular diffusivity and laminar layer pipe-flow expectations, but with the power-law formulation appearing to vary as the whim of the modeler. Comparisons of model predictions with field observations are difficult, because of the paucity of reliable field data. Van Aalst (1986) compared the predictions of many models and reported results varying over a range of more than an order of magnitude for three particle sizes — 1, 10 and 30 μm aerodynamic diameter.

In work many decades later, uncertainties about how models can best extend pipe-flow results to the open-air landscape persist. The discussion above of the differences between pipe-flow and open air studies assumes a simplification of a complicated set of contributing processes. For particle exchange, Katul et al. (2010) provide evidence that turbophoresis (a mechanism affecting eddy exchange) can, together with diffusiophoresis and thermophoresis, serve to fill in the well in the particle size dependence of the deposition velocity (as in Figure 41.1), The summary of recent field determinations of particle deposition velocities presented by Saylor et al. (2019) provides further evidence that (1) the pipe-flow studies cannot be extrapolated into all real-world circumstances with impunity and (2) major changes in sub-micrometer particle size have little effect on the relevant deposition velocity.

Field studies show that the 'well' in the deposition velocity versus particle size curve (as in Figure 41.1) becomes shallower as the surface becomes rougher (see Sehmel et al.,1973; Sehmel and Hodgson, 1978). Natural surfaces are rarely smooth. Moreover, many plant species have foliage with exceedingly complicated microscale surface roughness features. Earlier in this work, Figure 34.1 showed leaf hairs present on a plant species unknown and unwelcome in the author's garden. Leaf hairs have been shown to promote particle deposition and retention (q.v. Chamberlain, 1967) and their influence in natural circumstances could be considerable. Other factors, such as electrical charges and leaf stickiness, have been investigated. In every case, there is evidence that a natural canopy might be considerably different from the simplified surfaces usually preferred for wind tunnel and laboratory studies.

In analogy with the Reynolds number derived from studies of laminar flow and its transition to turbulence in pipes, micrometeorologists now favor the use of the roughness Reynolds number ($Re_* \equiv u_* z_0 / v$) as a scaling parameter. Upon distant reconsideration (the author having promoted the relevance of Re_* in early papers), Re_* as commonly evaluated combines aspects of SBL turbulence derived from measurements made well above the surface (u_*) with molecular-scale kinematic viscosity (v) and an extrapolation of observed wind profiles (z_0), hence challenging intuition. It is possible that there is more magic in this topic than hard science.

An intriguing feature of pipe-flow and water channel studies is that the intermittency now well recognized in the case of air-surface exchange with forests (e.g., Katul et al. 1997; 2004) is also a feature of exchange with the internal wall surfaces affecting channel flow (see Kline et al. 1967).

Conclusion

Extrapolate laboratory results with caution.

Chapter 42

Deposition to Buildings and Structures

"Statues remind us of history, also of air pollution" — Noel E. Ridere

The despoiling of buildings by exposure to air pollution has been a historic problem for those wishing to maintain a good life in cities and to preserve valuable antiquities. Removing the consequences of exposure to air pollutants is now a profitable enterprise, with positive outcomes visible in many major cities (such as New York, Philadelphia and London). Figure 42.1 shows how air pollution caused by local reliance on small furnaces burning high-sulfur coal and on the burning of animal waste (both causing high concentrations of SO_2 and NO_x in surface air) has affected the marble finish of the Taj Mahal. Rain washes deposited pollutants from the surfaces that it impacts but protected (underhanging) surfaces are not so fortunate.

Figure 42.1. A view of the exposed exterior of the Taj Mahal, showing the accumulated consequences of dry deposition of air pollutants to the surfaces protected from precipitation and the relatively clean surfaces where rainfall serves as a cleansing agent.

The transfer of some atmospheric constituent to a specific obstacle, or surface, exposed within a polluted flow field will certainly be the result of variety of processes, among which the most important might be well hidden. A simple example of surface deposition is the lead candle used in some of the earliest experiments — a tube of elemental lead mounted vertically with the intent to indicate the presence of sulfur dioxide in the air and to give some information of the wind directions mostly associated with it. Lead is blackened by the formation of lead sulfide when exposed to SO_2. In the days of the early industrial revolution, when the city of London emitted a pall of pollution, it was suspected that the emissions of sulfur dioxide could have contributed to the fading of dyes used in the silk industry of southern France. Results obtained using a network of lead candles succeeded in identifying the source of the pollution. The blackened candles pointed directly at London. (It is interesting that this experiment was reputably the personal brainchild of the head of the French Academy of Sciences, Louis Pasteur, whose work subsequently identified the major causes of dye fading as being microbial.) In general, the discoloration of buildings and exposed surfaces of any kind is a good indication of the presence of pollution and its air-surface exchange.

Early investigations of the deposition of airborne trace gases to exposed surfaces were usually in support of issues arising in building construction and maintenance. Exposure methods were developed to compare the sensitivity of different paints and other surface coatings to weather and air pollution. Standardized racks are often used, thus helping ensure that alternative materials are equally exposed. In urban surroundings, with rigid bluff bodies dominating the surface, it would appear a good first step to address the matter of deposition from the air in terms of such examinations of test surfaces. However, there remains considerable uncertainty about the optimal way to interpret the results of such information.

As was emphasized in the days of 'acid rain' concern (the 1970s and 1980s) it is not always the direct deposition of offending quantities that is responsible for the problems that arise. As an example, consider the decay of the ironwork of bridges. Oxidation of metalwork is a key offender. The rate of rusting is exacerbated by the deposition of molecules like SO_2 and HNO_3, especially to exposed areas that are wetted. Joints and cracks serve to prolong the resulting insult. Add to this mix the structural consequences associated with the freeze-thaw cycle and it becomes apparent that clean air helps protect infrastructure. An immediate inference is that surface complexity must rise in importance as the causes of environmental damage to exposed structures become better understood.

In the days of the National Acid Precipitation Assessment Program of the 1980s (NAPAP; q.v. Irving, 1990), there was considerable attention to the problem of despoiling of buildings and monuments. Intensive studies were conducted, involving landmarks like the Stature of Liberty but extending to examination and comparisons among tombstones, memorials and other structures. The intent was not only to derive understanding of the processes involved, but also to assess changes with time. Of particular interest were comparisons of the lifetime histories of similar structures in polluted cities versus cleaner ones. Sherwood (1992; 1995) explored this concept, using often-replicated bronze statues of 'the Hiker' (T. A. R. Kitson, sculptor) as a basis for comparisons. More than sixty copies of 'the Hiker' are distributed across the USA, most set up in the 1920s. Detailed diagnosis of the deterioration seen in these statues is confounded by the variations in siting and orientation, but the message is clear that the deposition of air pollutants is an issue worthy of attention by historians and civic authorities alike.

The investigations of damages to materials resulted in several conclusions of general relevance. It was shown, for example, that the presence of liquid water promoted deposition which, once present, accelerated further deposition. In this regard, deposition to corners, cracks and crevices was seen as a key contributor to eventual structural damage (as was discussed above). Decisions regarding the maintenance of bridges and other structures in a working and safe conditions necessarily entail consideration of the air pollution environment in which such things are located. Cleaning the atmosphere is certainly a step towards ensuring the longevity of physical and especially mechanical infrastructure of all kinds.

Conclusion

In construction, avoid sharp corners and crevices.

Chapter 43

Enter, Micrometeorology

Modern technology wins out.

Extreme examples of the consequences of air-surface exchange of trace gases (and particles) too large for the environment to accommodate are readily evident in (for example) views of the historic landscape around Sudbury, Ontario; Copper Hill, Tennessee; Broken Hill, Australia; and countless others. By the 1970s, the sensitivity of plants to different air pollution species (especially to ozone and sulfur dioxide) had been well studied in laboratory circumstances (e.g., Hill and Thomas, 1933; Mudd and Kozlowski, 1975). However, information regarding deposition to a landscape vegetated with different species of vegetation awaited the development of trace gas concentration measuring devices with precision and/or fast enough response time to use micrometeorological methods — at first gradient methodologies (e.g., Galbally, 1971, for ozone). Garland et al. (1973; 1974) and Fowler and Unsworth (1974) extended the micrometeorological methodology to the case of SO_2. Ray et al. (1986) presented an early chemiluminescent method for ozone measurement with response fast enough for use with eddy covariance. Field utilization of fast response O_3 sensors in eddy covariance applications followed (e.g., Wesely et al. 1978). Subsequently, results of field experiments converged on the understanding that V_d varies greatly according to the chemical species considered. In all cases it is necessary to distinguish between day (when stomata are open and atmospheric eddy diffusivity is permitting) and night (when stomata are closed and eddy diffusivities are minimized).

The development of fast-response SO_2 sensors in the early 1980s (initially flame photometers, q.v. Tanner et al. 1980) enabled direct measurement of gaseous and particulate sulfur over micrometeorological sites. Experimental programs making use of the flame photometric capability for measuring the eddy covariance of SO_2 to extensive areas of pasture and forests (Hicks et al. 1982a,b) yielded a wide range of values for V_d. The experimental variability was often interpreted as evidence of experimental error, but studies eventually revealed that this wide variability is an indication of how a simple quantity like V_d is an oversimplification of a complicated chain of processes. Fast response sulfur sensors were quickly adapted to differentiate between sulfur present as SO_2 and that present as particulate $SO_4^{=}$. Field studies over a forest (Hicks et al. 1982a) showed that while the flux of particulate sulfur was always downwards, that of smaller nucleation-size particles was often upwards, out of the canopy. Studies of the chemistry of biogenic emissions being made contemporaneously at the same location (a pine plantation in North Carolina) verified the presence of isoprene and related hydrocarbons, from which oxidation by interaction with ozone generated the very small particles that constituted the observed upflux. There was nothing new about this. More than twenty years beforehand, Went (1960) had explained the presence of the blue haze that characterizes many of the world's mountain forests. Those of us who grew up in areas influenced by the blue haze phenomenon and its pleasantly aromatic hydrocarbons have long accepted its ubiquity.

Subsequent field programs have shown that many physical, chemical and biological factors contribute to the overall rate of exchange of SO_2 from the air to whatever constitutes the eventual sink. In particular —

- The moist mesophyllic tissue inside stomata serves as a near-perfect sink for the SO_2 molecules that enter open stomatal cavities. Since it is these same mesophyllic cells that allow CO_2 to be made available for the photosynthetic process and to permit water vapor to be lost to the air (both via stomatal aperture), it is expected that the net resulting V_d must be greatest for quickly-growing annual plant species.
- There is little convincing evidence that deposition to the skin of leaves plays more than a minor role in SO_2 exchange. However, the variety of characteristics of vegetation leaves the conceptual door open. Consider cacti, for example.
- When incoming radiation is not sufficient to support photosynthesis, i.e., when stomata are closed, V_d for SO_2 will reduce to a minimal value that is perhaps indicative of a cuticular component. On the whole, it is commonly assumed that V_d will then be minimal (probably < 0.1 cm s^{-1}).
- When plant surfaces are wet (following precipitation or dewfall) very high SO_2 deposition velocities could be relevant, although probably limited by atmospheric stability and the consequent limitation on turbulent exchange.

In contrast, some atmospheric trace gases are deposited immediately they come into contact with a surface; e.g., HNO_3 which is both highly reactive and molecularly polar. Such gases offer analogy with the transfer of sensible heat, for which it is contact with leaf cuticle constitutes (in the present context) the effective source. For HNO_3, available data indicate that this analogy is closely appropriate (see Huebert and Robert, 1985). Recent studies indicate that Cl_2 behaves similarly (Spicer et al. 2021).

In the case of ozone deposition, it was originally assumed that the insolubility of the gas would limit its deposition to vegetation, however laboratory studies of bean leaves indicated that stomata were, once again, the controlling consideration (Rich et al. 1970) and this was subsequently confirmed by field studies (e.g., Wesely et al. 1978). In this particular case, it was concluded that the reactivity of ozone caused it to be quickly destroyed when encountering the organic-enriched environment within stomatal cavities. These and other related matters are discussed at length in the State of Science documentation of the National Acid Precipitation Assessment Program (Irving, 1990).

The accumulated results of studies of the deposition of airborne gases like H_2O, CO_2, SO_2, O_3, HNO_3, Cl_2 and a variety of oxides of nitrogen (NO_x) have resulted in an understanding that now permits rather confident extrapolation to other chemical species. However, two over-riding realizations have often been overlooked in simulations.

- Deposition is essentially a daytime phenomenon. In the distant past, measurements made in daytime were assumed to be representative of all times, but this is simply not the case. If one needs to estimate an appropriate weekly average, for example, then the relevant average deposition velocity will be much less than predictions for what daytime because R_a(night) >> R_a(daytime). If the gas being transferred is stomatally influenced, then stomatal closure at night must necessarily enhance this inequality.

- A high-quality study over a specific biome yields no more than results appropriate for that biome in the circumstances of the study. As always, extrapolation is risky.

Studies of the dry exchange of ammonia have risen to importance in recent years, because of its association with agriculture and more recently with its emergence as a factor in the energy sector. When present in trace quantities in air passing over surfaces not alkaline, ammonia is readily deposited. The concept of an NH_3 'compensation point' has received considerable attention, this being the central factor to determine whether the site-specific ammonia flux measured by (for example) eddy covariance is upwards or downwards. Farquhar et al. (1980) addressed the issue at the level of individual plants. Sutton et al. (1993) extended the compensation point thinking to landscapes. Asman et al. (1998) review the complexities that arise and offer recommendations about how the many areas of uncertainty could be addressed. Subsequent research has chipped away at these areas of relevant ignorance, with success apparently influenced by the biological and chemical characteristics of the areas selected for experimentation as well as the history of ammonia treatments or exposure in the surrounding region. In the usual experimental circumstances of spatially limited uniform surfaces, new innovative methods have arisen. Chamber methods have been used to measure the flux from the surface but are incapable of addressing the flux to the soil. Lichiheb et al. (2021) apply the concepts of chamber methodologies to develop a 'virtual chamber' approach suitable for extracting ammonia emission flux information data collected in strongly stable conditions over a swamp (also see Hicks et al. 2021). It remains to be seen how methods now evolving will help answer the many questions posed by Asman et al. (1998).

In similar vein, consider the air-surface exchange of the various components of oxidized nitrogen — nitric oxide (NO), nitrous oxide (N_2O) and nitrogen dioxide (NO_2). These (together with nitrogen pentoxide, NO_5) are collectively referred to as NO_x. In the atmosphere, NO_2 is the dominant species, a product of fossil fuel combustion. The reduced form NO is present in smaller amounts, arising as a trace emission from soil and vegetation. Based on an eddy covariance experiment conducted over a field of soybeans using a fast-response NO_x detector, Wesely et al. (1982) concluded that "as much as 50% of the NO_2 originally removed from the atmosphere during the daytime could have been released as NO by the soybeans."

From a study of vegetation in the Amazon basin, de Sena et al. (2007) report more than an order of magnitude variation between dry season and wet season stomatal resistances for some plant species, illustrating the importance of water stress and indicating that seasonal cycles that could well be of central importance to atmospheric simulations.

The discussion above leads to the inevitable conclusion that the matter of exchange of trace gases between the atmosphere and the surface is too complicated for easy definitive formulation. Considerations of small particles involves similarly complicated considerations. For example, the distinction between 'large' and 'small' particles remains indistinct, because the matter is confused by the roles of such considerations as particle shape and density. However, the particulate chemicals of major environmental and human health concern are in the accumulation size range, considerably smaller than the 'bricks' that fall by gravity.

Conclusion

Accept new methods but verify.

Chapter 44

The Multiple Resistance Analogy: Trace Gases

Reconsidering the Big Leaf

The multiple resistance analogy presents a simple view of how the processes contributing to the exchange between the air and the surface combine to cause the net exchange rate per unit area — the relevant flux. In the present discussion of trace gas and particle exchange, new considerations arise and it is appropriate to consider the overall exchange processes in more detail. A major contribution to the overall complexity is that the deposition of trace gases requires consideration of the role of biological processes at the plant physiological level and these are different for different chemical and biological species. An obvious starting point is the case of water vapor exchange and its formulation in terms of resistances to transfer (see Chapter 5). The situation being addressed here is that of a trace gas present in low enough concentrations that there is no limit on the capability of an exposed surface to capture the molecules impacting it. In other words, the number of available sites for chemical reaction is sufficient to accommodate all impacting molecules. The situation of the near field, in which concentrations are sometimes high enough to invoke a surface chemical reaction that depends on the concentration itself will be examined later.

> The present treatment differs from earlier work by the present author. There is no intent to show that one approach is preferred to another. It is simply requested that readers relate alternative views and conclusions to the applications that are of importance to them.

In the case of dry deposition of airborne pollutants, convention has accepted the multiple-resistance electrical analogy (based on the agricultural meteorology applications considered in Chapter 5), with different resistances being associated with specific contributors to the overall exchange. As in earlier contexts in this report, the flux of a trace gas from a specific level of reference in the SBL to the surface is expressed in simple terms as

$$F_c = C_z/(R_a + R_b + R_c) \tag{44.1}$$

where R_a represents the resistivity (an inverse conductance) to transport associated with turbulence, R_c expresses the resistance to uptake once the gas molecules being deposited contact the sink of relevance (the leaf cuticle and/or stomatal interior, the mesophilic tissue) and R_b is the integrated consequence of any and all surface boundary layer resistances limiting transport through the air in the immediate vicinity of the individual surface sinks. Upon consideration of the examinations of such matters as the apparent inequality of roughness lengths for momentum and sensible heat fluxes (Chapter 32) and the effective interfacial equivalent temperature T_e (Chapter 36) it is tempting to equate such concepts and their origins with the property R_b. All of these embody the realization that transport and mixing beneath the level relevant for determining R_a, i.e., below the familiar zero plane, needs attention. Given the present focus on aerosol deposition and the models used in contemporary research, it is appropriate to reconsider the various resistances as they relate to trace gas exchange and later, separately, to the particle flux issue.

These three resistances are analogous to transfer processes conceptualized for the case of a flate-plate. The model that results (initially referred to as the Dry Deposition Inferential Model – DDIM; Hicks et al. 1987) is intended for purposes in which areal estimates are desired. It lacks the detail necessary to address specific situations for which measurements of the fundamental plant physiological properties are available. Instead, averages of these quantities intended to represent the spectrum of plant species are used. To address areas where more detailed information is available and to provide a diagnostic tool for use by researchers, a Multi-Layer Model (MLM) was developed by Meyers et al. (1988); this has become a favored model for use in site-specific assessments of atmospheric chemical deposition. Experiments that explore the details of MLM are now common. In the mind of the present author, these serve to emphasize the complexity of the problem more than to reveal some generalization that might be widely appropriate. In particular, the assumption that fluxes relate to gradients in the subcanopy environment appears vulnerable. The matter will arise again, when recent results are examined in more detail. For the present, note that Figure 44.1 reflects the widespread interest in multi-layer sub-canopy approaches.

Fig. 44.1 illustrates the two approaches. The simplified symbolism reflects the intent to differentiate between the overall resistances to air-surface exchange as seen from the perspective of the atmosphere above (upper case symbols) and the individual components associated with different layers of the canopy (lower case symbols).

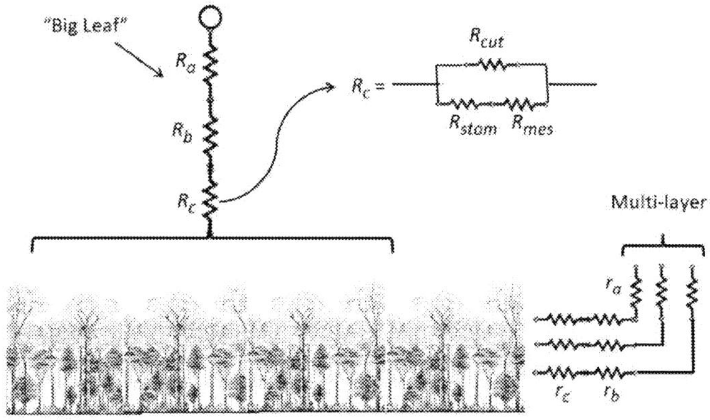

Figure 44.1. The pathways of trace gas deposition as simulated in the familiar multiple resistance models, depicting the fundamental differences between the whole-canopy 'big leaf' approach and the more detailed multiple-layer canopy models. In both cases, the effective canopy resistance is made up of cuticular, stomatal and mesophilic components.

Whereas the intent of the big leaf multiple resistance model is to simplify the way in which relevant chemical fluxes across areas are described, the alternative multi-level methods are suited for more diagnostic purposes. In the case of a multi-layer description of deposition characteristics of vegetation (e.g., Meyers et al. 1998), the subcanopy exchange is formulated in terms of a resistance varying with height, yielding an integrated value for use in the context of an air pollution model (e.g., Saylor et al. 2014). However, in the case of a big leaf view of the surface serving as a lower boundary to the atmosphere, the development has related R_b to be the single value paralleling the more detailed formalism of exchange in a wind tunnel or pipe but considering it to be indicative of the canopy as a whole. The differing formalisms involved have been well examined elsewhere (e.g., Chen et al. 2010; Saylor et al. 2014; Hicks et al. 2018) but clearly it remains challenging to show that the integral of the layer-by-layer quantifications is the same as the result obtained from micrometeorological measurements above the same surface. It is equally challenging to extract from the many analyses a consistent formalism that reflects the basic difference between pipe-flow studies and a natural landscape (as discussed above, in Chapter 41).

R_a: The Aerodynamic Resistance

R_a is conveniently expressed as

$$R_a = (ku_*)^{-1}[\ln(Z/z_0) - \psi_m] \tag{44.2}$$

At this point, the considerations already presented for the case of the surface energy budget and its interpretation can be repeated without reservation, save to reconsider the relevance of the friction velocity, u_*, in daytime conditions when dry deposition is of most importance. If it is accepted that u_* is not a controlling variable except in the stability regime near neutral, then for the unstable case, the free convection description of the transfer process yields a formulation depending on the sensible heat exchange alone, such that.

$$R_a(\text{free convection}) = 1.1\cdot[\theta/(gZ\cdot\overline{w'T'})]^{1/3} \tag{44.3}$$

where the factor 1.1 arises as $H^{*1/3}$ (see Equation 17.3). Including the property $(ku_*)^{-1}$ in the relationship, so as to parallel Equation 44.1, yields that

$$R_a(\text{free convection}) = (ku_*)^{-1}[u_*^3\theta/(kgZ\cdot\overline{w'T'})]^{1/3}\cdot(1.1k^{-2/3}). \tag{44.4}$$

The property in the square brackets is the inverse of the MOST stability index, ζ. Hence, the aerodynamic resistance in daytime, convective conditions over land can be expressed as

$$R_a(\text{free convection}) \approx 2(ku_*)^{-1}(-\zeta)^{-1/3}. \tag{44.5}$$

If the intent is to quantify V_d, the other resistances (R_b and R_c) in the conceptual models add to this value of R_a and hence the maximum possible deposition velocity in daytime is

$$V_d \approx 0.2u_*\,(-\zeta)^{1/3} \tag{44.6}$$

It is obvious, but worth emphasizing, that the resulting indication that V_d/u_* can be considered a function of stability alone is due entirely to the way the relationships are formulated. The derivation above already assumed (as data indicated) that in daytime u_* need not enter consideration, hence scaling by it is immaterial. The vertical transport in unstable conditions is dominated by convection, unless conditions are such that they can be considered to be near neutral (i.e., $0 > \zeta > -0.03$ or some other preferred value). However, as the surface itself is approached stability necessarily approaches neutrality and hence there is good physical reason to consider the role of u_* in a different context — as it contributes to the 'quasi-laminar' layer resistance, R_b.

R_b: The Quasi-Laminar Link between the Air and the Surface

During the 1970s and 1980s, field experiments using eddy covariance and a wide range of trace gas and particle sensors revealed that R_a alone (based on momentum exchange, z_0 and d) was not enough to describe the total in-air resistance to exchange of other properties. Principal among these other quantities was the sensible heat flux, subsequent studies of which generated the interest in the property kB^{-1} as was extensively considered in previous Chapters. Paul Frenzen (in a discussion at Argonne National Laboratory in 1976) proposed that this extension of laboratory thinking to the real world should be referred to as the 'quasi-laminar layer resistance,' a term that has stuck: R_b. At that early time, the similarity with the sensible heat situation that led to interest in kB^{-1} was recognized but was not explored — the thinking cap was worn above a different lobe.

For the case of a natural landscape, knowledge of R_b has arisen mainly from studies of the sensible heat flux, the temperature gradient and the relationship between them. In recapitulation of what was presented earlier, extrapolation of wind speed measured above a natural surface defines a roughness length, z_0. A different roughness length, z_{0H}, is determined by extrapolation of the temperature gradient to the level where the air temperature is that which is controlling heat exchange — commonly the 'surface temperature' (q.v. Garratt and Hicks, 1973). It is assumed that this temperature is the same as is seen by a downward-looking infrared thermometer but this topic has been explored further with the conclusion that an alternative works better — the 'effective interfacial temperature', T_e, that can be visualized as the average level of initiation of convective heat eddies (Hicks and Eash, 2021). While an analogy with the case of the wind profile might define the roughness length z_{0H} as indicated by extrapolation downwards from profiles measured in near-neutral conditions, this near-neutral requirement means that there is little temperature gradient to extrapolate and uncertainties correspondingly abound. In contemporary literature, quantifications of z_{0H} are usually derived from measurements of air temperature and infrared surface temperature in daytime conditions, this necessarily assuming that it is indeed the temperature of the actual surfaces that is relevant. This parallels the visualization associated with the transfer of many trace gases and particles to the surface.

The result originally proposed by Garratt and Hicks (1973) and subsequently confirmed and embellished by others (*q.v.* Figure 30.2) denied the assumption that Re_* is an important variable — the field data appear to be unchanging as Re_* varies. Indeed, examination of the quantity Re_* raises interesting intellectual questions. Neither z_0 nor z_{0H} are actual physical quantities (both are constants of integration determined by distant extrapolation, see Verheof et al. 1997), whereas v is a physical property at the molecular level and u_* depends on atmospheric turbulence. Bringing these variables together into a single index intended to have physical consequences presents a challenge to the imagination, usually and conveniently ignored.

Following the methods of conventional micrometeorology, the analysis shows that the surface quasi-laminar transfer resistance (R_b) relevant to the transfer of heat can be written as

$$R_b(\text{heat}) = (u_*B)^{-1} = (u_*k)^{-1}\ln(z_0/z_{0H}). \tag{43.7}$$

Data from field experiments yield further insight. A selection of results was shown in Figure 32.1. Upon accepting (a) that the dependence on Re_* is minimal and (b) the role of molecular or Brownian diffusivity is as derived from laboratory studies with the key factor being the Schmidt number raised to the ⅔ power (as is the conventional assumption with which the arguments of Chapter 41 disagree), the appropriate value of R_b for some other atmospheric trace constituent would be expected to be

$$R_b(\text{C}) = R_b(\text{heat})\cdot Le^{2/3} \tag{43.8}$$

in which the Lewis number $Le \equiv \kappa/D$ (where κ is thermal diffusivity and D is the diffusivity of the aerosol of interest) is introduced because the derivation above is based on thermal properties and adjustment to it should be referenced to thermal diffusivity and not to kinematic viscosity. Elsewhere, the Prandtl number ($Pr \equiv \nu/\kappa$)
kinematic to thermal diffusivity — $Le = Sc{\cdot}Pr$. In practice, the difference between Sc and Le is likely to be trivial and this even assuming that the formalism stands the test of time (which is questionable, considering its common extrapolation from the scale of molecular to eddy diffusion — typically more than four orders of magnitude). Most simulations assume that the pipe-flow and wind tunnel dependence on the Schmidt number carry over (much as above) to the atmospheric case. Other simulations apply the Schmidt (Lewis) number correction to the entire result of combining the contributing resistances R_a, R_b and R_c (as below) and this is clearly at odds with the assumptions made in the development (see Wesely et al. 2007). Not all models make any such adjustment and there are few data with which to test the alternatives. However it is clear that understanding of these (and associated) matters is dominated by experience over sites of micrometeorological acceptability and the assumption that wind tunnel and pipe-flow results provide relevant guidance. In particular, Friedlander's (1977) theoretical examination of pipe-flow observations leads to the conclusion that more appropriate expressions relevant to open-air real-world situations would replace the –⅔ exponent with –1 where it appears above. The search for data with which to test these alternatives continues. Extrapolating to the real world remains an issue of concern (Malhi, 1996).

Rc: The Surface Resistance

When present in large concentrations, formulating the deposition of a gas mixed with the air to a specific surface entails consideration of the ability of the surface to finalize the fine-scale collection and sequestration process. There is need to consider that the ability of an exposed surface to capture molecules that come in contact with it will diminish as the number of chemically effective capture sites is reduced (Spicer et al. 2021). In this case, the rate of deposition from the gaseous medium to the surface will be according to the chemical reactivity of the gas of interest with the surface(s) involved. The ability of deposition sites to sequester molecules impacting them will diminish with time, as the most chemically reactive sites are depleted first, leaving a dominance of less reactive sites for which the reaction coefficient is unrelated to the concentration. In concept, a central consideration is the number of sites of relevant chemical reactivity that populate the surface and the saturation of them as deposition proceeds, thereby reducing their number density. In the end, all available sites will be defunct and the surface itself will be saturated, i.e., in equilibrium with the concentration of the gas to which it is exposed. Equilibrium is then established between the polluted air and the exposed surface(s). As concentrations in air decrease following a period of unusually high exposure, the surfaces will likely outgas so as to maintain or return to such an equilibrium, which will then be changing with time. In much contemporary literature, the equilibrium state is commonly referred to as a 'compensation point' (particularly as in the case of ammonia, q.v. Farquhar et al. 1980; Flechard et al. 2013).

In the more familiar air pollution case, when the substance in question is present in low concentrations in the air, the number density of surface reaction sites is sufficient to accommodate all depositing chemicals and first and second order reaction expectations are no longer appropriate. The operational question is not so much how quickly the surface will become saturated as how efficient the capture process is. The concept of a saturated surface is then rarely relevant, although with obvious exceptions in some cases (such as ammonia).

Obviously, a reactive and polar pollutant like HNO_3 is likely to be removed immediately upon contact with any part of the surface, provided the surface is not exceedingly acidic. The central concept is then the resistance to deposition to the skin of leaves — R_{cut} or r_{cut}, depending on the way in which the canopy is being simulated. In this particular case, the end of the deposition process is the plant surface itself, the cuticle, and overall similarity with the case of heat exchange is expected. Data tend to support this, but precise and definitive verification is beyond current technological capabilities. Other pollutants are not so easily considered. SO_2, for example, is soluble in water, but is poorly reactive in comparison to HNO_3. Its final pathway is therefore expected to be via stomata (involving a resistance R_{stom}, or in the multiple-layered context, r_{stom}) with subsequent association with the moist mesophyllic tissue inside stomatal cavities (involving R_{mes}, or r_{mes}), or alternatively to any part of a plant (or other object) that is wet and not overly acidic. Convention has focused on the anticipation that stomata are a preferred route by which SO_2 and like gases are removed by air-surface exchange, but clearly the spectrum of biological species populating a landscape is not likely to be easily characterized by any single set of such variables. For many trace gases, plant physiological characteristics are known quantities, derived from detailed studies of leaves or bunches of leaves. However, it is clear that all such matters are at the mercy of site-specific considerations.

Conclusion

R_c is likely to have chemical and biological contributions. Equate R_c to zero with caution and avoid near-field situations in which concentrations are high.

Chapter 45

The Multiple Resistance Model: Particles

"The national average deposition of aluminum is dominated by crashing aircraft."
— Noel E. Ridere.

The progress of measurement methods for quantifying the flux of airborne particles to the surface will not be addressed here – it is documented elsewhere (e.g., Hicks et al. 2016). Instead, the discussion that follows centers on contemporary understanding and on the many remaining uncertainties. In line with convention, main attention will be on the deposition velocity, V_d, used to derive the surface flux of the particles in question (F_c) from knowledge of the concentration in air $\tilde{X}$ at some appropriate height (z) above the ground.

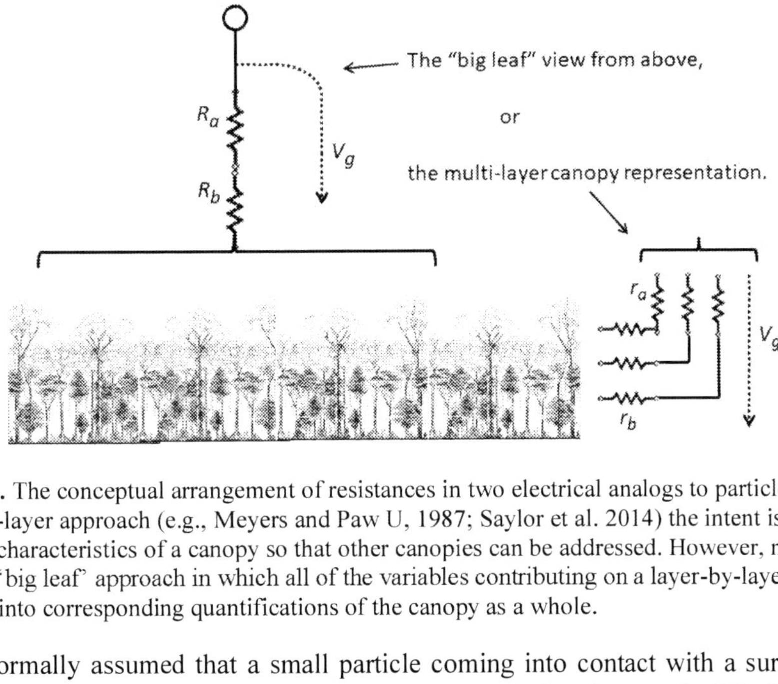

Figure 45.1. The conceptual arrangement of resistances in two electrical analogs to particle deposition. In the multi-layer approach (e.g., Meyers and Paw U, 1987; Saylor et al. 2014) the intent is to describe the capture characteristics of a canopy so that other canopies can be addressed. However, many models employ the 'big leaf' approach in which all of the variables contributing on a layer-by-layer level are assimilated into corresponding quantifications of the canopy as a whole.

It is normally assumed that a small particle coming into contact with a surface will be captured by the surface unless the particle is large enough that the van der Waal's (and other cohering) forces are over-ridden by gravity or some other mechanism. Davidson and Friedlander (1978) have summarized the different approaches used in models addressing the transfer of airborne particles to smooth surfaces. Friedlander and Johnstone (1957), for example, assume that particles are carried with atmospheric turbulence until they are within one stopping distance (as determined by inertia and the turbulent velocity) of the surface, whereas Sehmel (1970) assumes an effective sink at one particle radius from the surface. It is on the basis of such assumptions that very low deposition velocities to smooth surfaces have been predicted. Wind tunnel data have tended to support the prediction of low fluxes to smooth surfaces.

There is an area of uncertainty associated with all considerations of particle settling and deposition – their shape. The matter has received recent attention, due to the potential influence of carbonaceous particles on the global radiation balance. Soot particles reported by Wentzel et al. (2003) are well described by a fractal dimension from about 1.7 to 2.0 (± about 10%), depending on their origin. Xiong and Friedlander (2001) report that the fractal dimension of particles tends to increase as agglomeration proceeds, starting at about unity and eventually increasing to more than two, again for carbonaceous particles. Earlier, Rogak et al. (1993) associated particle structure with mobility, a matter that is conventionally accommodated by relying on the aerosol aerodynamic diameter (AAD) as a central property. AAD is the diameter of a spherical particle of unit specific gravity that would have the same settling speed as the particle in question. Most related work has focused on very small particles, however Schmitt and Heymsfield (2010) have shown that similar fractal dimensions are associated with larger atmospheric particles, especially ice crystals. At time progresses, it seems likely that there will be increased emphasis on particle morphology and the ways in which it influences the processes of interest here. For the present, we will step forward with confidence that such matters will rise to importance well after the demise of contemporary simulations (or simulators).

The multiple-resistance model shown in Figure 45.1 is commonly used as a starting point for discussion of particle deposition processes. The big-leaf and multi-layered approaches are illustrated separately. The surface capture resistance is conventionally taken to be zero and the contribution of gravitational settling (depending on particle dimensions and with settling speed V_g) is often included as an additive velocity component as in

$$V_d = (R_a + R_b)^{-1} + V_g. \tag{45.1}$$

However, Venkatram and Pleim (1996) point out that Equation 45.1 does not satisfy the constraint of conservation of mass; they propose a modified form:

$$V_d = (R_a + R_b + R_a \cdot R_b \cdot V_g)^{-1} + V_g \tag{45.2}$$

that has since gained wide acceptance. (It remains uncertain as to how the modeling community entertains the concept of the Lewis number correction as discussed above). However, there are other processes that intervene. Elsewhere (e.g., Katul et al. 2010; Hicks et al. 2016) there has been extensive deliberation over the roles of thermophoresis, diffusiophoresis, electrophoresis, turbophoresis and Stefan flow, all of which contribute forces to which small particles will respond. For the present, it is sufficient to show, in Figure 45.2, the range of predictions of various models of the deposition velocity processes (as consolidated by Saylor et al. 2019) for the case of transfer to forests and the agreement with field observations.

The field data shown in Figure 45.2 derive from eddy correlation measurements using fast-response particle detectors, gradient studies, and most recently eddy accumulation. Available evidence supports the theoretical expectation that in almost all situations a particle will be captured once it contacts the surface. For large particles (say > 10 μm diameter), deposition is dominated by gravity. For very small particles (say < 0.1 μm diameter) the dominant mechanism is Brownian diffusion. Unfortunately, and as already discussed, it is the intermediate size range that is usually of most concern, since this contains the so-called accumulation size range of particles carrying environmentally hazardous pollutants (sulfates and nitrates, primarily, in the range of about 0.2 to 3 μm). This is the size range that is of potential health concern — the inhalation size range. It also includes the range of the wavelength of visible light and hence is central in considerations of visibility. In the acid rain era of the 1980s, considerable effort was devoted to the development of instrumentation to measure concentrations of particles carrying sulfur and nitrogen, as a function of particle size. The results of numerous field studies then conducted are reported in the State of Science documentation of NAPAP (see Irving, 1990).

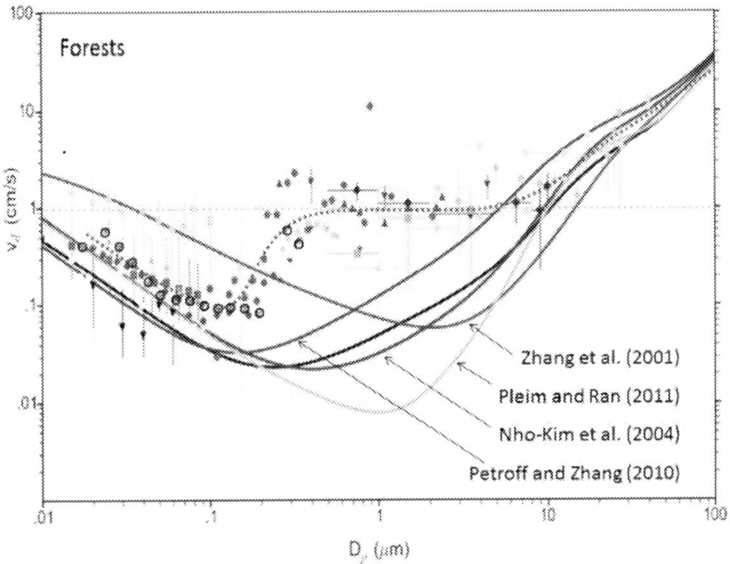

Figure 45.2. Deposition velocities as a function of aerodynamic diameter, showing the large difference between experiments and model expectations. The black dotted curve is drawn by eye. The black solid line represents the early prediction by Slinn (1982); more recent proposals are as identified. The experimental data plotted are as documented by Saylor et al. (2019). The data points circled (Gordon et al. 2011) represent total conductance rather than deposition velocity (see text). This diagram is based on various presentations by Rick Saylor.

It is tempting to manipulate the relationships so as to produce something close to the dotted line of Figure 45.2, however to do so would be a stretch. Expansions and enhancements are obvious, but their utility remains unclear. For example, the analysis by Hicks et al. (2018) indicates that worrying about the consequences of stability might be unnecessary. The reason is simple. Regardless of the approach taken, stability effects will enter through assumptions of the applicability of the micrometeorological ψ-functions and these are based on an integration of in-air gradients down to the level of the relevant sink. This is common practice, even though the procedure is conceptually incorrect, especially in the present case. So, why bother?

The biological characteristics of the plant species (or other objects) that make up the surface canopy present an additional level of complexity that could constitute a death knell to attempts to generate a universal multi-layer simulation. Figure 41.2 presented a photograph of a plant (*domesticus weedus*) growing in the author's garden. Leaf hairs of the kind shown are not uncommon and would constitute a problem for those trying to simulate particle deposition at the single leaf level. The concern is not new. Chamberlain (1967) tested the roles of leaf stickiness and hairiness in his wind tunnel tests. He concluded that "with the large particles (32 and 19 μm) the velocity of deposition to the sticky artificial grass was greater than to the real grass, but with those of 5 μm or less it was the other way around, thus confirming . . . that hairiness is more important than stickiness for the capture of smaller particles." The matter seems ripe for re-examination using modern technologies.

Following initial publication of the work on which Figure 45.2 was based, it was pointed out that some of the data had been plotted erroneously. Gordon et al. (2011) reported conductances associated with the upwards flux of small particles rather than the deposition velocities that would have been found if the direction of the flux was reversed. The similarity between the circled data points of Figure 45.2 with the others is intriguing and potentially important, because it indicates a similarity of the effective bulk conductances associated with small particle exchange, regardless of whether the forest serves as a source or a sink. There is no reason to dispute the equivalence of the turbulent diffusive mechanisms that cause exchange through the air; that is, R_a is the same regardless of the direction of the flux. In the conceptual model presently advocated (with apologies to Venkatram and Pleim, 1999, who argue that in its simple form the resistance model is not appropriate), R_b is an overall consequence of factors associated with small-scale transfer, primarily within the canopy. Since the velocities plotted in Figure 45.2 are always equivalent to $(R_a + R_b)^{-1}$, an obvious conclusion arising from the similarity among the data plotted (upwards fluxes versus downwards) is that R_b is the same, regardless of whether surfaces within the canopy are sources or sinks.

This leads to some conjectures that invite experimental examination.

1. It is known that a forest canopy will readily exude small particles resulting from chemical reactions among emitted trace gases and their interaction with atmospheric constituents (e.g., Held et al. 2004). If the causative reactions were distributed throughout a canopy it would be reasonable to then conclude that the relevant quantity kB^{-1} (as discussed above) would be similar to that appropriate for defining the deposition of the same particles to elements of the same canopy. In this case, it would be of little surprise to find equality between the bulk conductivities defining the transfer, upwards and downwards, of small particles by eddies.
2. The equality so postulated would apply only for the size range of particles generated within a canopy from gaseous precursors, but not to the case of the larger accumulation size range. For these larger particles, there is no doubt that measured conductances indeed relate to deposition, i.e., to cases when the flux is downwards, towards the canopy.
3. In the case of the finer particles it must be accepted that the effective fluxes as measured at some point above a forest canopy will be influenced by both emission and deposition (Held et al. 2004). The step decrease of apparent conductivity occurring at about 0.2 μm diameter (see Figure 45.2) corresponds closely to the classical demarcation between the two size regimes. For particles larger than this, exchange will then be dominated by deposition. For smaller particles, a mix of upward and downward transfer will be observed, but with similar conductances referring to each.
4. Figure 4 of Saylor et al. (2019) shows that the situation for grassland is different. Instead of the sudden and distinctive jump in deposition velocity occurring at about 0.2 μm diameter (as seen in the forest case, their Figure 3), the grassland data of their

> Figure 4 indicate little more than great scatter for all diameters less than about 10 μm. There is no order discernible; in comparison, the forest data are well behaved.

It is concluded that the step change in conductance/deposition velocity almost precisely at the particle size normally taken as demarcation between aerosols in the gas-to-particle conversion size range and the agglomeration-driven accumulation mode is a direct consequence of the natural competition between emissions and deposition. A field measurement will detect the net consequence of the combination without any apportionment.

The discussion of trace gas deposition, above, introduced the roughness Reynolds number (Re_*) and its role in defining the excess resistance to transfer, the quasi-laminar layer resistance, R_b. Figure 32.1 showed that there is a considerable difference between the Re_* dependence of $\ln(z_0/z_{0H})$ between flexible, fibrous surfaces and rigid, bluff bodies. Belot and Gauthier (1975) concluded that, for particle capture, "Elements of a fibrous shape are likely to be more efficient than those of a bluff shape." Further evidence of the importance of microscale surface roughness was also presented: "The deposits on leaves have a tendency to form aggregates." That is, particles tend to be deposited near previously deposited particles. The mechanism involved remains uncertain, but the prospect of easier deposition at locations where the otherwise-limiting laminar boundary layer is disrupted is quite attractive.

There have been many recent embellishments of the relationships describing V_d for particles, most intended to introduce other factors known to influence V_d but omitted in the development so far, such as turbophoresis (Katul et al. 2010). The consequences of these adjustments are to nudge the predicted values closer to the observations of Figure 45.2. Specific examples have been reported, but a general solution remains elusive.

Conclusion

The matter addressed by micrometeorological methods is the bidirectional particle flux. Equating the results to particle deposition involves an assumption that will always need substantiation.

Chapter 46

Dense Canopies and Complex Terrain

"Illustrations of complexity far outweigh understanding of the consequences" — (?)

In the case of a vegetated surface, the question arises as to whether known values of r_c can be combined according to the known biomass distribution, so as to approximate R_c. Most often, the property most relevant is the leaf area index, *LAI*. It is frequently proposed that if r_c is the surface 'capture' resistance per unit area of foliage, then R_c can be approximated as

$$R_c = r_c/LAI \tag{46.1}$$

(see Chapter 34). Note that *LAI* is sometimes quoted on the basis of single-sided leaf area, due initially to its use in the context of water vapor exchange which involves stomata – largely occurring on the upper (sunlit) sides of leaves. But in the case of HNO_3 or other trace gases that are highly reactive the more appropriate concept is probably to consider the two-sided *LAI*. It is not always clear which of these two alternatives is being considered in published papers.

There is an implicit assumption in the application of the *LAI* as in Equation 46.1 — that the foliage is sufficiently sparse that all layers of the canopy will be similarly available for deposition. However, if the vegetated surface constitutes a thick mat of impenetrable foliage, then Equation 46.1 is not likely to be appropriate. The matter was discussed earlier, in Chapter 34, with the conclusion that most likely spatial average is likely to be

$$R_c(\text{generalized}) = \bar{r}_c/\text{LAI}^{1/2}. \tag{46.2}$$

The question then arises as to how these considerations could influence the formulation of air-surface exchange processes in complex terrain. If interest relates to a specific site within a hilly area, with local slopes varying widely, the concept of a vertical aerodynamic resistance to transfer would need careful consideration. Regardless of other factors, formulating momentum exchange imposes the need to consider local streamlines and the $\overline{u'w'}$ plane described by them. In contrast, convection aligns with gravity; covariances $\overline{w'T'}$ and fluxes of trace gases and small particles should follow suit. In stable stratification at night and in the lack of a dominant buoyant force, the driving force is the momentum exchange and relevant covariances must then be normal to the plane of the streamlines (or close to it; there will always be a small restoring force affecting the covariances other than that of momentum, but generally this is diminishingly small). In this second case, the eventual vertical fluxes could be estimated by simple geometric adjustment of the axes (or neglected if the slope is small enough).

It is hard to imagine a circumstance in which a purist would avoid the need to allow axes to vary according to the situation. At night, most likely would be transfer across the plane of the streamlines, affected (of course) by whatever katabatic flow might be present. As height above the surface increases, the association with a vertically oriented coordinate system relevant to larger spatial scales will grow, imposing needs to consider advection and changes with time too complicated to be considered here. At some height, association of the fluxes with the slope of individual components of the surface will diminish and it is then that the complex terrain could best be seen as an area of variability (described perhaps by a fractal dimension) above which formalism returns to Cartesian expectations.

A central consideration affecting forested landscapes is whether the sub-canopy is dynamically opaque or transparent. In the former case, it could well be that flow regimes in hilly surroundings could be described in terms of familiar micrometeorological quantities (e.g., z_0 and d, the value of C_f, or some combination of Prandtl and Schmidt numbers, depending on where one went to school). The issue then would be to combine estimates of local gravity flow with the friction velocity formalism of convention and then to merge the axes from the near field to the gravity-controlled environment aloft.

In the latter case, local katabatic flow could well be confined to within the sub-canopy airspace, in which case the construct envisioned above could be extended by the simple pragmatic approach of adjusting d (and also, maybe, z_0). As I write this, the forest around me is of neither kind; the flow within its canopy is not as well defined as such a simple expectation imagines.

A final step (to generate a relevant spatial average) would entail consideration of the spectrum of like sites populating the area (a grid cell) in question. The no-stone-unturned approach favored by those with powerful computers is common. An alternative that builds upon specification of what is meant by 'complex terrain' will be explored later.

Conclusion

The easiest solution —move people away from shorelines and complex terrain.

Chapter 47

Suspension and Resuspension

"The simple believes everything, but the prudent gives thought to his steps."
— Proverbs 14:15

Particles residing on the surface can be inserted into the atmosphere by many mechanisms, such as the scavenging of mobile particles by the wind (sandstorms and dust devils, for example) and the utilization of mechanical systems to disrupt the surface (such as ploughing and street cleaning). In the era of radioactive fallout, related interest focused on the ways in which radioactive particles, once deposited, could re-enter the wind field and potentially affect areas further downwind. This resuspension was studied extensively, with main attention to areas affected by the detritus of weapons tests — a mix of fission products, containment and supporting materials, and soil particles. The tests were in uninhabited, arid and treeless terrain. Research identified the major contributor to the resuspension of deposited radioactive particles to be saltation, in which the wind causes sand (and/or soil) grains to be blown across the surface and to cause smaller particles they strike to be injected into the air above. The causative sand grains (or soil particles, depending on the circumstances) are themselves too large to be diffused into the air stream. They serve as the causative mechanism, not as a direct contributor to downwind exposure or risk. Descriptions of the processes involved and parameterizations of them have been provided by several specialists, such as Anspaugh et al. (1974), Sehmel and Lloyd (1974) and Horst (1975). The simulations derived are relevant to the situations of their observations — treeless and arid. Reviews of the science as was then known were presented by Sehmel (1980b) and Nicholson (1988)

In a verdant countryside, forested or otherwise, deposited particles will be captured and retained. There is little chance that the saltation mechanism will be in play, because (a) wind speed is attenuated near the surface and is inadequate to allow desert-like saltation to occur, (b) the surface drag is inadequate to break the bonding of particles already deposited to the surface and sequestered by many mechanisms (e.g., electrostatics, van der Waals forces, liquid bridges) to be overcome and (c) if a large particle is broken loose it has little chance of striking a downwind particle carrying a pollutant of importance.

In summary, there are two ways in which the consequences of resuspension can be considered. The first uses the relationships derived from studies of radioactive surface contamination which apply, mostly, to areas where people do not live but which describe the consequences in great detail. The second relates to areas where people actually live and where resuspension is conceptually a non-issue.

Conclusion

Suspension and resuspension can be important issues in arid areas with few people. In urban settings, traffic and street maintenance likely combine to generate an issue worth exploring.

Chapter 48

Indirect Monitoring of Dry Deposition

"Science is in support of regulations" — An EPA Administrator of the 1980s.

In the early days of concern about acid rain, the comparative ease of measurement of wet deposition and the importance of deposition as a whole (both wet and dry) in the context of environmental regulations led to a strong desire to construct a dry deposition network paralleling the wet deposition programs then in place in North America — the US National Atmospheric Deposition Program (NADP, see Lamb and Bowersox, 2000) and the Canadian Air and Precipitation Monitoring Network (CAPMoN, q.v. https://www.canada.ca/en/environment-climate-change/services/air-pollution/monitoring-networks-data/canadian-air-precipitation.html). Both networks monitored the deposition in rain of chemical species like sulfate, but CAPMoN also measured the air concentrations of the contributing chemicals – in this case gaseous sulfur dioxide and particulate sulfate. NADP was (and remains) a collaborative organization of deposition researchers, initially focused on wet deposition and relying on the wet bucket collections of samplers like the HASL wet/dry collector (Bogen et al. 1980; see Chapter 40).

The need for dry deposition information to be tabulated alongside the two-weekly averages of wet deposition provided by NADP was recognized, but a suitable methodology was lacking. The air concentrations measured by CAPMoN enabled estimation of dry deposition rates, provided relevant estimates of V_d were available. These estimates of V_d were derived from extrapolation of the rare field studies then reported, or on the expectations of models constructed from them. In either case, the products were data-poor. In response to the perceived need for an improved methodology, NADP set up a committee to advise on steps forward. This committee recommended (a) that use of the dry bucket measurements be discontinued, (b) that a program like CAPMoN should be inaugurated, but expanded to integrate routine measurements of variables known to influence V_d so that site-specific chemical fluxes could be estimated and (c) that filterpack methods should be used for concentration measurement but with protection against the consequences of high humidities. The last requirement arose following field tests in which the pressure drop across the sampling system was enough to allow liquid water to affect some of the filters. Deployments in the southern USA were especially susceptible to this phenomenon, which necessarily led to SO_2 being deposited on filter membranes intended to collect particulate sulfate. Hence $SO_4^=$ concentrations would be overestimated at the expense of SO_2. As protection against the effects of wetted filters, one filterpack configuration imposed a small heating of the incoming air and was employed in the Atmospheric Integrated Research Monitoring Network (AIRMoN), starting in 1984 (for details, see Hicks et al. 1991; Meyers et al. 1991). Sites were selected to closely satisfy micrometeorological requirements, so that the inferential methodology then being tested (q.v. Hicks et al. 1987) would have best chances of success. Distance from prominent sources was a major consideration because the desire to derive averages to parallel the wet deposition program of NADP imposed the need for air concentrations to be invariant (or near to it) with time of day. To test this, a test site was identified downwind of a major coal-burning power plant, the Bull Run steam plant in eastern Tennessee. It was anticipated that correlation between concentrations and time of day (and hence between C and V_d) would negate the calculation using the inferential methodology then being promoted. However it was found that nature was forgiving. Comparisons made with eddy covariance methods indicated that the inferential method underestimated the average value of V_d by about 20% (Matt et al. 1987) and underestimated long-term averages of sulfur deposition by a similar amount (Matt and Meyers, 1993).

A focus on the deposition of nitrogen species caused the pre-heating of the AIRMoN programs to be removed in newer network programs (particularly the US EPA Clean Air Status and Trends Network, CASTNet). In contemporary times, the histories of these programs (AIRMoN and CASTNet) have largely been lost and the activities have been assimilated into the National Atmospheric Deposition Program (NADP, see Lamb and Bowersox, 2000) and its National Trends Network. The need for comparison between the results of the two programs persists, especially since the programs for quantifying V_d have evolved (see Meyers et al. 1998; Clarke et al. 1997). A key (and very welcome) departure of CASTNet from the original NADP guidelines has been a focus on ozone.

CASTNet has evolved into mainly a chemical program (as illustrated by the claim that the precision of dry deposition data has been assessed, q.v. Sickles and Shadwick, 2002; whereas the relevant material refers only to the measurement of air chemistry). In fact, field tests using micrometeorology have yielded the expected results (e.g., Finkelstein et al. 2000), with the routine monitoring program (CASTNet, in this case) yielding average rates of deposition of sulfur to a forested landscape about half of what was measured using micrometeorology. The difference could be no more than the consequences of unprotected filter wetting, as already mentioned. For ozone, the agreement has been better.

The NAPAP State of Science and Technology documentation (Irving, 1990) includes an assessment of the national sulfur budget, based on monitoring and routine observations made by the NADP and AIRMoN programs. Data from the EPA CASTNet program were not available at that time. A two-year annual average revealed an influx into the continental United States of 0.07 Tg S (Boatman et al. 1989), to which 7.70 Tg were added by industrial emissions (Placet, 1990). Natural emissions (e.g., fumaroles, Placet, 1990) contributed a further 0.08 Tg, so that the source term amounted to about 7.85 Tg per year. The annual deposition rates quantified on the basis of NADP wet deposition 1.92 ± 0.14 Tg year^{-1}, Mountain Cloud Chemistry droplet deposition (0.17 Tg year^{-1}) and the AIRMoN dry deposition program (1.81 ± 0.24 Tg year^{-1}) totaled 3.8 ± 0.23 Tg year^{-1}. Independently, Galloway and Whelpdale (1987) estimated riverine outflow to the Atlantic to be about 3.2 Tg year^{-1}. The imbalance was therefore about 0.75 Tg year^{-1} which represents a first (and unverified) estimate of the rate at which the Atlantic Ocean might be affected as polluted air moves across it. A small part of this outflow from North America could reach Europe. In the current intellectual environment, similar computations commonly address the issue of carbon dioxide.

An increasingly common regulatory application of dry deposition estimates from networks like are described above is in the context of critical loads. A critical load is defined as the average deposition rate of some pollutant that is known to have a harmful consequence. The relevant load is defined as the sum of dry and wet components, the latter being readily measured via collection and analysis of precipitation. Having originated in Scandinavian situations involving inland lakes that suffered from sulfur deposition (along with the deposition of other acidifying species, like NOx), the concept has become popular among ecologists and especially regulators.

However, the following considerations are relevant.

1. Determination of an appropriate critical load is not straightforward. In many instances, actual data are lacking and the relevant values can only be obtained by extrapolation from some situation elsewhere that offers actual observations, or delphically. When asked directly about how the critical loads community handled the resulting uncertainties, a leading European advocate offered that the issue was not very important because they took their best estimate and applied a factor of ten.
2. It is the total deposition that is important, the sum of wet and dry. As discussed above, determination of the dry deposition component is demanding and the results uncertain. In the case of an inland lake, dry deposition to the actual water surface would be expected to be considerably smaller than that to the surrounding countryside, due to the relative smoothness of the water surface and the resulting low levels of turbulence in air near it. Turbulent exchange processes are then less efficient. Moreover, the stability regimes over inland lakes are likely to be different from those over the surrounding landscape, because in daytime the water surface will be cooler than the air and this will further curtail turbulent exchange. Nighttime situations remain uncertain. In general, estimates of dry deposition made at a nearby shoreline location are probably not indicative of what affects the water itself.
3. Run-off and ground water seepage need to be considered as relevant vectors. However, early studies using radioactive tracers showed that soil ion-exchange mechanisms quickly scavenged relevant ions from the water seeping through it. (Personal experience in this regard employed cosmogenic ^{35}S and ^{7}Be as tracers.) Clearly, the cases of hydrocarbon damage to ground water resources and the repercussions of large spills or containments illustrate that this generality cannot extend to cases in which the groundwater is overloaded with chemicals from outside sources, but in the trace quantities resulting from deposition processes ion exchange remains a dominant consideration.

In summary, critical loads have the attraction of being an easily understood basis for the imposition of regulations or other control measures. However, they cannot be enumerated with confidence (error margins are not specified) and the potentially offending deposition rates cannot be well quantified. They will not be considered further, here.

Conclusion

Strategies proposed for desired environmental protections often require regulations more readily expressed in concept than achievable in practice.

Chapter 49

Concentrated Gas Deposition and Canopy Retention

"Look at the near-field before the far" — Noel E. Ridere

The conventional considerations of dry deposition of gases relate to situations of gases being present in trace quantities, such as in situations distant from sources. This situation has been the focus of discussion so far. In these far-field conditions, it is usual to equate the surface resistance R_c to zero in the case of small particles and gases whose capture by the surface is efficient. In simple terms, there are then enough reaction sites available on the collecting surfaces to capture whatever substance is being deposited. For trace gases, the case of HNO_3 is a notable example.

In the event of a release (either intentional or otherwise) of a hazardous substance into the surface boundary layer, a first question relates to the identification of safe zones to which people could be relocated. A second initial concern is the identification of safe corridors by which emergency responders could access and mitigate the source. The answers are provided either by officials familiar with the characteristics of the material released, the prevailing meteorology and the surroundings of the release (not necessarily in that order), or by using some predictive model. Reliance on a model output is favored, mainly, it is suspected, because it isolates emergency responders from responsibility for making incorrect decisions.

The direction taken by the resulting plume or the path of an offending puff are of immediate concern to first responders. There is a lot of uncertainty involved, since local topography and obstacles cannot be ignored. There is nothing better than onsite real-time evidence to guide the decision-maker. A dispersion meteorologist is at risk of making an incorrect assumption that the wind is the only controlling factor, with obvious unwelcome consequences when topography or obstructions are involved. Results of past experiments are often relevant. However, these past experiments have suffered from the same affliction that has offended micrometeorology in general — they have been conducted in situations such that the simplest physical processes can be assumed to be relevant. Figure 49.1 illustrates the surface-related complexities confronting a responder after an accidental release of liquid chlorine from a ruptured containment vessel. Any surfaces or obstacles encountered by the flowing liquid would cause a phase change, resulting in rapid flow across the surface (as in the famous Liedenfrost effect). The upper part of the moving gas/liquid/aerosol cloud would be scavenged by whatever atmospheric turbulence might be relevant. Once this mixture of chlorine (in the case of the present example) escaped the confines of any surface concavity or vegetative canopy that might be present, generating the source term appropriate for initializing subsequent atmospheric dispersion can be estimated. There will probably be significant dilution involved.

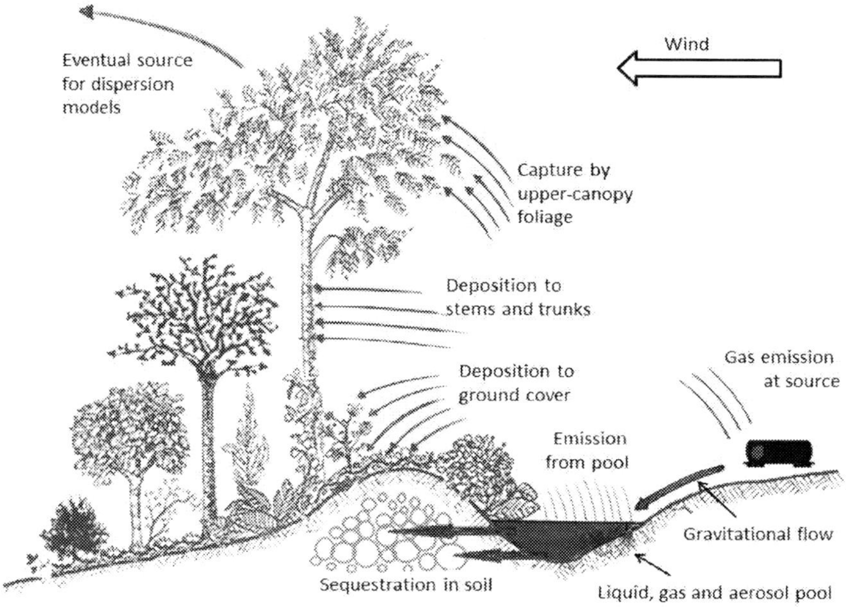

Figure 49.1. A schematic view of the processes contributing to the source term specification of numerical dispersion models. The material initially emitted is assumed to be liquid chlorine.

In the context of emergency response, predictions of dispersion are usually focused on the quantification of expected concentrations at specific places and times. There are many processes that must be taken into account in any model that is used; it is the formulation of such processes and the mechanisms of interaction among them that distinguish one such model from another. In practice, the ability to predict downwind concentrations is limited by the aleatoric variability of the key contributing processes — most notably the wind speed, atmospheric diffusivity (in all three dimensions) and the rate of removal by whatever surface is encountered. The formulations used in dispersion models are usually descriptions of behaviors expected of averages based on few observations, usually made elsewhere. The choice of which of many alternative parameterizations to employ tends to depend on the institution where a modeler was indoctrinated. In any event, the results produced by an assembly of the many contributing considerations are not likely to be normally distributed and specification of the uncertainties associated with the final products (concentrations) is therefore challenging. Modern computers enable many alternative methods to be employed (Monte Carlo, bootstrapping, etc.). While the results of such analyses can be revealing, they ultimately represent the consequences of assumptions made in the model they depend on. It is obvious that predictions cannot achieve point-by-point agreement with observations. All that must be expected is to get the most likely value right and to quantify the statistical distribution around it.

Irrespective of the characteristics of a receptor surface, it will eventually come into equilibrium with the concentration to which it is exposed. Immediately after exposure to constant concentrations of a specific gas in air, deposition will be the most rapid. The initial rate of transfer will be an indication of the areal number density of the available sites for capture of molecules impacting the surface. As sites become saturated, the remaining areal density of capturing sites will diminish. From the perspective of an observer, the surface resistance to exchange will increase. A chemical species-specific rate of accommodation with exposures can be identified experimentally, in straightforward laboratory experiments (e,g, Spicer and Fox, 2020; Spicer et al. 2021). For chemically reactive heavier-than-air gases, the matter is of concern in that hazardous chemicals (specifically including chlorine) can be sequestered by vegetation and soil, often with later release back into the atmosphere (as was demonstrated for chlorine in the Jack Rabbit experiments: Koopman et al. 2019). The source term description used in dispersion models must take such factors into account.

The earlier discussion of the association between fluxes and gradients in the case of a forest canopy has already intimated that there is a difficulty involved. In conventional simulations, downwind atmospheric dispersion following an injection of a dense gas commences at either a point source at some low height above the vegetation or an areal average at such a height. However a dense trace gas residing within a vegetation canopy will be somewhat (if not completely) isolated from the wind field above the canopy. The situation in mind is as illustrated in Figure 49.1. Beneath the canopy, topography and local surface features will tend to control dense gas spreading, until turbulence and wind gust incursions become influential.

The transfer of air (and the trace gasses accompanying it) from the sub-canopy airspace into the wind field above the canopy can be either continuously diffusive or irregularly intermittent (most likely a mix of these). The conventional view of the roles of 'sweeps' and 'ejections' (see Katul et al. 1997) is shown schematically in Figure 49.2. From the perspective of above-canopy dispersion modeling, there is need to account for a dilution caused by this exchange process. Elsewhere, but yet to be reported, it has been estimated that the dilution imposed by transfer through the canopy amounts to a factor of between 4 and 20, depending on the prevailing conditions. Central considerations are the density/transparency of the vegetation canopy and the presence or otherwise of convective mixing.

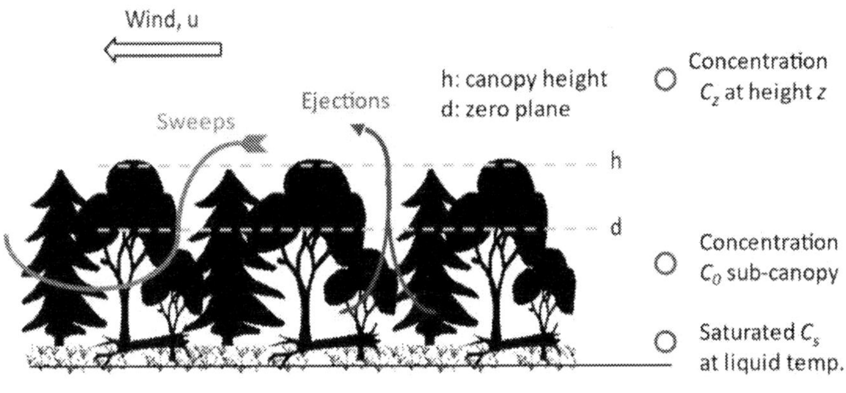

Figure 49.2. A schematic view of a forest canopy showing momentum-driven sweeps and thermal ejections, each penetrating through the level of the displacement height d and resulting in a well-mixed sub-canopy layer with concentration C_0, intermediate between the surface (C_s) and the air (C_a).

Thus, there are two corrections that need to be applied in order to estimate the concentrations intended to initialize conventional air pollution dispersion models — near-field sub-canopy deposition and trans-canopy exchange. Combining these into a single optimal formulation for operational application remains a topic for research, with dedicated field trials appearing a necessary component. In the meantime, it is clear that the matter of sub-canopy and/or near-field deposition requires attention to the relevant first-order (and second-order) chemical reactions occurring at exposed surfaces, whereas downwind far-field considerations do not.

As this is written, the author is sitting comfortably observing the leaves of a surrounding forest move in response to air motions. The forest is largely of mixed deciduous species, with an undercover of varying density in sloping terrain. It is apparent that this forest does not agree with much of the simplification expressed above. There is need to step beyond what has been learned from studies of homogeneous and managed plantations in micrometeorologically acceptable terrain.

Conclusion

If sub-canopy influences are ignored, models familiar to response communities will likely provide overestimates of downwind risks.

Chapter 50

Dispersion — Prairie Grass etc.

"If in doubt, avoid being downwind" — Noel E. Ridere

A trace gas injected into the atmosphere at some time and location will be transported with the wind and will diffuse as turbulence determines. Concentrations in the air will diminish as affected parcels of air are carried downwind. If the emitted substance is hazardous, then there is need to consider the exposures of downwind populations to the resulting dispersion (dispersion = transport plus diffusion). Verifying the reality of a perceived threat and assessing its subsequent consequences are specialties that draw heavily on the micrometeorological understanding summarized above. However, the focus of most micrometeorological research has been on understanding the contributing processes and especially their deterministic interactions. Dispersion represents the net consequence of all of the processes affecting the atmosphere, whether deterministic or otherwise. The development of the relevant skills has often departed from the lines of progression of micrometeorological understanding. However, much of what has been learned about micrometeorology has derived from studies of dispersion, in its various embodiments.

Here, the matter of local, near-field dispersion will be considered first. In this regard, 'local' dispersion is considered to be that which occurs (usually within the SBL) over a time scale small in comparison with the diurnal cycle. The Chapter to follow will focus on urban dispersion. The third Chapter in this dispersion sequence will address the extension to larger scales, as is commonly applied to assess risks over exposure periods longer than several days or at distances far removed from sources.

If all of the relationships contributing to the dispersion of trace gases or particles were known and if the diffusive spread of pollutants within the atmosphere were deterministic, then prediction of the consequences could be computed with precision. However, the micrometeorological formalisms discussed earlier in this document are no more than descriptions of those parts of the overall atmospheric environment to which such deterministic thinking applies. In practice, the movement of trace gases (or particles) through the atmosphere involves a range of motions that includes many lying outside the scope of the formulations now available. As an obvious example of how this complicates the overall issue of dispersion, remember that the magnitude of the correlation coefficient involved in the transfer of momentum $(\rho \overline{u'w'}$

specified in terms of the momentum flux or the corresponding friction velocity. That is, less than 20% of the turbulence influencing the velocity field above a surface can be one-on-one attributed to the surface itself. The extrapolation of standard micrometeorological expressions to address issues of dispersion is therefore not fully warranted, since the dispersion represents the net consequence of all atmospheric motions whether deterministic or not.

There has been a long history of dispersion studies involving the intentional release of known quantities of some readily measured tracer into the air. Many of these have been conducted in support of military applications and results of them remain largely unavailable.

Within the civilian community, the search for near-field understanding was led by the 1953 Great Plains program (Lettau and Davidson, 1957) which was a forerunner of the monumental 1956 Prairie Grass experiment (Barad, 1958; 1959), conducted over a large uniform field of stubble wheat near O'Neill, Nebraska. These experiments remain unequalled in the related sciences. They involved a large contingent of researchers (~ 60 people) over an experimental period exceeding a month. Their results have been the foundation of much of current understanding about what controls the dimensions of the plume resulting from source emissions at ground level, out to a distance of nearly 1 km. At a distance of 100 m, several towers were arranged to provide concentration measurements at heights of 0.5, 1.0. 1.5, 2.5, 4.5, 7.5, 10.5, 13.5 and 17.5 m above the ground. Examination of the Prairie Grass concentration profiles reveals that the plume at 100 m distance was most often ground hugging, much as expected since the gas released was sulfur dioxide, considerably heavier than air. The data invite further examination in the light of more contemporary understanding.

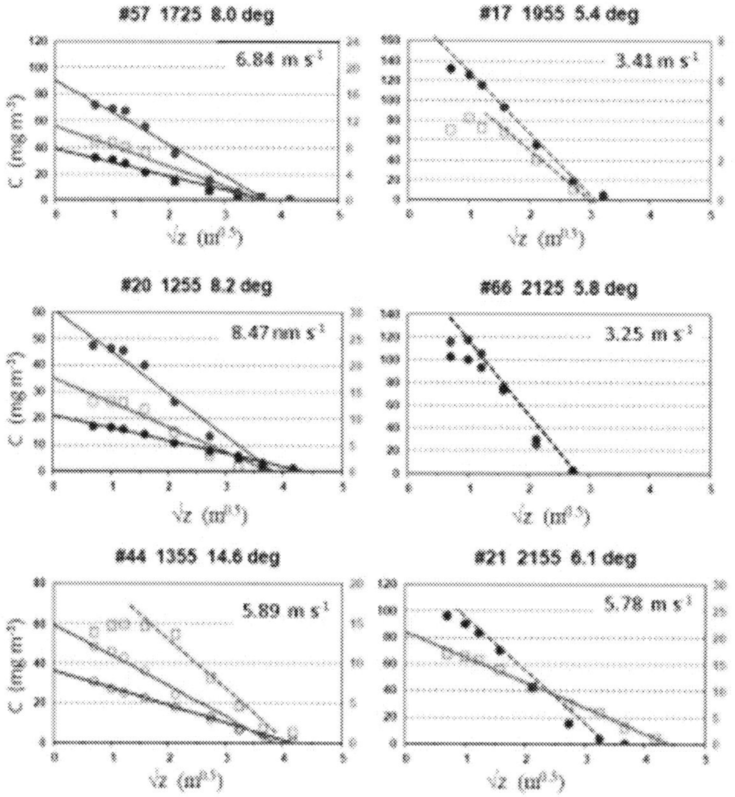

Figure 50.1. Results of six of the 68 Prairie Grass trace gas releases at ~0.5 m height, showing how concentrations observed at 100 m downwind varied with height at different locations on the arc. Solid lines are results of regressions. Dashed lines have been drawn by eye. Concentrations are plotted according to the left ordinate, except for the data plotted as red boxes which make use of the ordinate to the right.

Figure 50.1 provides an example of concentration profile data from the Prairie Grass 100-m arc. A review of the body of data indicates that the concentrations decreased approximately according to the square root of the height and hence Figure 50.1 has been arranged to derive from the documented data the most likely vertical reach of the gas cloud when the plume reached 100 m. The identifications of the diagrams indicate the experimental run number, the time of the release and the prevailing standard deviation of the wind direction (in degrees). In each diagram, the relevant wind speed is noted. The diagrams to the left relate to unstable (daytime) situations; to the right, to stable (nighttime) conditions.

The diagrams of Figure 50.1 indicate that scaling of observed concentrations according to the square root of the height is an acceptable approximation for the near-field case (out to 100 m or so). The x-intercepts of the lines drawn provide a measure of the vertical growth of the plume, extending to about 16 m in daytime but only to about 9 m at night. There is no obvious influence of either the wind speed or the prevailing turbulence regime as indicated by the measured values of $\bar{u}$ and σ_θ. However, it is obvious that these results cannot be used in other circumstances without careful consideration of the role of trace gas density and its impact on mixing at the point of release. A similar ground-hugging plume downwind of a release of (say) ammonia does not seem a likely occurrence, ammonia being lighter than air. This is not to negate the general applicability of the Prairie Grass results in other contexts, since it is always assumed that a tracer in low concentrations will be diffused without need for consideration of its basic physical properties once it is mixed well with the air and that this mixing is usually rapid. However, the Prairie Grass data seem to indicate that such mixing had often not occurred until the SO_2 plume had extended beyond the 100 m arc.

The Great Plains study of 1953 and the Prairie Grass experiment of 1956 were the first major field deployments of sonic anemometers (q.v. Suomi, 1957; Businger and Suomi, 1958). These early dispersion studies followed the same experimental philosophy as had been favored by micrometeorology — first, reduce the problems to addressable levels by ensuring that the consequences of terrain complexity and time stationarity could be minimized. In the years after Prairie Grass, European research extended the inventory of experiments with studies often concentrating on stable conditions and elevated sources (Högström, 1964). In the U.S.A. issues related to the dispersion of radioactive particles and gases caused attention to be given to the surroundings of reactor (and similar) installations and to the spread of the cloud of contaminants that would follow a release of radioactive material. Islitzer (1961) reported on a series of experimental programs studying dispersion at the National Reactor Testing Station in Idaho (now, Idaho National Laboratory). The need for related research was underlined by a review (Islitzer, 1962) of observations following an accidental release (3 January 1961) of I^{131} from a reactor in southeast Idaho. This emphasized the limitations on dispersion due to the proximity of the Rocky Mountains to the west. Studies of transport and diffusion affected by complex terrain quickly followed, with targeted investigations of the diffusion of trace gases in the terrain of the test area (Islitzer and Dumbauld, 1963). The influence of the upwind mountains on the overall flow characteristics of the reactor surroundings (the plain of the Snake River) was investigated first by watching the trajectories of smoke releases, by tracking zero-lift tetroons (Angell, 1964a; Angell et al. 1968; Dickson and Angell, 1968) and later by the use of remote probing.

The field observations of the 1950s and 1960s demonstrated the complexity of the near-field dispersion problem. Figure 50.2 provides two illustrations of the issues confronting those attempting to predict concentrations downwind. The influence of velocity variation with height is demonstrated in Figure 50.2a, where releases from a low level are seen to move in a direction differing from greater heights — observations reminiscent of the KISS phenomenon as discussed above (Chapter 27). However, the cross-sectional dimensions of the three plumes shown are similar, supporting the conclusion obtained from recent micrometeorological studies that there is not a great change of σ_θ or σ_ϕ with height. Figure 50.2b shows how meander enters into consideration. The plume path evident in the photograph emphasizes that tests of model predictions of concentrations downwind must consider both the diffusion of material in turbulent packets of air and the migration of these packets as influenced by velocity variability. The concept of packets of turbulence causing the diffusion of material within parcels of gas moving with the wind (as initially introduced by Inoue, 1950) corresponds well with the view of more modern micrometeorology, but clearly more investigation would be a good idea.

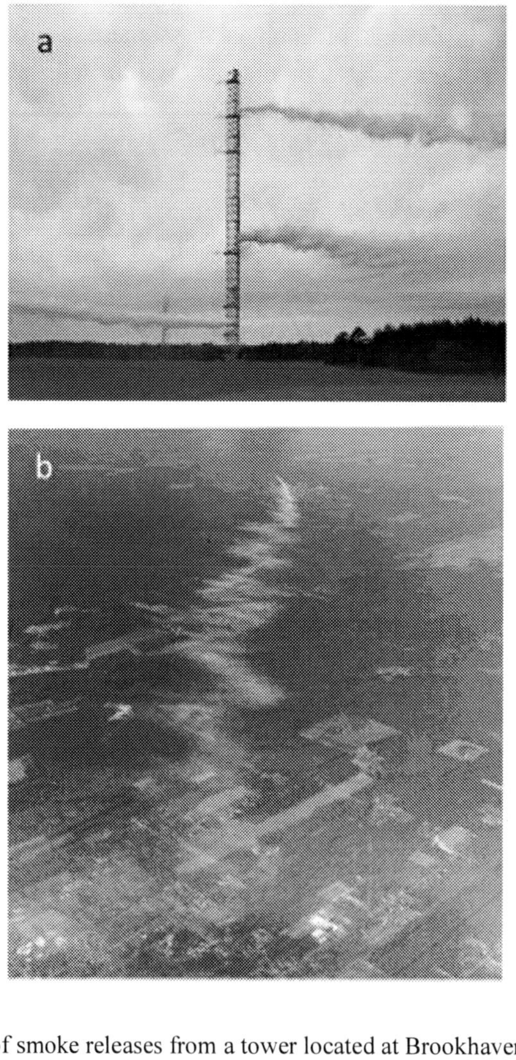

Figure 50.2. Two views of smoke releases from a tower located at Brookhaven National Laboratory in the 1970s. The source of these images is uncertain. The photographs were rescued from garbage left by others as they departed.

Elsewhere, field studies (see Irwin, 1983) started to address the matters of terrain complexity and heavy gases, in recognition that the risk to populations could be elevated when dispersion is directed by preferred wind directions and affected by limited diffusion due to the prevailing topography. Valleys and mountain canyons became the focus of dispersion studies beyond the flat-earth terrain of studies like Prairie Grass. In the U.S.A. the Atmospheric Studies of Complex Terrain program (ASCOT, q.v. Orgill and Schreck, 1985; Clements et al. 1989) provided a multi-organizational focus for experimentation and model development. Studies made use of atmospheric tracers like oil smoke, sulfur hexafluoride and perfluorocarbons (Lovelock and Ferber, 1982). Field tests were conducted to reveal details of both upslope and downslope flows and with attention primarily to well-defined and accessible topographic features of the Rocky Mountain foothills.

The tracer gas used in the Prairie Grass experiment was sulfur dioxide, SO_2, which is denser than air. The behavior of the gas, immediately upon its release, was doubtlessly affected by this density difference. In general, atmospheric dispersion models assume that dilution is quickly sufficient to eliminate concerns about effects of gravity, such as on buoyancy. But this does not negate concerns about behavior very near the source. The matter of what constitutes 'very near' depends on the circumstances and has resulted in the production of many simulations intended to help understand how heavier-than-air trace gases (and particles) disperse in the near field. Many field studies have addressed issues related to the plume evolution downwind of a release of such chemicals as ammonia, propane, chlorine, and many other gases. Within defense establishments, the purpose has often been to simulate the spread of nerve agents like sarin and cyclosarin. Additional concerns relate to the consequences of releasing at different heights, leading to many field studies of releases from (for example) smokestacks and the roofs of buildings. There is no doubt that each such study contributes to the overall understanding of the issues central to the problems of immediate concern; it is equally clear that the magnitude of this library obscures the extraction of meaningful and generalizable understanding. Recently, accidents involving chlorine-carrying railcars (e.g., Buckley et al. 2007) have caused a coupled experimental and model developmental focus on the need to take the special chemical and physical characteristics of liquid and gaseous chlorine into account (Fox and Storwold, 2011).

There have been at least two distinct paths followed by model developers. First, there has been the intellectual need to test how well particular sets of dispersion observations can be explained, for which purposes complex (diagnostic) models are regularly proposed. Second, there is a societal need for dispersion models simple enough to permit rapid utilization in difficult circumstances while providing data adequate for guiding responders and protecting people. These two paths rarely lead to the same products. However, the basic considerations of direction, speed, and rate of spreading are common to all dispersion problems. Surface deposition, precipitation scavenging and chemical transformation can also become important. Awareness of specific concerns has provided a continuing incentive for the development of simulations more detailed than legacy dispersion models. As time has progressed, models have been increasingly 'improved' by the incorporation of previously ignored complexities. An optimal situation would have all of the important input variables specified or accurately formulated for all places and times of relevance. It is clear that while modeling skill will depend upon how close we can get to this optimal situation; we will always be far from the optimum. For example, wind direction and speed, both critically important parameters, can be monitored using relatively straightforward instrumentation, but (as was emphasized earlier) no single wind-measurement station can hope to represent more than its immediate surroundings, except in very unusual circumstances of horizontal homogeneity, flat terrain and time stationarity (Nappo, 2002; Nappo et al. 2002). And even then chaos will intervene to confound the deterministic predictions. The contributing uncertainties compound to yield a resulting potential error in the final product that remains difficult to address (Irwin and Hanna, 2005), irrespective of the methodology utilized.

Modern computers have enabled a continuing evolution of simulations, permitting more attention to be given to the processes that cause each event to be unique in some way. Comparisons among different models serve little purpose other than to test the extent to which researchers and model developers adhere to a common understanding. In practice, the outputs of these new models can only be verified by comparing predictions to observations in situations of obvious relevance. In this regard, the dispersion studies of the 1950s (discussed above) serve as data sources yet to be repeated or generalized. The retrospective consideration of the Prairie Grass dataset (above) indicates that current model development and 'improvement' will likely require a field study of similar extent and complexity to satisfy more modern modelling questions. In the meantime, it remains unclear as to whether increased model complexity results in a better model prediction and how to demonstrate this if it is indeed so.

The simplest dispersion model currently popular is undoubtedly the Gaussian plume approach (Pasquill, 1961; Gifford 1961). This integrates all of the uncertainties associated with particular events into overall descriptions that are then relevant to assessments (i.e., to averages of many contributing situations) rather than to any particular episode. In an application addressing some specific event, predictions concerning averages are not particularly useful. Yet it is these specific and site-specific applications that are the applications of societal importance for a modern emergency response program.

Conclusion

Probabilistic predictions are likely unavoidable.

Chapter 51

Dispersion in Cities and Urban Areas

"If dispersion model predictions agree with observations, get better observations"
— Noel E. Ridere

During the last stages of WW-II, awareness of a potential attack on allied cities by an adversary making use of airborne biological agents generated field studies conducted in several of the allied countries. With the beginning of the cold war in the 1950s, field studies continued but conducted from an adversarial posture rather than defensive. In North America, biological weapons tests (BW) were conducted over cities that were like target areas in the USSR. Various tracer materials, culminating with the use of fine fluorescing ZnCdS particles (Leighton et al. 1963) and *Bacillus globigii* spores, were spread into the air upwind of more than 30 cities in the U.S.A. and Canada, with associated tests in Great Britain (q.v. U.S. Army 'Operation Dew', Operation LAC). Following international agreement, studies related to the use of BW agents ended in the early 1970s. Much of the related information has since been lost within the military bureaucracy, but concerns about possible adverse health impacts caused the National Academy of Sciences to review the program and summarize its potential impacts (NAS, 1997). It is clear that a lot of information relevant to today's concerns was gathered in these early studies and it remains possible that the relevant meteorological data exist in archives that have still to be located and explored.

The use of constant-level super-pressure tetroons with radar tracking was emphasized above, in Chapter 50. After proving the utility of the methodology, the capability was repeatedly used to study the flow modifications and dispersion characteristics associated with cities and urban areas, mainly in the U.S.A. but with one major departure — Cardington in the U.K. (Angell, 1964b). Among the cities/areas investigated by the tetroon team (with contributors Jim Angell, Gil Ferber, Ray Dickson, Walt Hoeker, Don Pack, Ken Peterson, Lester Machta and Dave Slade, all of ARL) were Atlantic City, Cincinatti, Columbus, Los Angeles, New York, Oklahoma City and St. Louis. In later years, the methods have been successfully used in studies elsewhere, e.g., as a component of the SIESTA program in Switzerland (Hu and Vogt, 1986) and in a study of the climate of Karlsruhe, Germany (Vogt and Thomas, 1988).

The development of perfluorocarbon trace gases (PFCs), detectable in minute quantities (Lovelock and Ferber, 1982) enabled a series of new dispersion studies, of longer duration and over greater distances than was previously possible. A short-duration demonstration was conducted in Oklahoma City in 1980 (Ferber et al. 1981). Subsequently, the Metropolitan Tracer Experiment (METREX, 1983-1984) was conducted in the Washington DC area, involving six-hour releases every 36 hours over a one-year period (Draxler, 1985). Three different release locations were employed, so as to allow consideration of local topographic differences (which proved to be significant). The results demonstrated the complexity of the situation more than its resolvability.

Studies of the dispersion in the SBL rose to a new level of importance following the World Trade Center incident of 11 September 2001 and the subsequent anthrax attack on US officials. An accelerated campaign of urban meteorology studies resulted, including a considerable effort to improve relevant numerical simulations. Arnfield (2003) and Kanda et al. (2007) have reviewed progress, with the latter pointing out the underlying similarity with the case of forest meteorology. Indeed, Oke et al. (1989) had previously emphasized the parallels by writing about 'the urban forest.' Some of the basic understandings of forest meteorology transfer to the urban case quite well. For example, Kanda et al. (2007) report values of the appropriate zero plane displacement and roughness length that are much the same as for forests: d/h typically in the vicinity of 0.6 and z_0/h about 0.1, where in this case h is the average height of buildings. The number of numerical models developed to address specific circumstances continues to grow, this in itself being illustrative of the complexity of the problem and the site-specific nature of any simulation.

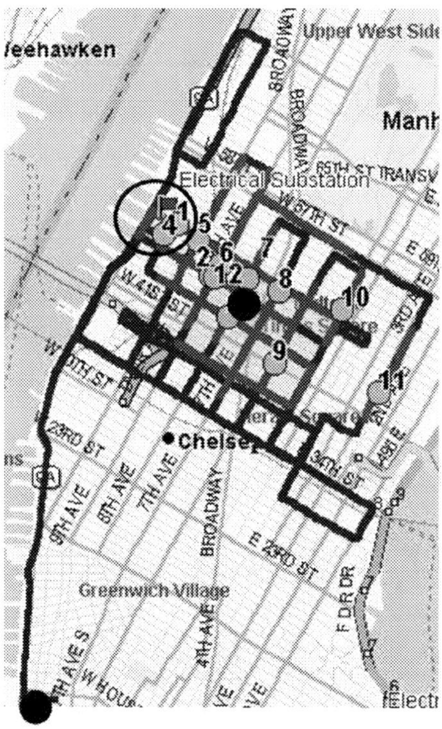

Figure 51.1. Results obtained in a survey of SF_6 concentrations in air at nose-level in the street canyons of New York City, in 2004. SF_6 emissions resulted from leakage from a power substation located as circled. The solid black circles indicate the starting and ending points of the street-level exploration.

In the case of an event like that considered here, all decisions are urgent. Waiting for guidance from some central facility is likely to be an unattractive option when simple capabilities using local information are immediately available to the response team. Regardless of the data input requirements of relatively simple dispersion models (there are too many of them to be considered here), the central considerations remain (a) the direction of plume migration and (b) its rate of spreading or dilution. In recognition of this, a research monitoring program was initiated in 2001 to explore how modern, reliable anemometry could be deployed in a cost-effective yet scientifically adequate manner to provide site-specific and real-time model input data for emergency responders. The resulting DCNet program (Hicks et al. 2012, 2014a) concentrated on the Washington, DC, situation, although with a secondary focus on New York City. The DCNet program has already been discussed, from the viewpoint of atmospheric physics and weather forecasting (in Chapter 28). Here, the dispersion aspects of the study will be addressed.

As mentioned previously, an immediate DCNet conclusion related to the inadequacy of data provided by conventional weather monitoring stations. In the DC case, local meteorological data are normally derived from instruments at the Reagan National Airport, some 5 km distant from the Capital and on the bank of the Potomac River. Wind roses shown in Figure 28.2 demonstrate that the airport wind data are certainly not always representative of the downtown environment. This is not surprising, since the airport data are intended to assist airport operations, for which application they serve well.

Based on measurements mostly made 10 m above the tops of selected buildings, the DCNet data indicate that from the perspective of σ_θ the downtown DC area is ergodic. That is, the variance of σ_θ with time at any single location is like the variance of σ_θ in space at a single time (Hicks et al. 2012). On the whole, however, wind speeds and friction velocities measured at any one of the DCNet sites in central Washington and its surrounding environs do not appear to differ greatly from measurements at other locations (see Figure 28.3).

In practice, streets and street canyons will provide a preferred route for the movement of any airborne hazard that might be present. DePaul and Sheih (1985; 1986) reported the results of tracer studies conducted in Chicago that verified the expectation (based on many wind-tunnel experiments) that the dimensions of street canyons are a controlling consideration. The street form-factor (≡ depth/width) is a key factor — a low value (< about 1) and winds above will quickly scavenge pollutants from street level; too high (> about unity) and pollutants will tend to remain at low levels within the canyon until mixing into the circulations induced by the wind above (see Britter and Hanna, 2003). A field survey studying the movement of sulfur hexafluoride (SF_6) in New York City in 2004 yielded illustrative data (Hicks, 2005). SF_6 is commonly used in studies of dispersion but is also used as an insulator in high-voltage urban power substations. Although efforts are made to minimize the loss of SF_6 from these installations, leakage always occurs. Obviously, old installations have greater leakage rates. The possibility arises to use leaks of SF_6 in studies 'of opportunity' of dispersion. Figure 51.1 shows results obtained in the New York street survey of 2004. Measurements were made using a portable analyzer, carried in a small van to survey the streets in the general vicinity of the source. Pockets of high concentration were frequently detected, at locations indicated by the red circles. The observations indicate the splitting of a continuous release from the substation as the resulting plume encountered intersections while being confined in the low levels of the streets.

A subsequent survey was conducted in Washington, DC, where there are substations near the White House (see Figure 51.2a) and around the National Capitol (Figure 51.2b). As expected on the basis of the New York City observations, street-level 'hot spots' were detected with greater frequency in the narrow streets of the east capital district of Washington (Figure 51.2b) than in the broader boulevards near the White House. It is likely that this difference was also influenced by differences in the leakage rates of SF_6, with some installations being older than others and hence (presumably) leaking more.

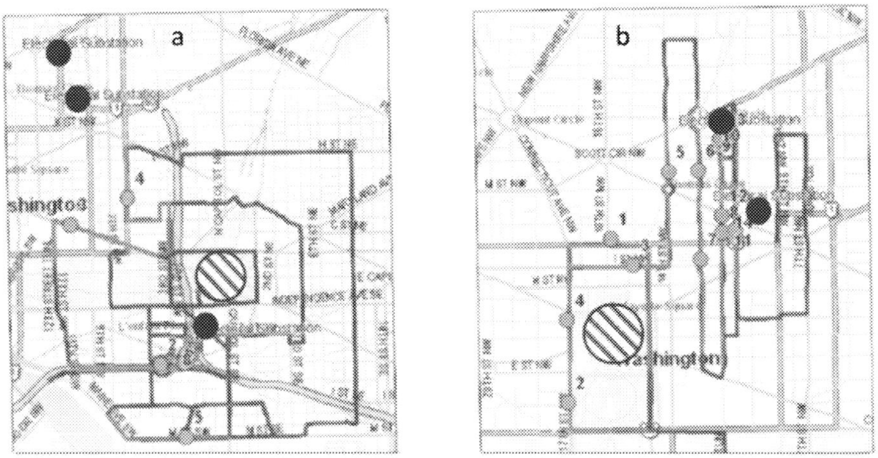

Figure 51.2. Results obtained in a 2004 survey of SF_6 concentrations in air at head height in the streets of Washington, DC, (a) near the White House and (b) in the area of the National Capitol. SF_6 emissions resulted from leakage from a power substation located as indicated by the large solid circles. Smaller numbered circles identify where SF_6 "hot spots" were encountered.

The density of SF_6 is a relevant consideration, in that the gas will tend to remain near the street surface unless (or until) disturbed by wind gusts or traffic. The New York and Washington SF_6 survey results indicate that the conventional model grid-cell focus is not well oriented towards the street-level concerns confronting first responders — protecting the public and accessing the source. Both of these require attention to the configuration and dimensions of streets, as was the case in the well-reported studies of Chicago street canyons conducted in the 1970s and early 1980s (DePaul and Sheih, 1985; 1986).

Conclusion

Every city has its own characteristics, to be considered individually.

Chapter 52

Dispersion Affecting the Far Field

"Experience is simply the name we give our mistakes" — Oscar Wilde.

The smoke-plume and tetroon experiments of the 1960s served well for studies out to a few tens of km. The search for a tracer suitable for use over greater distances initially focused on the use of radioactive emissions, from weapons tests or leaked from nuclear facilities. The familiar radioactive fallout provided a capability to explore dispersion at global scales: injections were generally at high altitudes and often stratospheric. In a first examination of long-range (>1000 km) dispersion from a low-altitude source, Draxler (1982) used measurements of ^{85}Kr, a noble gas released during the plutonium extraction process. (In more recent times, the emissions associated with reactor accidents have provided equivalent opportunities.) However, the need for a suitable and unique tracer was answered by Jim Lovelock (working with ARL at that time), who provided the capability to detect perfluorocarbons in trace quantities. After several successful tests of his electron-capture gas-chromatography method, the new tracers were employed in the 1983 Cross-Appalachian Tracer Experiment (CAPTEX; Ferber et al. 1986) in which ground-level releases were from sites in the vicinities of Dayton (Ohio) and Sudbury (Ontario), with the resulting dispersion monitored using a network of surface-air detectors stretching as far as the Atlantic coastline. The choice of these release sites was no doubt influenced by the acid rain concerns then dominating the atmospheric sciences.

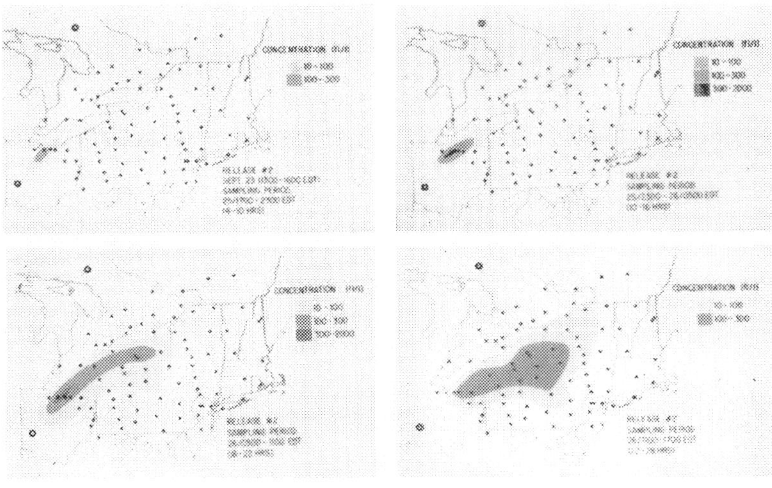

Figure 52.1. Fields of surface concentrations observed following a release of perfluorocarbons from a site in Dayton, Ohio, from 1305 to 1605 EDT on 25 September1983 (Ferber et al. 1986). The plots show the growth of the impacted area in six-hour intervals. Note that the times indicated in the diagrams are Eastern Daylight Time, lagging GMT by 4 hrs. After one day the footprint covered many of the Midwestern states.

Seven releases were made, five from Dayton, two from Sudbury. Releases were of 3-hr duration (except for one 30-min release from Dayton), occurring about once every week through September and October 1983. Release rates were about 60 kg hr^{-1}. Figure 52.1 shows examples of the growth of the cloud resulting from an Ohio release in September. The shape of the cloud is far from simple and imposes the need to consider its configuration as a component of any overall prediction scheme.

Several aspects of the sequence of plume dimensions indicated in Figure 52.1 are worthy of particular attention. First, the footprint becomes elongated as the released material becomes dispersed to greater heights and is influenced by the higher wind speeds there. Second, there is little resemblance of the surface-layer footprint to any coherent geometric shape as would likely be derived from models. Third, the surface data indicate the presence of small areas of high concentrations occurring in the datasets for the pre- and post-dawn periods following the release. Discussion earlier in the present document of the role of nocturnal intermittency is potentially relevant, in that some mechanism needs to be invoked to transport concentrations from altitudes well above the surface so that they could be detected by the ground-level sampler array. Classical fumigation doubtlessly contributed. Moreover, the area in question is notorious for the presence of low-level jets and the episodes of turbulence associated with them. There is no substantial evidence that would permit any association of these with the observed ground level concentration regimes.

Several aircraft were equipped with samplers and were used to explore the vertical cross-sections of the plume at various times and locations. Using the aircraft data, Figure 52.2 (based on Figure 21 of Ferber et al. 1986) shows that after 24 hrs of transportation the plume centroid was at a height of about 1.1 km agl. Obviously, the surface samplers did not reflect the complete picture of plume passage. However, at an age of one day and a distance of about 700 km, the elevated plume had a cross-sectional width of at least 200 km. This is similar to the conclusion based on surface concentration sampling. In considering the interpretation of the data obtained, there is an obvious need to determine which of two alternative views is most appropriate in the context of relevance — either the ground-level footprint which is what would be an appropriate measure when health risks and surface consequences are considered, or PBL plume descriptors which would be preferred if emphasis were on distant effects.

A longer-range tracer study using PFCs followed: the Across North America Tracer Experiment (ANATEX) of 1987 (Draxler and Heffter, 1989; Stunder and Draxler, 1989). ANATEX involved the released of tracer gas at two locations: St. Cloud, Minnesota, and Glasgow, Montana. Releases were every 2.5 days, alternating between 0500-0800 and 1700-2000 GMT. Data obtained in ANATEX have been accumulated and combined with data from CAPTEX in a database (DATEM) described by Ngan and Stein (2017). Results from these two studies have served as the base for many model assessments and subsequent developments, e.g., Haagenson et al. (1990) and Sykes et al. (1993).

The METREX, CAPTEX and ANATEX tracer studies shared a characteristic focus on multiple releases, so that the influence of changing meteorology was revealed. A similar experiment, also using PFC tracer gas, was conducted in Europe in 1994, designed to assist the development of dispersion models for emergency response application. This European Tracer Experiment (ETEX, Nodop, et al. 1998; van Dop et al. 1999) involved two tracer releases (c.f. CAPTEX 7, ANATEX 33). The ETEX surface footprint was measured with finer resolution than CAPTEX and ANATEX; the ETEX surface network consisted of 168 surface stations out to a distance of about 1500 km from a release point in Monterfil, France. The CAPTEX surface network was of 86 locations, out to 1100 km from the source. ANATEX used 77 surface sites distributed out to about 1500 km (but with several additional 'remote' stations). All of the studies made use of aircraft to investigate variations with height. ETEX was a multi-national effort designed to enable evaluations of different emergency response dispersion models as favored by the contributing countries. Comparisons among models immediately revealed the dominant need to get the wind direction right. A variety of pragmatic responses ensued, most intended to eliminate the effects of meteorological wind direction variability (meander) from consideration, so as to permit a comparison on the basis of the diffusion aspects alone. The conclusions reached from ETEX deliberations did not differ substantially from those arising from the METREX and ANATEX programs.

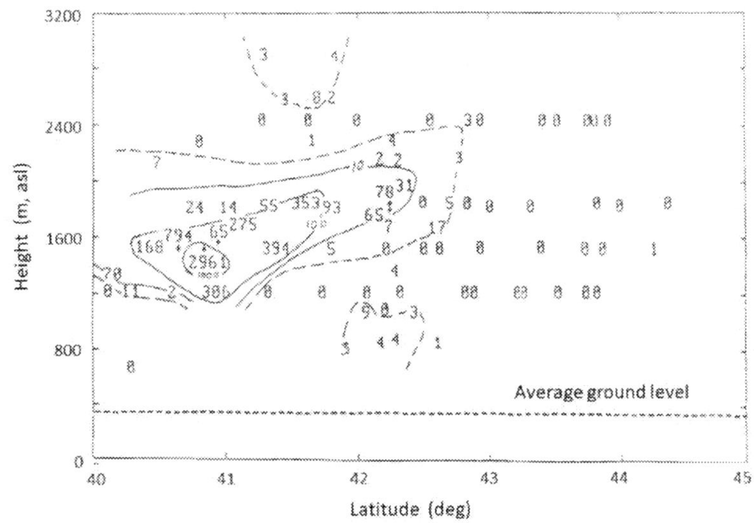

Figure 52.2. A latitudinal cross-section of the elevated ANATEX plume, one day after release at Dayton, Ohio, on 2 October 1983. The diagram is based on Figure 21 of Ferber et al. 1986, which presented an integrated product of measurements made by several aircraft operating within an hour of 2100 Z (1700 EDT). Concentration units are fl l^{-1}.

Using ETEX data, tests of a large number of models have been reported (e.g., D'amours, 1998; Stohl et al. 1998). As already mentioned, meteorological variability tended to be a major contributor (e.g., Straume et al. 1998; Sofiev et al. 2006), but even when this was eliminated from consideration (as in the case of many of the ETEX applications) it remained evident that predictions of concentrations in surface air were not well verified by data. In practice, a plot of predicted concentrations against observed will necessarily occupy only one quadrant of a Cartesian domain, no concentrations being less than zero. A statistical examination of the results so presented would then be at the mercy of whatever decision is made about how to take the many zero axis-values into account (i.e., those results for when predictions were of a finite concentration but no tracer was detected and vice versa). If such axis-values were eliminated from consideration, then the remaining 'fly-speck' depiction of the results would yield a correlation coefficient that might well seem encouraging whereas the reality is quite different.

The perfect tracer experiment does not yet exist. Moreover, the search for such a study will necessarily be unrewarding since the issues of topography (deployment of sensors at sea appears an unachievable requirement), synoptics and unresolved variance will dominate any consideration of a specific experimental venue or time. On the other hand, there are recurring instances of tracers of opportunity resulting from volcanic eruptions, for example. Each such event provides an opportunity to test models, to assess related uncertainties and to incorporate improvements — e.g., as following the Etna (Scollo, et al. 2007) and Nabro (Fairlie et al. 2014) volcanoes, the Bhopal industrial accident (Singh and Ghosh, 1987), the Chernobyl reactor failure (ApSimon et al. 1986; Pudykiewicz, 1989) and other incidents such as the Fukushima event (Mészáros et al. 2016). All such occasions, while being regrettable disasters, present opportunities for researchers and model developers. It is obvious, however, that a perfect dispersion model is unlikely to be developed. All that can be done is to estimate concentrations downwind and, more meaningfully, to quantify the probability that some perceived level of concern might be exceeded.

An opportunity for model testing arose after retreating Iraqi forces ignited oil wells in Kuwait in 1991. Figure 52.3 is a view of the plumes from the two major oil fields involved, one north of Kuwait City and the other to the south of it. Inspection of the photograph reveals a feature of interest from the perspective of turbulence periodicity and/or intermittency: the edges of the plumes indicate an irregularity (as circled in the diagram) at approximately 1 km intervals. An overview of the Kuwait oil fire experience is given by Hicks (1992) and Husain (1995). Descriptions of the plumes, the modeling of them and their health consequences are presented by (e.g.) Draxler et al. (1994), Husain (1994), Heller (2011).

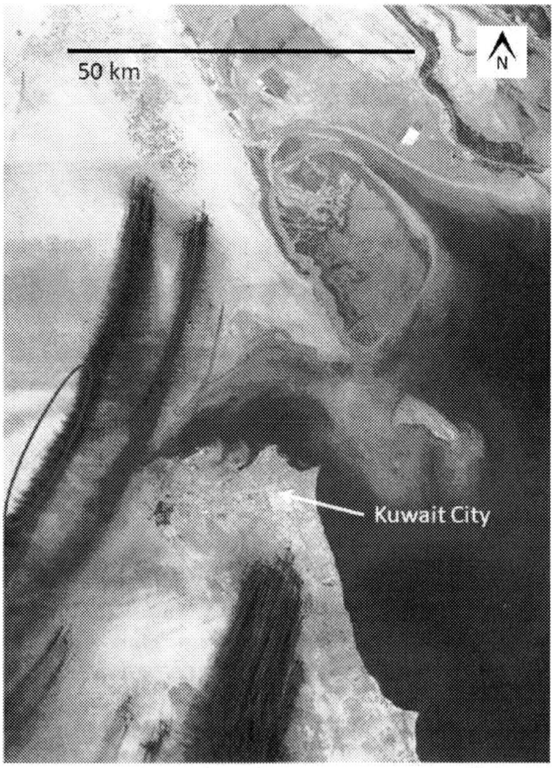

Figure 52.3. A sample image of the oil fire smoke plumes over Kuwait in 1991.

The probability distributions of the final predictions of concentrations can be revealing. If some specific location is indeed affected by the passage of a cloud or plume of pollutants, then the predicted concentrations cannot be negative and at the other extreme they could be large. The most likely statistical distribution is therefore log-normal, or close to it (e.g., perhaps a Gamma distribution with the upper limit being associated with factors such as the transit time). On the other hand, the probability of a specific location being affected by the migrating cloud of pollutants is also not necessarily well constrained, even though in the near field it is classically considered to be determined by a Gaussian spread. The problem is complicated and it is clear that there is need to associate considerable importance to the aleatoric uncertainty — that which is due to atmospheric variability that cannot be addressed using deterministic models regardless of their refinement.

There is no doubt that associating levels of uncertainty with predictions of any model presents difficulties, since most often it is the model itself that forms the basis for error assessment. Following the methodology outlined by Gumbel (1958), if a log-normal approximation to the appropriate distribution is assumed (as verified by many, e.g., Sykes et al. 1993, using ANATEX data), then the coefficient of variation $\check{V}$ will be about unity and the standard deviation of ln(x) will be close to $(1 + \check{V}^2)^{0.5}$. Using a standard table for a normally distributed variate (but allowing for the log transformation), it follows that with 95% confidence the 'true' value will fall within a range corresponding to a factor of about 9.6 above or below the predicted level. In this regard, recall the conventional admonition to dispersion modelers that their predictions should be interpreted as relating to truth within a factor of ten. Irwin and Hanna (2005) used Monte Carlo simulations of long records to show that the total standard deviation associated with surface concentration estimates (expressed relative to log-normal distribution) decreases as the number of replications (or the length of the time period addressed) increases, trending to a value of 1.46 whereas the back-of-the-envelope discussion above indicates an expected value of $(1 + \check{V}^2)^{0.5} = 1.41$.

Considerations such as are outlined above assume that the direction of transport is known. If this is erroneous, then the concentration probability distributions will be dominated by zeros and statistical examination becomes far more complicated. The estimations made above could be improved, however the level of uncertainty associated with the relative roles of the processes involved is such that the benefit of any such additional effort is (in the mind of the present author) probably unjustified unless doing so improves the career prospects of its advocates.

The matter of intermittency affects all of the issues considered here. It is useful to separate, in concept if not otherwise, the process of diffusion within an air parcel and the meander of the wind field carrying it. The Gaussian methodology addressed briefly above applies to a case in which diffusion dominates along a straight-line path, such that meander is not an issue. However, whenever the distance scale exceeds a few kilometers, meander and changes in the wind field due to synoptics and topography become an issue. The probability of some individual being exposed to a fatal dose of some hazardous substance could well be more determined by the wind field streamlines than by the diffusion and dilution modifying concentrations within the parcel of air being transported. There is an underlying challenge to those studying dispersion in its many forms: Does a proposed change in modeling represent an improvement upon what could be derived from assuming a cloud being moved by the wind with Gaussian diffusive spread?

The rate of such spreading remains somewhat contentious. The discussion above relates mainly to diffusion within the SBL. Extrapolation to the PBL cannot be by the simple expedient of extending SBL relationships beyond their range of applicability. In daytime, the depth of the convectively mixed boundary layer enters as an important consideration. Using measurements reported by Izumi and Caughey (1976), Lenschow et al. (1980) showed that the quantities σ_u, σ_v and σ_w are not well ordered with height within the PBL. In their study, the velocity statistics were normalized according to the PBL convective velocity scale, $w_* \equiv [(g/T)z_i\overline{w'T'}]^{1/3}$. Hicks (1985) considered the same dataset in conjunction with observations made using aircraft over the Coral Sea (reported by Warner, 1972), in this case also normalizing the velocity statistics by the surface turbulence velocity scale u_* and substantiating the lack of consistent order according to height within the PBL but above the SBL. Figures 52.4a and 52.4b show how the Coral Sea measurements are scattered through the PBL depth, just as was observed in the terrestrial case and reported by Lenschow et al. (1980). Also plotted is the expectation of Pantha and Clayton (1980) which decrees a consistent decrease of the turbulence velocity standard deviation ratios as height increases through the PBL, trending to zero at the height $z = z_i$. The difference is striking; consider that if the velocity standard deviations were zero at $z = z_i$ then the momentum flux through the layer of air at that altitude would also be zero and hence the air above would accelerate without constraint. Nevertheless, the assumption that the rate of spread of a diffusing cloud decreases as height increases through the PBL appears to have permeated through the dispersion modeling community.

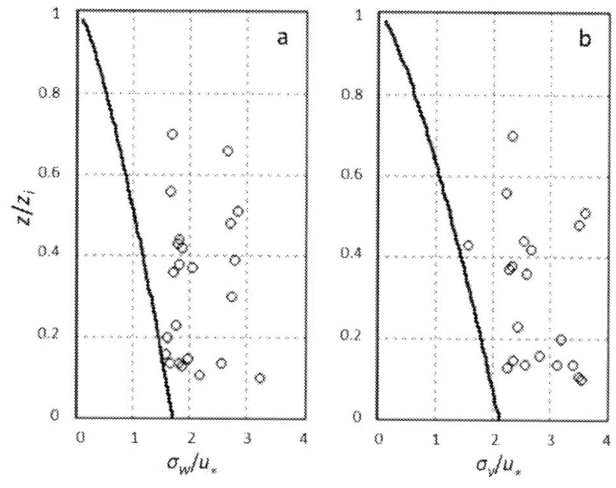

Figure 52.4. The variation of normalized velocity statistics with height within the PBL, derived from the tabulation by Hicks (1985) of observations reported by Warner (1970) over the Coral Sea.

Regardless of the way the velocity turbulence statistics are normalized (either by w_* or by u_*) there is no known observational evidence that the normalized ratios change with height within the PBL. To compare the contributions of the PBL-controlled process and the familiar SBL contribution to the overall velocity variance above the SBL, Hicks (1985) examined the terrestrial and oceanographic data as mentioned above and concluded that a best fit to the observations then available was

$$\sigma_w^2 = 1.2u_*^2 + 0.35w_*^2 \tag{52.1}$$

and

$$\sigma_v^2 = 3.6u_*^2 + 0.35w_*^2. \tag{52.2}$$

Trivial algebra reveals that Eqs, (52.1) and (52.2) are equivalent to

$$\sigma_w^2 = 1.2u_*^2\,(1 + 0.16(-z_i/L)^{2/3}) \tag{52.3}$$

and

$$\sigma_v^2 = 3.6u_*^2\,(1 + 0.16(-z_i/L)^{2/3}). \tag{52.4}$$

The matter of intermittency arises once again. Observations of fair-weather 'pop-corn' cumulus immediately reveal the consequences of convective cells sporadically reaching the top of the mixed layer (or the condensation level, perhaps). Severe convective storms exemplify the capacity of buoyant cells to combine as they ascend and to inject themselves into the troposphere well above the level z_i indicated by previous measurements. At the same time as this penetration occurs, the returning downdraft will bring with it the characteristics of the velocity regime aloft and if this regime carries wind shear then that shear will be transferred into the PBL. The statistics associated with the occurrence of tornadoes illustrate that the horizontal distance scales and the time scales involved in the buoyancy-induced mixing can be long — certainly longer than the time step of a model trying to predict the occurrence. The dilemma remains as a matter for further consideration.

In the meantime, the anecdotes of pilots and the archives of the aircraft community provide a gold mine of related information, limited at the severe-episode end of the spectrum by the lack of evidence from pilots who did not survive.

Conclusion

The distinction between predictable and unpredictable requires more attention.

Part VI: Looking Forward

Chapter 53

Flat-Earth Micrometeorology

"You don't need to see the whole staircase, just take the first step" — Martin Luther King Jr.

Legacy micrometeorology relies on the assumption that the goal is to describe key properties in terms of their departure from neutral. In practice, neutral is a rare phenomenon, often not found during the duration of a field experiment. On the other hand, the controlling property in daytime over land is certainly the sensible heat flux (actually the virtual heat flux, but let's avoid too many complications). At night, the dominant influence is what stratification can impose — resulting from the cooling of the surface and the influence of gravity as well as friction. After many decades of actual observation, it is proposed that it is better to concentrate on the stable and unstable periods of the SBL and let neutral serve as an ephemeral intermediary.

Several aspects of modern micrometeorology remain difficult to understand. For example, it is obvious that impressive curves can be drawn using relatively poor data if the MOST framework is used — i.e., by expressing dimensionless gradients as functions of stability formulated as in $\zeta = Z/L$ and then plotted logarithmically. It is suspected, in distant retrospect, that one of the early attractions of MOST was its ability to generate convincing diagrams from otherwise unconvincing observations. The rebellion of those of us with a background in statistics eventually resulted in awareness of the shared variable syndrome, which affects a large quantity of published literature.

Even in the earliest days of MOST it was evident that description of dimensionless profiles according to *Ri* in unstable conditions provided a more attractive depiction than dependence on $\zeta = Z/L$. One reason is obvious. The former (*Ri*) involves the combination of two variables ($\partial\bar{\theta}/\partial z$ and $\partial\bar{u}/\partial z$). The latter involves three such variables ($\overline{w'u'}$, $\overline{w'T'}$ and $(z - d)$). Each of these three quantities is also involved in the computation of ϕ_H. In this more familiar ζ-case, the variance introduced by considering measurements made at different heights is an important consideration: it is not the errors of measurement that are then important, but the dynamic range of the variable. Figure 18.1 is an example of how *Ri*-dependence looks better than ζ-dependence over land and in daytime.

In unstable conditions, the dominant driving force is buoyancy. Early free convection theory proved sufficient to explain the relationship between the sensible heat flux and the potential temperature gradient in the surface boundary layer for the stability range extending from $\zeta \approx -0.03$ (or thereabouts) to a distant negative value not yet well demonstrated. This is substantiated by results from many of the experiments considered in this document. There is no known set of observations that demonstrates that for mass and energy exchange the friction velocity is a controlling property in unstable conditions. As an example, the figures of Chapter 16 demonstrate that u_* is not a central property when instability is outside the near-neutral range. The friction velocity appears to be important because the dimensional requirements of MOST impose a dependence.

In unstable conditions, the empirical observation that $\phi_m = \phi_H^{1/2}$ is compatible with simplistic thinking — at any instability, ϕ_m will lie somewhere between ϕ_H and neutral, with the most likely situation being such that at any stability ϕ_m will average to about the geometric mean between these two extremes. QED. This is not a winning argument so much as an indication that we should not be surprised by the finding for unstable stratification that $\phi_m = \phi_H^{1/2}$ and hence that Ri and ζ are numerically the same. However, the observations now available indicate that beyond $\zeta \approx -1$ there is a continuing decrease of ϕ_H whereas ϕ_m appears to trend back towards neutral (q.v. Figure 15.2). There is room for some careful experimental micrometeorology using modern instrumentation and extending the studies of the past.

For the case of stable stratification, the driving forces are then the momentum of the air above, the drag at the surface and gravity, the last becoming a key consideration if the terrain is not horizontal. Over even the slightest of slopes katabatic flow can be generated, evolving as a night proceeds and reaching heights determined by the location as well as by the prevailing synoptics. In the examples considered here (especially in Chapter 27), the stratified flow regime undercuts the neutral logarithmic profile that would otherwise prevail. The familiar log-linear wind profile is then a consequence of elevating the neutral profile, perhaps intermittently. The time-averaged situation can be described mathematically but awaits more experimental testing of the viewpoints promoted here. Such evaluation will necessarily need to pay careful attention to the matter of local terrain, vegetation and coordinate alignment.

Both unstable and stable situations are affected by intermittency, primarily resulting from convective initiation in the former case and a combination of surface and synoptic factors in the latter. In this context, ‘neutral’ is often a period during which the stable and unstable atmospheric characteristics balance out, with intermittency contributing a level of uncertainty in all analyses that are then conducted. The classical requirement for neutral stability (high wind speeds under heavy cloud cover during the dusk or dawn transition periods) is often replaced by a post-event search through archived data for periods when the intermittent periods of instability are accompanied by compensating periods of stability. Chapters 19 and 20 emphasize the issues that arise during many morning and evening transitions, with intermittency often dominating the SBL turbulence regime regardless of what time-averaged ‘neutrality’ might predict. The terminology ‘transitional stability’ appears to be a more appropriate descriptor than ‘neutral’.

The classical view of turbulence as a continuing homogeneous spectrum of fluctuations (such as is required for some Fourier analyses) is violated by intermittency. During periods when intermittency is a strong feature, a single observing instrument will report an irregular passage of packets of turbulence — exemplified by turbulence ‘solitons’ as described by Eniichi Inoue (in the 1960s and 1970s) in consideration of ‘honamis’ affecting crops and by the commonplace observation of coherent buoyant cells migrating with the wind across arid terrain such as in the early Australian experiments. It is not yet clear whether the descriptions of convective initiation derived from laboratory studies can be extrapolated to the open atmosphere, but the observations of Chapters 19 and 20 look promising and dedicated field experimentation seems a good next step.

An assumption underlining conventional thinking is that the roughness length describing some particular surface is constant. However, the literature contains many examples in which z_0 changes with the wind speed (e.g., waving wheat) or wind direction (a vineyard). What was known about z_0 was originally the result of examinations of wind profiles without measurements of the momentum flux. Now that such measurements can be made, z_0 and the displacement height d, can be investigated in ways not previously possible (see Chapters 25, 26 and 27). The present contention is that by plotting observations of the wind speed normalized by friction velocity (and exponentiated) against the height of measurement an improved quantification of z_0 and d is obtained, such that statistical error margins can be associated with both.

Early on, it was reported (by Garratt and Hicks, 1973), that measurements then available indicated roughness lengths for sensible heat differing by about an order of magnitude from what was known to describe momentum exchange (i.e., $z_{0H} \approx z_0/10$). This large difference has always been difficult to relate to the physical processes involved. It is proposed here that the reason for the large difference lies in the assumption that temperatures (T_0) measured by downward-looking infrared thermometry represent the zero-plane for convective exchange. However, comparison of T_0 with soil temperature T_s and air temperature T_a, permits quantification of the equilibrium, or effective interfacial temperature T_e, defined as that which explains the apportionment of incoming energy between fluxes into the soil and into the air. In the heat of the day, initial data indicate that T_e is less than T_0 by about 5 °C. This is enough to account for the legacy result already discussed, that $z_0/z_{0H} \approx 10$. An alternative theoretical examination of the matter by Zilitinkevich et al. (2008) and Zilitinkevich (2009) appears to have been accepted by the modeling community.

Measurements of T_0 permit a refinement of the conventional BREB heat-energy apportionment methodology so that the combined influence of canopy heat storage and photosynthesis can be accounted for (ABRA: Chapters 32–35), although not on a run-to-run basis. In this regard it is changes in T_0 that are the relevant variable, not its absolute value.

At the time of the first publication of the familiar expression for the dimensionless temperature profiles (Dyer and Hicks, 1970), the authors were well aware that their curve ($\phi_H = (1 - 16\zeta)^{-1/2}$) was a close approximation to Priestley's free convection expectations when constrained to trend to unity as stability approached neutral. The equation was arranged to trend to unity at neutral because no mechanism was recognized for a near-neutral departure. Figure 15.2 herein shows support for this when modern instrumentation is used. However, Figure 17.1 shows that the distribution of data points supports the free convection approach better than the familiar expression for ϕ_H, with the point at which the free convection approach converges with neutral being about $\zeta = -0.03$. In practice, any such trend is hidden by the scatter resulting from the many considerations already mentioned — mainly intermittency as well as a variety of sampling and instrumentation inadequacies.

In the present document there is little mention of the integrated form of the dimensionless gradients ψ_H, ψ_m, etc. The reason is simple. The ψ-functions are intended to provide a method for comparing state variables ($\bar{u}$, $\bar{T}$, $\bar{q}$) between two levels within the SBL, not above the constant flux region or below the level at which the effects of individual surface roughness elements become apparent, i.e., within the height range approximated as $2h < z < d + |L|$. Extrapolating the ψ-functions beyond their region of applicability is an invitation to error. The most familiar expressions describing the dependence of ψ on stability represent an integration from height d to z_a, either by formal integration of the empirical ϕ-functions (Paulson, 1970) or by best-fitting raw data (Kanemasu et al. 1979). A recent examination of this issue results in the conclusion that disregarding the ψ-functions is as good an approach for the SBL as would be correcting for them (q.v. Hicks et al. 2018). An alternative methodology, replacing considerations involving z_0 and d and relying instead on the friction coefficient, C_f, and the

Stanton and Dalton coefficients, appears more in keeping with contemporary observations and attractively simple. In the case of the friction coefficient, Ocotillo data indicate that there is a strong and linear dependence of C_f on $\overline{w'T'}$ in unstable conditions, and no dependence in stable. This is illustrated in Figure 53.1, where diagrams a and c relate to the stable case, b and d to the unstable. All of the results have been computed using the 10 m observations. The lower pair of diagrams of Figure 53.1 show how $\overline{w'T'}$ influences the determination of ζ.

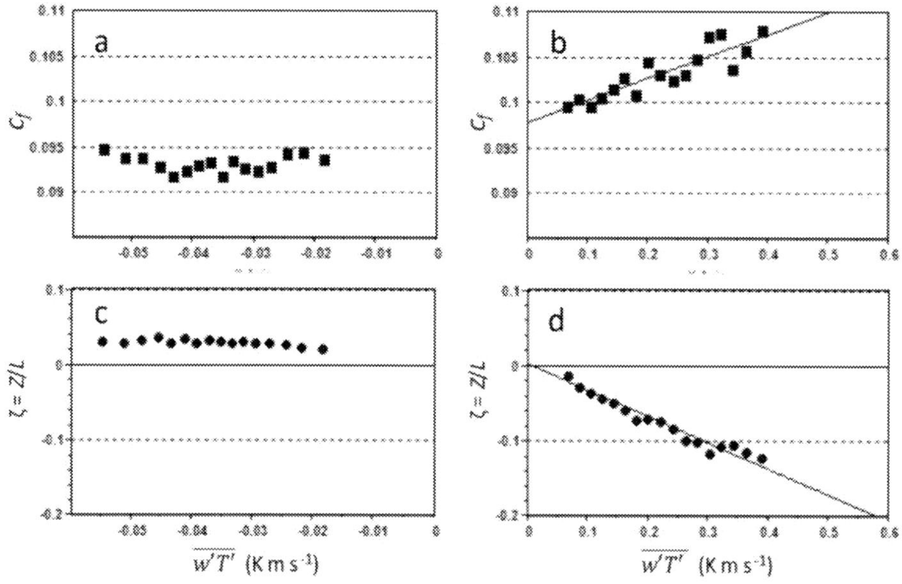

Figure 53.1. The dependence of Cf and ζ on $\overline{w'T'}$, as revealed by the Ocotillo (Texas) observations.

Questions about the energy balance at the surface have dominated a considerable portion of recent literature, mainly associated with studies of forests. The data available to the present author show no signs of any discrepancy, provided allowance is made for the roles of heat storage, advection and photosynthesis (see Chapter 36). It is suspected (but not yet tested) that inappropriate corrections applied to measured sensible and latent heat fluxes by coordinate rotation could be the cause for some of the reported imbalances, especially for areas of indicated discrepancy in sloping terrain.

Whereas covariances are strongly associated with gradients in the unstable case, the observations now reported indicate that this association cannot always be expected to extend into stable conditions. In particular, Figure 16.1 shows the expected strong relationship between fluxes and gradients in unstable conditions, but little correlation in stable. In this case, the observations were from a tower located in an area certainly influenced by local katabatic flow at night — on the plain of the Snake River in Idaho. Note, however, that these same observations appear to behave rather well when shown in the conventional form, as ϕ_m versus ζ in Figure 18.1a. But plotting against *Ri* (Figure 18.1b) reveals that all is not as well as might be concluded if only the ζ-dependence were considered. Other datasets reveal similar behavior, but the more complex terrain situation addressed in Chapter 27 indicates that the association between wind speed and the momentum flux remains strong within the regime that is elevated above the region of influence of local katabatic flow. It is tentatively concluded that the origin of the log-linear lore lies with the generation of local gravity flows and their interaction with the logarithmic wind profile that would otherwise prevail. A low-level gravity flow could well be difficult to detect. Experimental investigation would necessarily require measurements at low heights and detailed attention to fluxes and profiles both before and after coordinate rotation, using both flat and sloping sites.

Examination of observations from the transition periods around dawn and dusk reveal need to consider more than conventional micrometeorology suggests. In particular, changes with time (and presumably with space) then have considerable effect, amplified when there is a thick vegetation coverage and/or when terrain is not horizontal. Descriptions of air-surface exchange as a one-dimensional process may well prove satisfactory over accepted micrometeorological 'perfect' sites during the central daytime and nighttime hours but appears likely to have severe shortcomings otherwise.

Conclusion

Entrenched fallacies are hard to dislodge.

Chapter 54

Assembling the Pieces

"You don't know what you can get away with until you try." — Colin Powell

The dissatisfaction with MOST that initiated the preparation of this document grew from a recognition of the roles of the causative physical and biological processes. From the atmospheric/terrestrial perspective, the causative processes are the surface drag (i.e., the flux of momentum) and the convection occurring in daytime (mainly the sensible heat flux). By considering the interaction among different dimensionless quantities describing the gradients with which these fluxes are associated, we end up describing one symptom of the physical behavior in terms of another symptom. The inner physicist is unsatisfied. First, reduce the problem to its simplest considerations.

In daytime, the heat input from the sun is most often an obvious controlling factor. In the terrestrial case, the dominant nighttime feature is that the ground cools and the atmosphere in contact with it becomes stratified. It is during the intermediate and usually transitional cases near dusk and dawn that sensible heat exchange is minimal and if wind speed is high enough near-neutral conditions might be encountered. These considerations generate insecurity about the common acceptance of neutral stratification as a reference for turbulence characteristics in other stabilities. As already emphasized, a time average identified as neutral is usually a transitional state, made up of periods of stability and instability that are smoothed out when observations are expressed as time averages. Identification of a period that corresponds closely to neutrality is asking a lot of poorly-determined data, especially since no measure of either the sensible heat flux or the momentum flux at a single location can always be expected to be better that ± 10% (an approximation for 30–min averages).

Examination of currently available datasets indicates that the evening transition is not as problematic as the morning, because in the late afternoon the atmosphere slowly relaxes as radiation decreases and turbulence persists whereas there is a more sudden 'switch' at sunrise. As night approaches, convection decreases and eventually the mixed layer becomes stratified. Synoptics permitting, the lower atmosphere is often bounded by a jet generated either topographically or as a consequence of upstream meteorology. Near the surface, the velocity field responds to local topography and the wind profile above tends to be displaced, much as in an extension of the familiar consideration of a displacement height and zero plane. Eventually, the air starts to move in distinct stratified layers. Shallow sheets of air above the area of local influence slide across each other, moving largely independently. With the drag at the surface diminished, the air well above the surface accelerates in response to gravity and meso-scale pressure differences. As the resulting low-level jets grow, their imposed shear can cause busts of turbulence and/or waves that can have consequences that can be detected at the surface. For example, fast response traces of such indicators as CO_2 concentrations at heights of a few meters over land often show a slow increase of concentrations as emissions from the soil accumulate in the stratified lowest parts of the atmosphere, with sudden concentration decreases as sporadic packets of turbulence enable mixing with layers at greater heights.

In much of daytime, instability is such that free convection reigns. It does not matter if the air masses are moving; the common view that free convection initiates at some instability in the vicinity of $\zeta = -0.5$ arises largely as a consequence of injecting u_* into the relevant formulations. Field observations indicate that after the wind profile is normalized by u_* there is no remaining contribution of u_* to the unstable flux-gradient relationship when conditions are more unstable than about $\zeta = -0.03$. For the unstable case, an obvious next step is to consider the repercussions of substituting the free convective scaling velocity for u_* in the conventional deliberations. Re-writing Equation 12.1 shows that the temperature gradient is then given by

$$\partial\bar{\theta}/\partial z = (H^*)^{-2/3}(g/T)^{-1/3}(\overline{w'T'})^{2/3}z^{-4/3}. \quad (54.1)$$

Integrating this from the height of effective origin of the convection to a height of observation (z) we get that

$$\begin{aligned}\overline{\theta 0} - \overline{\theta z} &= 3\cdot(H^*)^{-2/3}(g/T)^{-1/3}(\overline{w'T'})^{2/3}(z_{0H}^{-1/3} - z^{-1/3}) \\ &= 2.52(g/T)^{-1/3}(\overline{w'T'})^{2/3}(z_{0H}^{-1/3} - z^{-1/3}). \quad (54.2)\end{aligned}$$

This is with full awareness that integration at the lower level assumes the applicability of the relationship all the way to the surface, which is not physically defensible but which nevertheless can be used (as frequently discussed earlier) to define an effective thermal "roughness length" z_{0H}, just as the familiar integration of the logarithmic wind profile leads to z_0. Equation 54.2 provides a ready quantification of the aerodynamic resistance associated with turbulent transfer of heat in unstable conditions dominated by free convection:

$$R_a(fc) = (\overline{\theta 0} - \overline{\theta z})/\overline{w'T'} = 2.52(g/T)^{-1/3}(\overline{w'T'}^{\,-1/3}(z_{0H}^{-1/3} - z^{-1/3}). \quad (54.3)$$

This expression could serve as an entry point for consideration of other fluxes, as has been considered in presentations elsewhere (e.g., Businger et al. 1971). However, any such application will necessarily invoke fears of the shared variable syndrome. For the present, further deliberation of the ramifications of Equations 54.2 and 54.3 will be left to others, with the caution that there is always the danger of partaking in 'mathematical manipulation for self–gratification' as was described, decades ago, by Bill Swinbank. A key step of unusual simplicity would be to describe an area of interest in terms of the average friction coefficient, which would then permit determination of the SBL wind profile by accepting that $\phi_m = \phi_H^{1/2}$. Merging with average temperature and velocity profiles above the SBL would then become a simple matter of adapting actual experimental observations to this new framework.

However, it is clear that the separation of mechanisms associated with a downward momentum flux from those imposed by an upward sensible heat flux (as in unstable stratification) invokes memories of the development of the O'KEYPS formulation. Sellers (1962) emphasizes the numerous assumptions underlying the development. In general, the present interpretation of the data now available indicates that such a separation is indeed warranted, but that in the extreme stabilities, both unstable and stable, the averaging period (15 min to 60 min) normally associated with micrometeorological research is populated with irregular strong events interspaced with periods of relative calm.

As I write this, I sit on my patio on a hot afternoon watching a nearby Bradford pear, in full plumage, semaphoring its experience with turbulence intermittency. The winds are classical light and variable. The surrounding area is forested. The tree responds well to the features of the turbulence that impacts it. The local gusts are not from any uniform direction and often those that affect me are of a completely different direction from those affecting the tree – about 20 m distant. Once again, the human eyeball serves as a discerning scientific instrument. Details of intermittency are obvious. The gusts have the expected characteristics of packets of turbulence moving across the surface, each providing greatly improved coupling between the air and the surface – honamis.

The role of intermittency presents a challenge for those who wish to describe flow fields and related properties (temperature and humidity, for example) in either extremely stable or extremely unstable conditions. The wind speed is then inevitably low and the occurrence of an intermittent event will largely determine the average fluxes during the time interval in which it occurs. However, the corresponding gradient will be comparatively unaffected. The basis for associating fluxes with gradients (as in the O'KEYPS derivation) is violated. Of course, the dimensionless wind gradient might itself be considered an inappropriate quantity, being fundamentally the wind gradient normalized by the momentum flux — a time average of a continuing flow divided by a consequence of episodic non-stationarity. The same characteristics are evident in the nighttime association of the temperature gradient with the sensible heat flux. In general, the data now considered suggest that the ability to associate fluxes with gradients in the stably stratified SBL can be quantified in terms of the prevailing net radiation, or perhaps better — the cloud cover. If the net radiation is more than about –30 W m^{-2} (admittedly site dependent), legacy flux-gradient relationships might well be used without great concern. In other nighttime conditions, all bets are off. Further generalization or formulation of these considerations requires more understanding and certainly more observations than are available to the present author. The attention of others is sought, with both hope and trepidation.

The situation following dusk appears to offer similar complexities. Intermittency is a common occurrence. The conclusion now drawn is that this nighttime intermittency is often (but certainly not always) of PBL origin rather than SBL, due to influences of upwind topography or of kinematic oscillations associated with the low-level jet. Intermittency is often evident following dawn; however the causative factor is then associated with the surface rather than the wind aloft. The onset of convective turbulence is mathematically predicted to involve a quasiperiodic occurrence of bursts of turbulence, as can be observed following sunrise. The deciding property appears to be the turbulence persisting through the night. If this level has dropped so that surface conditions at dawn are calm, then the onset of convection is observed to be intermittent in much the way that theory prescribes for flow impacting a flate-plate (Crutchfield and Huberman, 1980).

In extremely unstable conditions, when the heat flux is high and winds are light, time-averaged surface heat fluxes are made up of strong convective episodes, occurring sporadically. When representing a spatial average by a single-station measurement of $\overline{w'T'}$ consideration is the number of flux events that occur within a sampling period. There is no need to reconsider the way in which legacy analyses address this situation — standard depictions of cospectra agree with the view that as instability increases so does the low-frequency component of the cospectrum. Early in the field studies of fluxes and gradients, in Australia, when analysis made use of real-time analog computers, the contributing events were actually counted and the statistical certainty associated with the average was then computed and used in considering the analysis to follow. In modern times, the details addressable using digital methods contribute much to the physical understanding, but also tend to obscure the fundamental processes involved.

Muschinski et al. (2001) emphasize that conventional turbulence theory assumes a monochromatic characteristic of the field of turbulence being studied, whereas some of the processes now known to contribute to air-surface exchange are far from monochromatic. It is for this reason that there is no discussion of spectra or cospectra here. The early observations of intermittency during the field programs in Australia led to the opinion that applying an analytical technique like a fast Fourier transform to a time series lacking the requisite time uniformity was an exercise best left for others. Rather than think of standard spectral characteristics, attention was directed to aspects such as event amplitude and pulse repetition frequency, with the episodes visualized as packets of turbulence migrating with the wind. This view was supported by studies of honamis over rice paddies and other crops (Inoue, 1955a,b; Tani, 1960; Inoue, 1963) and turbulence footprints over barley crops, wheat fields and other flexible plant canopies (Finnigan and Mulhearn, 1977; Finnigan, 1985), from which reassurance can easily be drawn even though the causative factors are probably diverse.

In the SBL, surface drag will be intermittently increased as coupling between the surface and the air is enhanced by mixing, either due to convection (in daytime) or to topography and/or PBL kinematic and gravity-induced processes (mainly at night). The observations now considered indicate that the time-averaged momentum flux in strongly unstable conditions will be determined by the strength of the turbulent gusts that occur within the period and the frequency of them. The evidence presented in Figure 17.2 is relevant. Figure 17.2 (a) shows that for unstable conditions the logarithmic form of the wind profile is a good fit to observations provided $\overline{w'T'}$ does not exceed about 0.2 K m s^{-1}. The data also indicate that the measures of surface roughness do not change markedly over this near-neutral unstable region.

When a situation of interest is not as micrometeorologically desired, deployment of towers and the orientation of their instruments becomes an issue requiring careful consideration. Analysis would probably require attention to measured velocity components before any coordinate rotation is applied. It is the present contention that conventional coordinate rotation provides a method to adjust measurements of the momentum covariance so that the flux is normal to the plane of the streamlines. When the sensible heat flux is enough for buoyancy to be a dominant factor, mass and energy fluxes would seem better quantified using coordinates aligned according to gravity. The concept of a single height-independent coordinate system for all covariances near the surface seems a stretch. In practice, a strong wind speed will likely override all other considerations and hence the determination of how to deploy instruments or to interpret near-surface observations must remain an area for contemplation. Well above complex terrain, the relevant coordinate system should by determined by gravity. The open questions are how to define 'complex,' how to determine the height of relevance and how to account for effects of mountains of sufficient height that all considerations of this kind become irrelevant.

Quantification of spatial averages of z_0 and d has proved to be a demanding challenge for field experimentalists. The issues that commonly arise are mitigated by the use of the friction coefficient ($C_f \equiv u_*/\bar{u}$) referred to some standard height above the surface or above the vegetation. The checker-board surface problem has been addressed experimentally (e.g., Hicks and Wesely, 1977; Tsvang et al. 1991). The observations considered here indicate that when a model grid cell consists of large areas of different crops yielding separate friction coefficients C_{f1} and C_{f2}, the most likely areal average results from the product $\bar{u} \cdot u_*$ being about the same over the two surfaces. Hence, $\bar{u}_1/\bar{u}_2 \approx (C_{f2}/C_{f1})^{1/2} \approx u_{*2}/u_{*1}$. The extension of this thinking to more complex circumstances seems worth attention. The drop in wind speeds as hurricanes move from over water to over land seems amenable to such considerations.

The case of inland lakes is intriguing, since it must be expected that the water body will change its temperature slowly, so that the air will be warmer than the water in daytime and cooler at night. Therefore, it is expected that air above small inland lake will display a near-surface stability cycle opposite in phase from the surrounding land. A large lake will, of course, generate its own boundary layer, developing as described in many models that often lack the verification provided by extensive observations.

In the case of dispersion and deposition of some trace quantity carried by the wind, the conventional approach is to accept the results repeatedly yielded by pipe-flow studies that indicate the need to consider molecular-scale (or Brownian) diffusion with an exponent value of $-2/3$ whereas a -1 exponent would be more in place with classical flate-plate theory. The difference aligns with expectations based on the geometric constraints of pipe-flow. It is the present contention (outlined in Chapter 40) that the early suggestion (by Wesely and Hicks, 1977) leading to the wide adoption of the $-2/3$ exponentiation methodology was an unjustified extension to the open atmosphere of well-verified results obtained in constrained pipe-flow. Relevant data derived from open-air field studies are required.

The eastern North American forest coverage elevates the need for improved understanding of how such a complicated surface interacts with the air above it. In the event of some hazardous pollutant being introduced into the sub-canopy air mass, little guidance exists about the way in which this substance will enter the velocity field above. Two aspects of this exchange process appear important — the delay imposed by the transfer and the extent of dilution. Clearly, the matter depends on the density and transparency of a canopy (Lalic and Mihailovic, 2004). Once again, open-air experimentation is needed.

Conclusion

Tread carefully, but tread anyway.

Chapter 55

Extension to Modeling

> "With four parameters I can fit an elephant, and with five I can make him wiggle his trunk"
> — John von Neumann

The iconoclastic conclusions presented herein, based on actual observations, could easily provide an alternative foundation for simulating the surface boundary layer of the atmosphere. The steps involved are probably best considered by others; the present author has no capability to either construct a demonstrative model or test it. The models of relevance are usually Eulerian, operating on a Cartesian grid imposed by the proclivities of developers as much as by practical requirements. There are alternatives — non-Cartesian and data assimilative approaches have advantages (Bacon et al. 2000), but these models fly in the face of Cartesian dominance and intellectual inertia. Lagrangian models offer great advantages when local characteristics need to be emphasized, or when dispersion from known sources is being considered (Wilson and Sawford, 1996). In the present document, discussion of the relative merits of different modeling approaches has been intentionally avoided. The literature is full of examinations of alternatives.

It is quite possible that modifying the way the surface boundary layer is depicted in meso-scale meteorological models will have little effect on predictions of changes in the standard meteorological properties, as affect people directly. However, it is certain that matters such as air pollution, atmospheric deposition, dispersion and climate prediction will be more affected. In the climate case, the coupling of the atmosphere with the oceans is central. The present text addresses air-sea exchange only in passing, but it is clear that the sea surface differs from most terrestrial situations — the sea surface in merely an intermediary in the transfer of momentum from the air to an eventual sink at some poorly defined depth.

The is an obvious preference among modelers to adapt micrometeorological detail to their requirement for relevant depiction of the surface. As a result, the descriptors familiar to air-surface exchange investigators have been adapted for use on many numerical simulations, despite their complexity. This complexity (for example, specification of the surface in terms of roughness length, displacement height and leaf area index) is primarily developed to describe the characteristics of a specific surface under consideration, such as a crop or a forest. Expanding this kind of depiction into the landscape gid cells of numerical models then requires consideration of each grid cell in terms of the fundamental micrometeorological quantities, a challenging task given the biodiversity of typical areas. A simpler methodology is advocated, based on the use of the gross surface characteristics. Table 1 lists values of the friction coefficient ($C_f \equiv u_*/\bar{u}$), demonstrating the ability to describe the surface roughness in terms of a slowly moving quantity while avoiding the complexity introduced by focusing on z_0 and d. There are many tables of this kind. The values now listed are results obtained in the author's various field studies.

Table 1. Effective friction coefficients for different homogeneous surfaces

		h (m)	*d* (m)	z_0 (m)	$C_f(10)$
Open ocean –	5 m s^{-1}	0	0	0.00005	0.033
	10 m s^{-1}	0	0	0.0002	0.037
	15 m s^{-1}	0	0	0.0005	0.042
Sand		0	0	0.001	0.050
Mowed grass		0.01	0.01	0.01	0.070
Pasture		0.1	0.08	0.03	0.10
Soybeans		0.8	0.6	0.1	0.13
Maize (mature)		3	2.4	0.5	0.17
Shrubs (5 m)		15	10	1	0.20
Forest (25 m)		30	20	1.5	0.22

A reference height of 10 m above the level of the zero plane is assumed. Following familiar terminology, *h* is the average height of the canopy – the top of the vegetation (or other structures), *d* is the expected value of the displacement height and z_0 is the roughness length.

Given that C_f already normalizes the drag at the surface by the wind speed, the remaining determining quantity is the sensible heat flux (or better, the virtual heat flux). A straightforward way to address this dependence is therefore to consider how C_f varies with $\overline{w'T'}$, as in Figure 53.1. The plots (a) and (b) show that C_f is approximately constant through the night but varies consistently with $\overline{w'T'}$ in daytime. The variations of ζ with $\overline{w'T'}$ in diagrams (c) and (d) are in accord with these findings. Familiar micrometeorological depictions result mainly from the inclusion of u_* in the formulation of ζ — apparently an unnecessary complexity. The plotted averages were from the Ocotillo experiment, as already considered, using the 10–m $\bar{u}$ values to ensure compatibility with Table 1.

There is a tendency to view surface roughness as a characteristic and unchanging feature of a terrestrial surface. Clearly, there are large changes as the seasons progress, in some areas due to farming practices but also in areas where forests are deciduous. Figure 55.1 shows results from an intensive campaign to study the changes in an eastern Tennessee deciduous forest over a period of several years. The seasonal change in biomass (as indicated by changes in leaf area index) is striking. Without substantial supporting information, the observations of Figure 55.1 indicate that any assumptions about *d* must be questioned. The apportionment between *H* and L_wE will obviously change with season.

While eastern North America is largely forested, much of central North America is farmland, supporting crops of many varieties (but mainly maize, wheat and soybeans) that grow quickly and inject an obvious seasonal variability in any depiction of areal air surface exchange. Such seasonal cycles doubtlessly affect not only the air-surface exchange properties but also the magnitude of the fluxes of main concern in consideration of climate change — water vapor and carbon dioxide. A model that disregards such seasonality must be considered askance.

There are several modeling systems (as already discussed) intended to describe the characteristics of air-surface exchange in a general manner, so that site-specific details can be ingested into the calculations (e.g., the Noah system, Ek et al. 2003; SURFEX, Le Moigne et al. 2012). A review of a sampling of these models suggests that their assemblies of process descriptions are often unbalanced — some processes are addressed in more detail than others. These simulations are essentially assemblies of the detailed formulations derived from micrometeorological experiments. The consequence is that these schemes tend to require attention to specific details that are not available routinely or are not relevant to situations in which the surface is not homogeneous. It is a matter of widespread faith that the displacement height *d* is the same for all quantities being transferred. Clearly, the matter depends on the density and transparency of a canopy (Lalic and Mihailovic, 2004). In any event, it is usual to consider the relevant flux-gradient relationships at heights well above the surface, so that any differences in *d* become inconsequential. A simpler approach like the reliance on C_f suggested above would appear to offer substantial advantages.

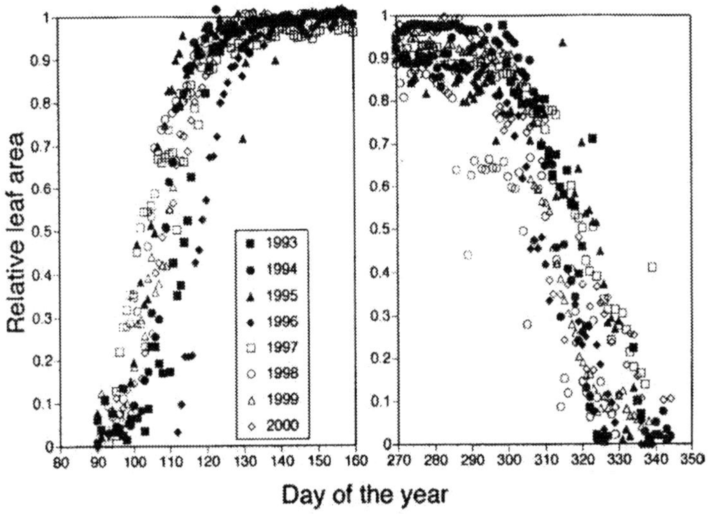

Figure 55.1. The variation of leaf area with time of year, as measured at the Walker Branch watershed in Oak Ridge, Tennessee. The forest grew during the seven years of this illustration. The data plotted are normalized to the maximum leaf area index measured during the growing season in question. The forest is mixed deciduous (q.v. Hutchison et al. 1986).

Such air-surface exchange schemes (e.g., as considered by Noilhan and Planton, 1989; Chen et al. 1996; Mitchell et al. 2005 and Chaney et al. 2016) build upon familiar MOST descriptions of the SBL to provide one-dimensional descriptions of what is basically a multi-dimensional problem. Whether the simplification is adequate remains to be tested. So far, tests appear to have been limited to situations like those that yielded the information used to construct the models (see the listing of 'validators' by Mitchell, 2005). A review of recent publications regarding such land-surface models reveals that (a) the formulations are based on the assumption that u_* is always a contributing variable, (b) relationships that clearly violate the shared variable constraint are commonly used, (c) leaf area indices are applied as divisor corrections imposed without consideration of canopy density effects and (d) various ratios of roughness lengths for heat and momentum are permitted (see the discussion above). Some of what is assembled into such grid-cell schemes is based on the outputs of other models.

Reduced to its basics, a numerical simulation is a detailed representation of what some investigator or team of investigators considers the optimal and sufficient suite of relevant processes. The historic focus on parsimony appears to have suffered from ever-increasing computer power. If the intent is to conduct an assessment of some situation over large time and space scales, then including details of processes contributing to variance at smaller scales will mainly satisfy those with unused computer capability. At the opposite extreme, if the goal is to predict some particular situation then detailed attention to smaller scales would seem appropriate. A first order conclusion arises immediately — models proven to be satisfactory for assessment purposes cannot automatically be accepted as relevant for addressing specific events. There is indeed a major distinction between models specifically designed to forecast weather and those intended to assess changes in climate.

It is the present author's opinion that we should step past the concepts of 'validation" and 'verification,' and move instead towards demonstrations that a model's outputs are adequate simulations of reality and appropriate for the intended application. Many successful and informative tests of models have focused on modeling dispersion in the SBL, where terrain complexity and lack of stationarity present obstacles to the clear-cut endorsement of any simulation. In the ASCOT program (Orgill and Schreck, 1985) atmospheric tracers were used to examine a selection of dispersion models. Luhar and Rao (1994) conclude that available complex terrain models can predict downwind concentrations in general terms, using hourly-averaged data. Rao and Hosker (1993) discuss procedures for comparing model performances, concluding that a spread of likely results should be envisioned, with the models providing an indication of the most likely expectations. In considerations of weather forecasting, 'skill scores' have provided a basis for model comparisons (Murphy and Epstein, 1989). Given that the models involved invariably rely on familiar SBL MOST expressions, there seems little point in considering this matter further.

The periods near sunrise and sunset are often avoided because of their complexity. The confounding issue is often little more than time stationarity. To assume that the PBL can be described as a time series of one-dimensional approximations challenges the imagination and ignores the results obtained in even the first studies of direct relevance (e.g., Lettau, 1951). The various data analyses presented in this document reveal that the time rate-of-change of surface temperature is sometimes the most highly correlated of all contributing variables affecting the sensible heat flux.

Multiple regression (and artificial intelligence) analysis of archived data has been a favored basis for estimating tomorrow's air quality. An alternative methodology for predicting adverse air quality is to extend the simulations used for weather forecasting to the air quality application. This necessitates a measure of point source emission rates of the pollutants of interest, deposition rates of these pollutants during atmospheric transport and allowance for chemical reactions affecting their concentrations in the air. The difficulties confronting weather forecasting are compounded by these new requirements and air-surface exchange is a more important concern. There are many large-scale air quality prediction efforts in recent development – such as the Australian Air Quality Forecasting System (AAQFS; Cope et al. 2004), Eta-CMAQ in the United States (Otte et al. 2005; Eder et al. 2006), 'Caliope' for the Iberian Peninsula (Baldasano et al. 2008) and 'PREV'AIR' in Europe (Rouil et al. 2009). In many cases, the focus is on prediction of surface concentrations of ozone, this being an irritant and a known environmental hazard.

Conclusion

No matter what a modeler might propose, there will be at least one experimentalist who will complain.

Chapter 56

The Need for New Studies

> "No amount of experimentation can ever prove me right; a single experiment can prove me wrong."— Albert Einstein.

Back in 1959, in his monograph, Priestley loudly advocated that it was well time that the research community stopped looking at near-perfect sites in increasing detail, but instead should start addressing issues of the imperfect world that surrounds us. His comment is as appropriate today as it was then. We now have instrumentation that is clearly capable of refreshing the legacy relationships developed fifty years ago, but moreover we are now profiting from observational capabilities that produce information in previously unimaginable quantity and detail. A major problem confronting today's science is the complexity of analysis when memory banks are swamped with relevant information. There is an obvious need for utilization of artificial intelligence (AI) methods, just as there was for digital computers when John von Neumann explored the complexity of weather forecasting (q.v. Charney et al. 1950).

Investing scarce resources on studying the causes of small departures from expectations is likely to continue. But a better investment would be on studying the situations where expectations are found to be at odds with observations. Hence, experimental focus would be most profitably focused on situations where predictions are known to be flawed. In the past, there was a strong desire to ask modelers to design experiments. The result was a spate of studies that contributed little more than support for what the models predicted. The lack of historic studies of circumstances where the models did not work continues to contribute to a void in common understanding. It is to the credit of the research community that modern experimental programs appear to have stepped past this balking point. There have been many recent experimental programs conducted specifically to extend understanding into situations where understanding was previously deficient. There are three specific aspects of these same considerations that lead to a modicum of guidance for future research programs.

1. Surface heterogeneity must be defined and quantified in such a way that its consequences can be included in simulations. In this regard, conventional 'complex terrain' constitutes an extreme example.
2. The role of atmospheric behaviors external to the conventional surface boundary layer must be studied in more detail and in more situations. This is clearly a requirement if high quality simulations of the nighttime SBL are desired.
3. The way in which the atmosphere is coupled with the surface must be adjusted so that it is no longer viewed as a one-dimensional problem. In particular, the consequences of changes with time must be incorporated into related formulations. The role of water must be acknowledged.

The experimental capabilities now available have attracted many development programs. Principal among these are the many efforts to optimize the use of satellite imagery, both passive (spectral radiometry) and active (radar). It is in this context that modern AI capabilities appear to offer special promise.

Surface-based remote probing systems are now common. Weather radar is a standard provider of information to guide local atmospheric models. Other remote-probing systems (SODAR, LIDAR) offer opportunities that seem best suited for intensive research applications, provided methods to extract desired information from their outputs can be arranged. Airborne research platforms have proved to be an asset in many experimental campaigns, but while providing a good spatial coverage over limited time periods they usually fail to yield details on time variability over specific terrain. Unmanned aircraft (drones) have been advocated as an answer to many pragmatic problems that arise, but experimentation using drones appears to have been dominated by a community infatuation with the drones themselves.

Studies of specific crops or other surface features require particular attention to the surface, rather than to the atmosphere as affected by it. In this context, the contemporary dominance of eddy covariance technology should not be allowed to relegate gradient methods of flux measurement to obscurity. The methods relying on interpretation of gradients and the surface heat balance offer unique advantages over standard eddy covariance (EC) methods: The Bowen Ratio Energy Balance (BREB) approach offers a capability to derive flux data from measurements made much closer to the surface than would be required for EC methods. The data obtained by BREB methodology are therefore less susceptible to issues of fetch inadequacy than are EC techniques. As has been shown in the present documentation, infrared surface temperature measurements can be used to help prove this point. Moreover, the availability of surface temperature data enables use of an Augmented Bowen Ratio Analysis (ABRA) introduced here, to correct for the consequences of heat storage in the energy balance relationship. Once a specific site is 'calibrated' for heat storage and the consequences of time-dependent photosynthesis, the results can then be used to derive carbon dioxide (or other) fluxes from concentration data measured alongside the necessary temperature and humidity sensors.

It is expected that the most informative field studies will be conducted in places and situations where nobody else has yet thought to go.

Conclusion

There is much need for new experiments conducted 'outside the box.' The central question is 'Who defines the box.'?

Part VII: ENVOI

"So many books, so little time." — Frank Zappa

In the introductory comments, it was emphasized that this is not a textbook. Nor is it finished. It represents the thinking of the author at the time of its writing and there are no revisions anticipated. As more data become available and especially as more analyses of them are permitted, new investigators will be able to test the opinions and conclusions presented here. It is not expected that there will be a loud cry for revision of the conventions to which I habitually object. For example, flux/gradient associations are well locked into the practice of our discipline, and it is not expected that new data will be able to displace this familiar dogma quickly. It has not been the intent to disparage any model or system of models. However, the discussions presented here indicate that many simulations suffer from an overenthusiastic acceptance of legacy relationships and tend to extrapolate them with unwarranted confidence. And if some of my colleagues decry that I have misrepresented their results, let me emphasize that I am quite capable of screwing up.

Textbooks have become our bibles; their teachings have replaced reality and have hindered additional investigation. It is well past time to step forward, (1) to accept that the atmosphere near the surface is chaotic enough to make precision forecasting an impossible dream, (2) to allow for the enhancement of mixing in the PBL due to turbulent intermittency, (3) to profit by using the advanced sensing systems now available to investigate the nature of the atmosphere and its coupling with the surface (whether terrestrial, marine or vegetative) so as to understand reality, however inconvenient, and (4) to recognize that life (not only that of mankind) is a factor affecting the future of this planet.

In their examination of the Gaia hypothesis Lovelock and Margulis (1974) pointed out that this planet's atmosphere carries the signature of life. The Earth is a living biosphere whose lifeforms have adjusted to their circumstances and, wherever possible, have adapted these circumstances to suit their own purposes. Mankind is surely no more than the most recent and perhaps most virulent of the planet's lifeforms bent on modifying the planet itself. In the end, this planet might look like Mars. As matters related to climate change become increasingly urgent, it is clear that describing changes in the atmosphere must involve consideration of terrain, the biosphere and the hydrosphere. The models must address the exchanges between the media if they are expected to represent any one of them with confidence. To this end, it seems that the days of Cartesian gridded numerical modeling may well be numbered. To my mind, future simulations would benefit from adopting the adjustable-grid methods already in place in other disciplines. But to do this would involve a departure from the accepted tried-and -true methodologies. As Santayana said, "Habit is stronger than logic."

There continues to be evidence that those who develop simulations suffer from inadequate field experience. I recommend spending a considerable part of individual research agendas on simply watching the environment that needs to be addressed. Human eyes are great scientific tools.

I will keep watching until the lights go out.

Appendix 1: Recurring Symbols

α	Charnock constant
β	Bowen ratio (H/L_wE)
γ	Psychrometric constant (c_p/L_w)
ε	Production rate of TKE
ζ	Stability (Z/L)
η	Enthalpy
θ	Potential temperature
θ_a	Air potential temperature
θ_v	Virtual potential temperature
θ_z	Potential temperature at height z
ρ	Air density
σ	Standard deviation
τ	Momentum flux
ϕ_H	Dimensionless temperature gradient
ϕ_m	Dimensionless wind gradient
ϕ_w	Dimensionless humidity gradient
ψ_H	Integrated departure from neutral of ϕ_H
ψ_m	Integrated departure from neutral of ϕ_m
ψ_w	Integrated departure from neutral of ϕ_w
A,a	Constants
B,b	Constants
C_d	Drag coefficient $(u_*/\overline{u})^2$
C_f	Friction coefficient $(u_*/\overline{u})$
c_p	Specific heat of air at constant pressure
C_z	Concentration at height z
CoV	Coefficient of Variation $(\sigma_x/\overline{x})$
d	Displacement height
e_a	Water vapor pressure in air
e_z	Water vapor pressure at height z
E	Flux of water vapor $(\rho\overline{w'q'})$
f_x,F_x	Fluxes of quantity x
g	Acceleration due to gravity
G	Heat flux into the ground
h	Height of a canopy
H	Sensible heat flux
h_m	Depth of the PBL mixed layer
k_l	Conductivity $(R_a + R_s)^{-1}$

k_2	Atmospheric conductivity (R_a^{-1})
L	Obukhov scale length
L_w	Latent heat of vaporization of water
LAI	Leaf Area Index
R_1	Total correlation
R_{12}	Correlation coefficient between X_1 and X_2.
$R_{12.3}$	Correlation X_1 with X_2 holding X_3 constant
R_a	Aerodynamic resistance
R_b	Quasi-laminar boundary layer resistance
R_c	Canopy uptake resistance
R_n	Net radiation
T_0	Surface temperature (infrared)
T_a	Air temperature
T_e	Effective interfacial temperature
T_s	Soil temperature
T_1, T_2	Temperatures at heights z_1, z_2
u	Horizontal along-wind velocity component
v	Horizontal cross-wind velocity component
w	Vertical wind component
X	Variable
x	Along-wind Cartesian coordinate
y	Cross-wind Cartesian coordinate
z	Height; Vertical Cartesian coordinate

Overbars indicate averages
Wavy overbars identify limited averaging
Primes indicate departures from averages.

Appendix 2: Acronyms

AAD	Aerosol Aerodynamic Diameter
ABRA	Augmented Bowen Ratio Analysis
AI	Artificial Intelligence
ANATEX	Across North America Tracer Experiment
ANL	Argonne National Laboratory
ARL	Air Resources Laboratory (of NOAA)
ATDD	Atmospheric Turbulence and Diffusion Division (of ARL)
BREB	Bowen Ratio Energy Balance
BUBBLE	Basel UrBan Boundary Layer Experiment
BW	Biological Warfare
CAPMoN	Canadian Air Pollution Monitoring Network
CAPTEX	Cross Appalachian Tracer Experiment
CASES–99	Cooperative Atmosphere–Surface Exchange Study—1999
CASTNet	Clean Air Status and Trends Network
CAT	Central African Time
CBLAST	Coupled Boundary Layers and Air-Sea Transfer study
CDABRA	Carbon Dioxide Augmented Bowen Ratio Analysis
CHESS	Chestnut Ridge Environmental Site
CST	Central Standard Time
DATEM	Data Archive of Tracer Experiments and Meteorology
DCNet	Washington DC research network
DDIM	Dry Deposition Inferential Model
EPA	Environmental Protection Agency (US)
EPRI	Electric Power Research Institute
EST	Eastern Standard Time (US)
ETEX	European Trasport Experiment
FLIP	Floating Instrument Platform
INL	Idaho National Laboratory
KISS	Kinematic Isolation in Stable Stratification
LT	Local Time
METREX	Metropolitan Tracer Experiment
MLN	Multi-Layer Model
NADP	National Atmospheric Deposition Program
NAPAP	National Acid Precipitation Assessment Program
NOAA	National Oceanic and Atmospheric Administration (US)
ORNL	Oak Ridge National Laboratory
PBL	Planetary Boundary Layer

RSL	Roughness Sub-Layer
SBL	Surface Boundary Layer
SIESTA	SF_6 International Experiments in Stagnant Air
SRNL	Savannah River National Laboratory
SURFEX	The Surface Externalis´ee system
TKE	Turbulent Kinetic Energy
UTC	Coordinated Universal Time

Appendix 3: Details of Sites at Which Observations Presented Here Were Obtained

		Latitude	Longitude	Elevation	Surface
Alabama	Belle Mina	34.6898 N	86.8809 W	180 m	Pasture (cattle)
Daly Waters	(Koorin #1) (Koorin #2)	16.2936 S 16.3006 S	133.3797 E 133.2247 E	220 m 245 m	Sparse forest Sparse forest
DCNet sites	AGU	38.9147 N	77.0452 W	28 m	Height above ground
	DOC	38.8937 N	77.0325 W	40 m	"
	DOE	38.8866 N	77.0251 W	40 m	"
	EMA	38.8540 N	76.9954 W	20 m	"
	HU	38.9217 N	77.0209 W	25 m	"
	NEA	38.9059 N	77.0362 W	35 m	"
	NRL	38.8212 N	77.0248 W	30 m	"
	NSF	38.8930 N	77.0478 W	25 m	"
	RFK	38.8895 N	76.9728 W	45 m	"
	SSG	38.9924 N	77.0303 W	60 m	"
	TSQ	40.7603 N	73.9837 W	125 m	"
	EML	40.7260 N	74.0076 W	36 m	"
Edithvale, Victoria		38.0222 S	145.1471 E	2 m	Pasture (cattle)
Hay (NSW) #1		34.6555 S	144.7949 E	90 m	Arid grassland (sheep)
#2		34.4943 S	144.9312 E	95 m	Arid grassland (sheep)
Idaho INL (tower site)		42.5833 N	112.9402 W	1530 m	Arid sagebrush
Illinois, Sangamon CY		39.5800 N	89.6900 W	220 m	Various maize/soybean
Dresden		41.3675 N	88.2819 W	160 m	Reactor cooling pond
Kerang, Victoria		35.8223 S	144.0306 E	80 m	Pasture (sheep)
Mt. Gambia, SA		37.8067 S	140.6682 E	65 m	Pine plantation
Ocotillo, Texas		32.1205 N	101.3755 W	640 m	Sparse sagebrush
Ohio (Mansfield)		40.6060 N	82.6740 W	426 m	Maize
Pinnacles		38.6177 N	78.3468 W	1017 m	Mixed deciduous
Sth. Carolina, Aiken, SRNL		33.2091 N	81.7410 W	45 m	Coniferous forest
Tennessee, Loudon		35.6958 N	84.3875 W	270 m	Maize
CHESS site		35.9311 N	84.3323 W	390 m	Deciduous forest
Walker Branch		37.9606 N	84.2910 W	360 m	Deciduous forest
Zimbabwe, Harare		17.7220 S	31.0209 E	1460 m	Maize, other crops

References

Abu–Hamdeh, N. H., 2003. Thermal properties of soils as affected by density and water content. *Biosystems Eng.* 86:97–102.

Allen, R. G. L. S. Pereira, D. Raes and M. Smith, 1998. *Crop evapotranspiration – Guidelines for computing crop water requirements,* 2nd ed. United Nations Food and Agriculture Organization, Rome.

Allen, R. G., W. O. Pruitt, J. A. Businger, L. J. Fritschen, M. E. Jensen and F. H. Quinn, 1996. Chapter 4 "Evaporation and Transpiration" in *ASCE Handbook of Hydrology.* New York, NY. pp. 125–252.

Allwine, K. J., J. H. Shinn, G. E. Streit, K. L. Clawson and M. Brown, 2002. Overview of URBAN 2000: A multiscale field study of dispersion through an urban environment. *Bull. Amer. Meteorol. Soc.* 83:521–536.

Alves, I., A. Perrier and L. S. Pereira, 1998. Aerodynamic and surface resistances of complete cover crops: How good is the "Big Leaf"? *Trans. ASAE* 98:345–351.

Andreas, E. L. and B. B. Hicks, 2002. Comments on "Critical Test of the Validity of Monin–Obukhov Similarity during Convective Conditions." *J. Atmos. Sci.* 59:2605–2710.

Angell, J. K., 1964a. Use of tetroons to investigate the kinematics and dynamics of the planetary boundary layer. *Mon. Wea. Rev.* 92:465–470.

Angell, J. K. 1964b. Measurements of lagrangian and eulerian properties of turbulence at a height of 2,500 ft. Quart. *J. Roy. Meteorol. Soc.* 90:57–71.

Angell, J. K. 1972. A comparison between circulations in transverse and longitudinal planes in an unstable planetary boundary layer. *J. Atmos. Sci.* 29:1252–1261.

Angell, J. K. D. H. Pack and C. R. Dickson, 1968. A Lagrangian study of helical circulations in the Planetary Boundary Layer. *J. Atmos. Sci.* 25:707–717.

Anguelova, M. D. and F. Webster, 2006. Whitecap coverage from satellite measurements: A first step toward modeling the variability of oceanic whitecaps. *J. Geophys. Res.* 111:C03017, doi:10.1029/2005 JC003158 2006.

Anspaugh, L. R., J. H. Shinn and D. W. Wilson, 1974. Lawrence Livermore Laboratory Report IAEA–SM–184/13 UCRL–75250, 17 pp.

ApSimon, H. and J. Wilson, 1986. Tracking the cloud from Chernobyl. New Scientist (London), 111:42-45.

Arnfield, J., 2003. Two decades of urban climate research: a review of turbulence, exchanges of energy and water, and the urban heat island. *Int. J. Climatol.* 23:1–26.

Asman, W. H., M. A. Sutton and J. K. Schjorring, 1998. Ammonia; emission, atmospheric transport and deposition. *New Phytol.* 139:27–48.

Auble, D. L. and T. P. Meyers, 1992. An open path, fast response infrared absorption gas analyzer for H_2O and CO_2. *Boundary-Layer Meteorol.* 59:243–256.

Baas P., G. J. Steeneveld, N. J. H. van de Wiel, A. A. M. Holtslag, 2006. Exploring self–correlation in flux–gradient relationships for stably stratified conditions. *J. Atmos. Sci.* 63:3045–3054.

Bacon, D. P., N. N. Ahmad, Z. Boybeyi, T. J. Dunn, M. S. Hall, P. C. S. Lee, R. A. Sarma and M. D. Turner, 2000. A dynamically adapting weather and sispersion model: the Operational Mulitscale Environment Model with Grid Adaptivity (OMEGA). *Monthly Weath. Rev.* 128:2044–2076.

Baldocchi, D. D., 1997. Flux footprints within and over forest canopies. *Boundary-Layer Meteorol.* 85:273–292.

Baldocchi, D. D., E. Falge, L. Gu, R. Olson, D. Hollinger, S. Running, P. Anthoni, Ch. Bernhofer, K. Davis, R. Evans, J. Fuentes, A. Goldstein, G. Katul, B. Law, X. Lee, Y. Malhi, T. Meyers, W. Munger, W. Oechel, K. T. Paw U, K. Pilegaard, H. P. Schmid, R. Valentini, S. Verma, T. Vesala, K. Wilson and S.

Wofsy, 2001. FLUXNET: A new tool to study the temporal and spatial variability of ecosystem–scale carbon dioxide, water vapor, and energy flux densities. *Bull. Amer. Meteorol. Soc.* 82:2415–2434.
Ball, K., 1956. The theory of strong katabatic winds. *Aust. J. Phys.* 9:373–396.
Banta, R. M., 2008. Stable–Boundary-Layer regimes from the perspective of the low–level–jet. *Acta Geophys.* 56:58–87.
Barad, M. L. (ed.), 1958. Project Prairie Grass: a field program in diffusion. Air Force Cambridge Research Center, *Geophysical Papers No. 59,* 299 pp.
Barr, S. and M. E. Orgill, 1989. Influence of external meteorology on nocturnal valley drainage winds. *J. Appl. Meteorol and Climatol.* 28:497–517.
Belcher, S. E., N. Jerram and J. C. R. Hunt, 2003. Adjustment of a turbulent boundary layer to a canopy of roughness elements. *J. Fluid Mech.* 488:369–398.
Beljaars, A. C. M. and A. A. M. Holtslag, 1991. Flux parameterization over land surfaces for atmospheric models. *J. Appl. Meteorol.* 30:327–341.
Belot, Y. and D. Gauthier, 1975. Transport of micronic particles from atmosphere to foliar surfaces. Pp 583–591 in *'Heat and Mass Transfer in the Biosphere'* (D. A. de Vries and N. H. Afgan eds.), Scripta, Wahington.
Best, A. C., 1935. Transfer of heat and momentum in the lowest layers of the atmosphere. *Geophys. Memoirs Meteorol.* Office, #65, 66 pp.
Black, P. G., E. A. D'Asaro, W. M. Drennan, J. R. French, P. P. Niiler, T. B. Sanford, E. J. Terrill, E. J. Walsh and J. A. Zhang, 2007. Air–sea exchange in hurricanes: Synthesis of observations from the coupled boundary layer air–sea transfer experiment. *Bull. Amer. Meteorol. Soc.* 88:357–374.
Blackadar, A. K., 1957. Boundary-Layer wind maxima and their significance for the growth of nocturnal inversions. *Bull. Amer. Meteorol. Soc.* 38, 283–290.
Blanchard, D. C., 1971. Whitecaps at sea. *J. Atmos. Sci.* 28:645.
Boatman, J. F., D. L. Wellman, C. C. Van Valin, R. L. Gunter, J. D. Ray, H. Sievering, Y. Kim, S. W. Wilkison and M. Luria, 1989. Airborne sampling of selected trace chemicals above the central United States. *J. Geophys. Res. Atmos.* 94:5081–5093.
Bodenschatz, E., J. R. de Bruyn, G. Ahlers and D. S. Cannell, 1991. Transitions between patterns in thermal convection. *Phys. Rev. Letters,* 67:3078–3081.
Bogen, D. C., S. J. Nagourney and C. Torquato, 1980. A field evaluation of the HASL wet–dry deposition collector. *Water Air Soil Pollut.* 13:453–458.
Bonner, W. D., 1968. Climatology of the low–level jet. *Mon. Weather Rev.* 96:83–85.
Bosenquet, C. H. and J. L. Pearson, 1936. The spread of smoke and gases from chimneys. *Trans. Faraday Soc.* 32:1249–1263.
Bowen, I. S., 1926. The ratio of heat losses by conduction and by evaporation from any water surface. *Phys. Rev.* 27:779–787.
Britter, R. E. and S. R. Hanna, 2003. Flow and dispersion in urban areas. *Annual Rev. Fluid Mechanics* 35:469–496.
Broecker, W. S. and T. H. Peng, 1974. Gas exchange rates between the air and the sea. *Tellus,* 26:21–35.
Brook, R. R., 1985. The Koorin nocturnal low–level jet. *Boundary-Layer Meteorol.* 32:133–154.
Brooks, I. M., M. J. Yelland, R. C. Upstill-Goddard, P. D. Nightingale, S. Archer, D. d'Asaro, R. Beale, C. Beatty, B. Blomquist, A. A. Bloom, B. J. Brooks, J. Cluderay, D. Coles, J. Dacey, M. DeGrandpre, J. Dixon, W. M. Drennan, J. Gabrielle, L. Goldson, N. Hardman-Mountford, M. K. Hill, M. Horn, P-C. Hsueh, B. Huebert, G. de Leeuw, T. G. Leighton, M. Liddicoat, J. J. N. Lingard, C. McNeil, J. B. McQuaid, B. I. Moat, G. Moore, C. Neill, S. J. Norris, S. O'Doherty, R. W. Pascal, J. Prytherch, M. Rebozo, E. Sahlee, M. Salter, U. Schuster, I. Skejelvan, H. Slagter, M. H. Smith, P. D. Smith, M. Srokosz, P. K. Taylor, M. Telszewski, R. Walsh, B. Ward, D. K. Woolf, D. Young and H. Zemmelink, 2009. Physical exchanges at the air-sea interface. *Bull. Amer. Meteorol. Soc.* 90:629–644.
Brower, C. and M. Heibloem, 1986. *Irrigation water management: Irrigation Water Needs.* U.N. Food and Agriculture Organization, Rome.
Brown, E. N., C. A. Friehe and D. H. Lenschow, 1983. The use of pressure fluctuations on the nose of an aircraft for measuring air motion. *J. Climate Appl. Meteorol.* 22:171–180.

Brutsaert W., 1975. A theory for local evaporation (or heat transfer) from rough and smooth surfaces at ground level. *Water Resour. Res.* 11:543–550.

Brutsaert W., 1982. *Evaporation into the atmosphere*. D. Reidel (New York) 299 pp.

Buck, A. L., 1976. The variable–path Lyman–alpha hygrometer and its operating characteristics. *Bull. Amer. Meteorol. Soc.* 57:1113–1118.

Buckley, R. L., C. H. Hunter, R. P. Addis and M. J. Parker, 2007. Modeling dispersion from toxic gas released after a rain collision in Graniteville, SC. *J. Air Waste Manage. Assoc.* 57:268–278.

Bunker, A. F., 1955. Turbulence and shearing stresses measured over the North Atlantic Ocean by an airplane–acceleration technique. *J. Meteorol.* 12:445–455.

Burba, G. G. and S. B. Verma, 2005. Seasonal and interannual variability in evapotranspiration of native tallgrass prairie and cultivates wheat ecosystems. *Agric. Forest Meteorol.* 135:190–201.

Businger, J. A., 1969. Note on the critical Richardson number. *Quart. J. Roy. Metetorol. Soc.* 95:653–654.

Businger, J. A., 1973. Turbulent transfer in the atmospheric surface layer. Pp. 67–100 in *"Workshop on Micrometeorology"* (D. A. Haugen, ed.), Amer. Meteorol. Soc. Boston.

Businger, J. A., 1988. A note on the Businger–Dyer profiles, *Boundary-Layer Meteorol.* 42:145–151.

Businger, J. A. and V. E. Suomi, 1958. Variance spectra of the vertical wind component derived from observations with the sonic anemometer at O'Neill, Nebraska in 1953. *Arch. Met. Geoph. Biokl. A.* 10:415–425.

Businger, J. A., J. C. Wyngaard, Y. Izumi and E. F. Bradley, 1971. Flux profile relationships in the atmospheric surface layer. *J. Atmos. Sci.* 28:181–189.

Calder, K. L., 1952. Some recent British work on the problem of diffusion in the lower atmosphere, pp 787–792 in *"Proc. U.S. Tech. Conf. Air Poll.,"* McGraw–Hill, New York.

Casati, B. L., J. Wilson, D. B. Stephenson, P. Nurmi, A. Ghelli, M. Pocernich, U. Dumrath, E. E. Ebert, B. G. Brown and S. Mason, 2008. Forecast verification: current status and future directions. *Meteorol. Appl.* 15:3–18.

Chamberlain, A. C., 1966. Transfer of gases to and from grass and grass–like surfaces. *Proc. Roy. Soc. A* 290:236–265.

Chamberlain, A. C., 1967. Transport of lycopodium spores and other small particles to rough surfaces. *Trans. Roy. Soc.* London, A, 296:45–70.

Chamberlain, A. C., 1968. Transfer of gases to and from surfaces with bluff and wave–like roughness elements. *Quart. J. Roy. Meteorol. Soc.* 94:318–332.

Chamberlain, A. C., 1983. Roughness length of sea, sand and snow. *Boundary-Layer Meteorol.* 25:405–409.

Chaney, N. W., J. D. Herman, M. B. Ek and E. F. Wood, 2016. Deriving global parameter estimates for the Noah land surface model using FLUXNET and machine learning. *J. Geophys. Res. Atmos.* 121:13–218.

Charlson, G. S. and R. L. Sani, 1971. On thermoconvective instability in a bounded cylindrical fluid layer. *Int. J. Heat Mass Transfer* 14:2157–2160.

Charnock, H., 1955. Wind stress on a water surface. *Quart. J. Roy. Meteorol. Soc.* 81:639–640.

Chaudhuri, P., W.-D. Lo, W.-Y. Loh and C.-C. Yang, 1995. Generalized regression trees. *Statistica Sinica* 5:641–666.

Chen, F., K. Mitchell, J. Schaake, Y. Xue, H. L. Pan, V. Koren, Q. Y. Duan, M. Ek and A. Betts, 1996. Modeling of land surface evaporation by four schemes and comparison with FIFE observations. *J. Geophys. Res: Atmos.* 101:7251–7268.

Chen, J., S. Hagos, H. Ciao, J. D. Fast and Z. Feng, 2020. Characterization of surface heterogeneity–induced convection using cluster analysis. *J. Geophys. Res: Atmospheres* 125: doi.org/10.102/2020jD032550.

Chen, Y. K. Yang, D. Dzou, J. Qin and X. Guo, 2010. Improving the Noah land surface model in arid regions with a appropriate parameterization of the thermal roughness length. *J. Hydrometeorol.* 11:995–1006.

Cheng, Y. and W. Brutsaert, 2005. Flux–profile relationships for wind speed and temperature in the stable atmospheric boundary layer, *Boundary-Layer Meteorol.* 114, 519 – 538.

Cheng, Y., M. B. Parlange and W. Brutsaert, 2005. Pathology of Monin–Obukhov similarity in the stable layer. *J. Geophys. Res.* 110:D06101.

Chilton, T. H. and A. P. Colburn, 1934. Mass transfer (absorption) coefficients prediction from data on heat transfer and fluid friction. *Ind. Eng. Chem.* 11:1183–1187.

Choi, H–D., T. J. Sharac and T. M. Holsen, 2008. Mercury deposition in the Adirondacks: a comparison between precipitation and throughfall. *Atmos. Environ.* 42:1818–1827.
Cionco, R. M., 1972. A wind–profile index for canopy flow. *Boundary-Layer Meteorol.* 3:255–263.
Clarke, J. F., E. S. Edgerton and B. E. Martin, 1997. Dry deposition calculations or the clean air status and trends network. *Atmos. Environ.* 31:3667–3678.
Clarke, R. H. and R. R. Brook, 1979. *The Koorin Expedition.* Australian Departrment of Science, Bureau of Meteorology. Australian Government Printing Service, Canberra, 359 pp.
Clawson, K. L., R. G. Carter, B. R. Reese, R. C. Johnson, N. F. Hukari and D. J. Lacroix, 2001. GAUNTLET SF_6 atmospheric tracer release and field test support. *NOAA Tech. Memorandum OAR ARL–249,* 45pp.
Clawson, K. L., R. M. Eckman, N. F. Hukari, J. D. Rich and N. R. Ricks, 2007. Climatography of the Idaho National Laboratory, 3rd edition. *NOAA Tech. Memo. OAR ARL–259,* Air Resources Laboratory, Idaho Falls, ID. 322 pp.
Clements, W. E., J. A. Archuleta and P. H. Gudiksen, 1989. Experimental design of the 1984 ASCOT field study. *J. Appl. Meteorol and Climatol.* 28:405–413.
Clough, W. S., 1973. Transport of particles to surfaces. J. *Aerosol Sci.* 4:227–234.
Collet, P., J.–P. Eckmann and H. Koch, 1981. Period doubling bifurcations for families of maps on R^n. *J. Stat. Phys.* 1:1–14.
Cope, M. E., G. D. Hess, S. Lee, K. Tory, M. Azzi, J. Carras, W. Lilley, P. C. Manins, P. Nelson, L. Ng, K. Puri, N. Wong, S. Walsh and M. Young, 2004. The Australian air quality forecsting system. Part I: Project description and early outcomes. *J. Appl. Meteorol.* 43:649–662.
Corrsin, S., 1949. *Extended applications of the hot–wire anemometer.* National Advisory Committee for Aeronautics Technical Note No. 1864, Washington, DC, 62 pp.
Coulter, R. L. and J. C. Doran, 2002. Spatial and temporal occurrences of intermittent turbulence during CASES–99. *Boundary-Layer Meteorol.* 105:329–349.
Cowan, I. R., 1965. Transport of water in the soil–plant–atmosphere system. *J. Appl. Ecology* 2:221–239.
Cramer, H. E. and F. A. Record, 1953. The variation with height of the vertical flux of heat and momentum. *J. Meteorol.* 10:219–226.
Crawford, T. L. and R. J. Dobosy, 1992. A sensitive fast–response probe to measure turbulence and heat flux from any airplane. *Boundary-Layer Meteorol.* 59:2257–278.
Crawford, T. L., R. T. McMillen and R. J. Dobosy, 1990: Development of a "generic" mobile flux platform with demonstration on a small airplane. *NOAA Tech. Memo. ERL ARL-184,* 81 pp.
Crutchfield, J. P. and B. A. Huberman, 1980. Fluctuations and the onset of chaos. *Phys. Letters* 77A:407–410.
Cunliffe, M. and J. C. Murrell, 2009. The sea–surface microlayer is a gelatinous biofilm. *ISME J.* 3:1001–1003.
Currier, E. L. J. B. Knox and T. V. Crawford, 1974. Cooling pond steam fog. *J. Air Poll. Control Assoc.* 24:860–864.
Dai, Y., R. E. Dickinson and Y. Wang, 2004. A two–big–leaf model for canopy temperature, photosynthesis, and stomatal conductance. *J. Climate* 17:2281–2299.
D'amours, R., 1998. Modeling the ETEX plume dispersion with the Canadian emergency response model. *Atmos. Environ.* 32:4335–4341.
Davidson, B., 1963. Some turbulence and wind variability observations in he lee of a mountain ridge. *J. Appl. Meteorol.* 2:463–472.
Davidson, B. and M. L. Barad, 1956. Some comments on the Deacon wind profile. *Eos Trans. Amer. Geophys. U.* 37:168–176.
Davidson, C. I. and S. K. Friedlander, 1978. A filtration model for aerosol dry deposition: Application to trace metal deposition from the atmosphere. *J. Geophys. Res.* 83:2343–2352.
Deacon, E. L., 1949. Vertical diffusion in the lowest layers of the atmosphere. *Quart. J. Roy. Meteorol. Soc.* 75:89–103.
Deacon, E. L., 1950. The over–estimation error of cup anemometers in fluctuating winds. *J. Sci. Instruments* 28:231–233.
Deacon E. L., 1953. Vertical profiles of the mean wind in the surface layers of the atmosphere. *Geophysical Memoirs,* No. 91, Air Ministry Meteorological Office, London.

Deacon, E. L., 1959. The measurement of turbulent transfer in the lower atmosphere. pp. 211–228 in *"Atmospheric Diffusion and Air Pollution"* (F. N. Frenkeil and P. A. Sheppard, eds.), Academic Press, London, 475 pp.

Deacon, E. L., 1977. Gas transfer to and across an air–water interface. *Tellus* 29:363–374.

Deacon, E. L., 1979. The role of coral mucus in reducing the wind drag over coral reefs. *Boundary-Layer Meteorol.* 17:517–521.

Deacon, E. L. and D. R. Samuel, 1957. A liner, temperature compensated hot wire anemometer. *J. Sci. Instr.* 34:24–25.

DePaul, F. T. and C. M. Sheih, 1985. A tracer study of dispersion in an urban street canyon. *Atmos. Environ.* 19:555–559.

DePaul, F. T. and C. M. Sheih. 1986. Measurements of wind velocities in a street canyon. *Atmos. Environ.* 20:455–459.

De Sena, J. O. A., H. A. Zaidan and P. R. de Camargo e Castro, 2007. Transpiration and stomatal resistance variations of perennial tropical crops under soil water availability conditions and water deficit. *Braz. Arch. Biol. Technol.* 50:225–230.

Dickson, C. R. and J. K. Angell, 1968. Eddy velocities in the planetary boundary layer as obtained from tetroon flights at Idaho Falls. *J. Appl. Meteorol and Climatol.* 7:986–993.

Dilley, A. C., 1968. On the computer evaluation of vapour pressure and specific humidity gradients from psychrometric data. *J. Appl. Meteorol.* 7:717–719.

Draxler, R. R., 1982. Measuring and modeling the transport and dispersion of Krypton-85 1500 km from a point source. *Atmos. Environ.* 16:2763–2776.

Draxler, R. R., 1985. Metropolitan tracer experiment (METREX). *NOAA Tech. Memo ERL ARL–140.* 24 pp.

Draxler, R. R., 1987. The accuracy of tajectories during ANATEX calculated using dynamic model analyses versus rawinsonde observations. *J. Appl. Meteorol.* 30:1446–1467.

Draxler, R. R. and J. L. Heffter (eds.), 1989. Across North America Tracer Experiment (ANATEX). *NOAA Tech. Memo ERL ARL–167.* 24 pp.

Draxler, R. R., J. T. McQueen and B. J. Stunder, 1994. An evaluation of air pollutant exposures due to the 1991 Kuwait oil fires using a Lagrangian model. *Atmos. Environ.* 28:2197–2210.

Dryden, H. L. and A. M. Kuethe, 1929. The measurement of fluctuations of air speed by the hot–wire anemometer. *National Bureau of Standards Report No. 320*, Washington, DC, 32 pp.

Duce, R. A. and P. S. Liss, 2002. The surface ocean-lower atmosphere study (SOLAS). *Atmos. Environ.* 36:5119—5120.

Dugas, W. A., L. J. Fritschen, L. W. Gay, A. A. Held, A. D. Matthias, D. C. Reicosky, P. Steduto and J. L. Steiner, 1991. Bowen ratio, eddy correlation, and portable chamber measurements of sensible and latent heat flux over irrigated spring wheat. *Agric. For. Meteor.* 56:1–20.

Dyer, A. J., 1961. Measurements of evaporation and heat transfer in the lower atmosphere by an automatic eddy–correlation technique. *Quart. J. Roy. Meteorol. Soc.* 87:401–412.

Dyer, A. J., 1965. The flux–gradient relation for turbulent heat transfer in the lower atmosphere. *Quart. J. Roy. Meteorol. Soc.* 91:151–157.

Dyer, A. J., 1967. The turbulent transport of heat and water vapour in an unstable atmosphere. *Quart. J. Roy. Meteorol. Soc.* 93:501–508.

Dyer, A. J., 1974. A review of flux–profile relationships. *Boundary-Layer Meteorol.* 7:363–372.

Dyer, A. J., 1981. Flow Distortion by Supporting Structures. *Boundary-Layer Meteorol.* 20:243–252.

Dyer, A. J. and E. F. Bradley, 1982. An alternative analysis of flux–gradient relationships ate thye 1976 ITCE. *Boundary-Layer Meteorol.* 22:3–19.

Dyer, A. J. and B. B. Hicks, 1970. Flux–gradient relationships in the constant flux layer. *Quart. J. Roy. Meteorol. Soc.* 96:715–721.

Dyer, A. J. and F. J. Maher, 1965. Automatic eddy–flux measurement with the Evapotron. *J. Appl. Meteorol.* 4:622–625

Dyer, A. J., B. B. Hicks and K. M. King, 1967. The Fluxatron – a revised approach to the meanurement of eddy fluxes in the lower atmosphere. *J. Appl. Meteorol.* 6:408–413.

Eastman, J. A. and D. H. Stedman, 1977. A fast response sensor for ozone eddy–correlation flux measurements. *Atmos. Environ.* 11:1209–1211.

Eckman, R. M., R. J. Dobosy, D. L. Auble, T. W. Strong and T. L. Crawford, 2007. A pressure–sphere anemometer for measuring turbulence and fluxes in hurricanes. *J. Atmos. Ocean. Tech.* 24:994–1007.

Eder, B., D. Kang, R. Mathur, S. Yu and K. Schere, 2006. An evaluation of the Eta–CMAQ air quality forecast model. *Atmos. Environ.* 40:4894–4905.

Edgerton, E. S., T. F. Lavery and R. P. Boksleitner, 1992. Preliminary data from the USEPA dry deposition network: 1989. *Environ. Pollut.* 75:145–156.

Edson, J., T. Crawford, J. Crescenti, T. Farrar, N. Frew, G. Gerbi, C. Helmis, T. Hristov, D. Khelif and A. Jessup, 2007. The coupled boundary layers and air–sea transfer experiment in low winds. *Bull. Amer. Meteorol. Soc.* 88:341–356.

Elagina, L. G., 1962. Optical device for measuring gthe turbulent pulsations of humidity. *Izv. Geophys. Ser.* 8:1100–1107.

Elliott, W. P., 1958. The growth of the atmospheric internal boundary layer. *Transactions, American Geophysical Union* 39:1048–1054.

Ellison, T. H., 1957. Turbulent transport of heat and momentum from an infinite rough plane. *J. Fluid Mech.* 2:456–466.

El–Madany, T. S., H. F. Duarte, D. J. Durden, B. Pass, M. J. Deventer, J. –Y. Juang, M. Y. Leclerc and O. Klemm, 2014. Low–level jets and above–canopy drainage as causes of turbulent exchange in the nocturnal boundary layer. *Biogeosciences* 11:4507–4519.

Engel, A. and L. Galgani, 2016. The organic sea-surface microlayer in the upwelling region off the coast of Peru and potential implications for air-sea exchange processes. *Biogeosciences* 13:989–1007.

EPA, 1988. *Clean Air Status and Trends Network (CASTNet)* Deposition Summary Report (1987–1995), US Environmental Protection Agency report EPA–600/R–98/027.

EPRI, 1983. *Overview, results and conclusions for the EPRI plume model validation and development project: plains site, EA–3074,* Final Report. Project 1616–1. TRC–Environmental Consultants Inc, East Hartford, CT.

Evans, J. H., 1972. Dimensional analysis and the Buckingham Pi theorem. *Amer. J. Phys.* 40:1815–1821.

Fairall, C. W., E. F. Bradley, J. E. Hare, A. A. Grachev and J. B. Edson, 2003. Bulk parameterization of air–sea fluxes: updates and verification for the COARE algorithm. *J. Climate* 16:571–591.

Fairall, C. W., J. E. Hare, J. B. Edson and W. McGillis, 2000. Parameterization and micrometeorological measurement of air-sea gas transfer. *Boundary-Layer Meteorol.* 96:63–105.

Fairall, C. W., A. B. White, J. B. Edson and J. E. Hare, 1997. Integrated shipboard measurements of the marine boundary layer. *J. Atmos. Oceanic Technol.* 14:338–359.

Fairlie, T. D., J. P. Vernier, M. Natarajan and K. M. Bedka, 2014. Dispersion of the Nabro volcanic plume and its relation to the Asian summer monsoon. *Atmos. Chem. and Phys.* 14:7045–7057.

Farquhar, G. D., P. M. Firth, R. Wetselaar and B. Weir, 1980. On the gaseous exchange of ammonia between leaves and the environment: determination of the ammonia compensation point. *Plant Physiol* 66:710–714.

Feigenbaum, M. J., 1979. The onset spectrum of turbulence. *Physics Letters,* 74A:375–378.

Feigenbaum, M. J., 1980. The transition to aperiodic behavior in turbulent systems. *Commun. Math. Phys.* 77:65–86.

Ferber, G. K., J. L. Heffter, R. R. Draxler, R. J. Lagomarsino, F. L. Thomas and R. N. Dietz, 1986. *Cross–Appachian tracer experiment* (CAPTEX–83). NOAA Technical Memo. ERL ARL–142. 60 pp.

Ferber, G., J. K. Telegardas, J. L. Heffter, C. R. Dickson, R. N. Dietz and P. W. Krey, 1981. Demonstration of a long–range tracer system using perfluorocarbons. *Environmental Protection Agency Tech. Report* EPA–600. 63 pp.

Ferber, G., J. K. Telegardas, J. K. Heffter and M. E. Smith, 1977. Air concentrations of Krypton–85 in the Midwest United States during January–May 1974. *Atmos. Environ.* 11:379–385.

Fichtl, G. H. and P. Kumar, 1974. The response of a propeller anemometer to turbulent flow with the mean wind vector perpendicular to the axis of rotation. *Boundary-Layer Meteorol.* 6:363–379.

Finkelstein, P. L., T. G. Ellestad, J. F. Clarke, T. P. Meyers, D. B. Schwede, E. O. Hebert and J. A. Neal, 2000. Ozone and sulfur dioxide dry deposition to forests: Observations and model evaluation. *J. Geophys, Res.: Atmos.* 105:15365–15377.

Finkelstein, P. L and P. F. Sims, 2001. Sampling error in eddy correlation measurements. *J. Geophys. Res.: Atmos.* 106:3503–3509.

Finn, D., K. L. Clawson, R. M. Eckman, H. Liu, E. S. Russell, Z. Gao and S. Brooks, 2016. Project Sagebrush: revisiting the value of the horizontal plume spread parameter σ_y. *J. Appl. Meteorol. and Clim.* 55:1305–1322.

Finn, D., B. Reese, B. Butler, N. Wagenbrenner, K. L. Clawson, J. D. Rich, E. Russell, Z. Gao and H. Liu, 2016. Evidence for gap flows in the Birch Creek Valley, Idaho. *J. Atmos. Sci.* 73:4973–4894.

Finnigan, J. J. 1985. Turbulent transport in flexible plant canopies. pp 443–480 in *"The Forest–Atmosphere Interaction"* (B. A. Hutchson and B. B. Hicks, eds.), Reidel, Dordrecht, 684 pp.

Finnigan, J. J. and P. J. Mulhearn, 1977. Modelling waving crops in a wind tunnel. *Boundary-Layer Meteorol.* 14:253–277.

Flechard, C. R., R.-S. Massad, B. Loubet, E. Personne, D. Simpson, J. O. Bash, E. J. Conter, E. Nemitz and M. A. Sutton, 2013. Advances in understanding, models and parameterizations of biosphere-atmosphere ammonia exchange. *Biogeosci.* 10:5183–5225.

Foken, T., 2006. 50 years of the Monin–Obukhov similarity theory. *Boundary-Layer Meteorol. 1*19:431–447.

Foken, T., F. Wimmer, M. Mauder, C. Thomas and C. Liebethal, 2006. Some aspects of the enegy balance closure problem. *Atmos. Chem. Phys.* 6:4395–4402.

Fowler, D. and M. H. Unsworth, 1974. Dry deposition of sulphur dioxide on wheat. *Nature* 249:389–390.

Fowler, D., J. N. Cape and M. H. Unsworth, 1989. Deposition of atmospheric pollutants on forests. *Phil. Trans. R. Soc. Lond. B* 324:247–265.

Fox, S. B. and D. Storwold, 2011. *Project Jack Rabbit: Field Tests.* U.S. Department of Homeland Security, Chemical Security Analysis Center, CSAC 11–006, 162 pp.

Frank, J. M., W. J. Massman and B. E. Ewers, 2013. Underestimates of sensible heat flux due to the vertical velocity measurement errors in non–orthogonal sonic anemometers. *Agric. For. Meteorol.* 171:72–81.

Frenzen, P., 1968. Fast–response cup anemometer features cosine response. *AEC–NASA Tech. Brief* 68–10202, 2 pp.

Frenzen, P., 1974. Swinbank on Boundary-Layer research. *Boundary-Layer Meteorol.* 7:249–255.

Frey, N. M., 1977. The role of organic films in air–sea gas exchange. pp. 121–172 in *'The Sea Surface and Global Change'* (P. S. Liss and R. A. Duce eds.), Cambridge University Press, 536 pp.

Friedlander, S. K., 1977. *Smoke, Dust and Haze.* John Wiley and Sons, New York, 317 pp.

Friedlander S. K. and H. F. Johnstone, 1957. Deposition of suspended particles from turbulent gas streams. *Ind. Eng. Chem.* 49:1151–1156.

Friedrich, B., D. Hoffmann, J. Renn, F. Schmaltz and M. Wolf (Eds.), 2007. *One Hundred Years of Chemical Warfare: Research, Deployment, Consequences.* Springer, Cham, Switzerland, 408 pp.

Frisch, U., 1985. Fully developed turbulence and intermittency. Pp. 71–88 of "Turbulence and Predictability in Geophysical Fluid Dynamics and Climate Dynamics," *Soc. Italiana di Fisica [Italian Society of Physics],* Bologna.

Fritschen, L. J., 1985. Characterization of boundary conditions affecting forest environmental phenomena. pp 3–23 in *'The Forest–Atmosphere Interaction'*
Dordrecht, 684 pp.

Frost, R., 1948. Atmospheric turbulence. *Quart. J. Roy. Meteorol. Soc.* 74:316–338.

Funk, J. P., 1959. Improved polythene–shielded net radiometer. *J. Sci. Instr.* 36:267–270.

Galbally, I. E., 1971. Ozone profiles and ozone fluxes in the atmospheric surface layer. *Quart. J. Roy. Meteorol. Soc.* 97:18–29.

Galloway, J. N. and D. M. Whelpdale, 1987. Watox–86 overview and western North Atlantic Ocean S and N atmospheric budgets. *Global Biogeochem. Cycles* 1:261–281.

Garland, J. A., D. H. F. Atkins, C. J. Readings and S. J. Caughey, 1974. Deposition of gaseous sulphur dioxide to the ground. *Atmos. Environ.* 8:75–79.

Garland, J. A., W. S. Clough and D. Fowler, 1973. Deposition of sulphur dioxide on grass. *Nature* 242:256–257.

Garratt, J. R., 1977. Review of drag coefficients over oceans and continents. *Monthly Weather Rev.* 105:915–929.

Garratt, J. R., 1979. Turbulent flux measurements. pp 55–60 in *"The Koorin Expedition: atmospheric boundary layer data over tropical savannah land"* (R. H. Clarke and R. R. Brook, eds.).

Garratt, J. R., 1980. Surface influence on vertical profiles in the atmospheric near–surface layer. *Quart. J. Roy. Meteorol. Soc.* 106:803–819.

Garratt, J. R., 1992. *The Atmospheric Boundary Layer.* Cambridge University Press, Cambridge, 316 pp.

Garratt, J. R., D. E. Angus and P. Holper, 1998. *Winds of Change.* CSIRO Publishing, Collingwood, Australia, 136 pp.

Garratt, J. R., R. J. Francey, I. C. McIlroy, A. J. Dyer, I. Helmond, E. F. Bradley and O. T. Denmead, 1980. The International Turbulence Comparison Experiment (Australia 1976) – meteorological support data. *CSIRO Div. Atmos. Phys. Tech.* Paper #37.

Garratt, J. R. and B. B. Hicks, 1973. Momentum, heat and water vapour transfer to and from natural and artificial surfaces, *Quart. J. Roy. Meteorol. Soc.* 99:680–687.

Garratt, J. R. and B. B. Hicks, 1990. Micrometeorological and PBL experiments in Australia. *Boundary-Layer Meteorol.* 50:11–29.

Garratt, J. R. and A. J. Prata, 1996. Downwelling long–wave fluxes at continental surfaces – a comparison of observations with GCM simulations and implications or the global land–surface radiation budget. *J. Climate* 9:646655.

Garratt, J. R. E. K. Webb and S. McCarthy, 2011. Charles Henry Brian Priestley 1915–1998. *Historical Records of Australian Sci.* 22:126–151.

Garrett, W. D. and J. D. Bultman, 1963. Capillary–wave damping by insoluble organic monolayers. *J. Colloid. Sci.* 18:798–801.

Geiger, R., 1965. *The Climate Near the Ground.* Harvard University Press, Cambridge, MA, 611 pp.

Gibson, C. H. and W. H. Thomas, 1995. Effects of turbulence intermittency on growth inhibition of a red tide dinoflagellate, Gonyaulax polyedra Stein. *J. Geophys. Res.* 100:24,841–24,846.

Gifford, F. A., 1961. Use of routine meteorological observations for estimating atmospheric dispersion. *Nuclear Safety* 2:44–57.

Gifford, F. A., 1967. Translation of *"The Principles of Dimensional Analysis"* by R. Saint–Guilhem, available as US Environmental Sciences Research Administration, Atmospheric Turbulence and Diffusion Laboratory Report IERTM–ATDL 3, 86 pp.

Gill, G. C., 1973. The helicoid anemometer. *Atmosphere* 11:145–155.

Gollub, J. P. and S. P. Benson, 1980. Many routes to turbulent convection. *J. Fluid Mech.* 100:449–470.

Gong J., G. Wahba, D. R. Johnson and J. Tribbia, 1998. Adaptive tuning of numerical weather prediction models: Simultaneous estimation of weighting, smoothing, and physical parameters. *Mon. Wea. Rev.* 126:210–231.

Gordon, M., R. M. Staebler, J. Liggio, A. Vlasenko, S. -M. Li and K. Hayden, 2011. Aerosol flux measurements above a mixed forest at Borden, Ontario. *Atmos. Chem. Phys.* 11:6773–6786.

Guan, C. and L. Xie, 2004. On the linear parameterization of drag coefficient over sea surface. *J. Phys. Oceanog.* 34:2847–2851.

Gumbel, E. J., 1958. *Statistics of extremes.* Columbia university press. 375 pp.

Haagenson, P. L., G. Kun and Y.-H. Kuo, 1990. Eva;uation of meteorological analyses, simulations, and long-range transport calculations using ANATEX surface tracer data. *J. Appl. Meteorol. and Climatol.* 29:1268–1283.

Hall, B. D. G. S. Dutton, D. J. Mondeel, J. D. Nance, M. Rigby, J. H. Butler, F. L, Moore, D. F. Hurst and J. W. Elkins, 2011. Improving measurements of SF_6 for the study of atmospheric transport and emissions. *Atmos. Meas. Tech.* 4:2441–2451.

Hall, D. O and K. K. Rao, 1987. *Photosynthesis.* Edward Arnold, London, 122 pp.

Hall, T. M., T. W. N. Haine and D. W. Waugh, 2002. Inferring the concentration of anthropogenic carbon in the ocean from tracers. *Global Biogeochem Cycles,* 16:78–1–78–15

Hall, T. M. and D. W. Waugh, 1998. Influence of non–local chemistry in tracer distributions: inferring the mean age of air from SF_6. J. *Geophys. Res.* 103:13,327–13,336.

Handford, P. M. and P. Bradshaw, 1989. The pulsed–wire anemometer. *Expts. in Fluids* 7:125–132.

Hanna, S. R., 1982. Applications in air pollution modelling. In *"Atmospheric Turbulence and Air Pollution Modelling"* (F. T. M. Nieuwstadt and H. D. van Dop, eds.), D. Reidel Publishing, Dordrecht, Holland.

Hanna, S. R. and J. C. Chang, 1992. Representativeness of wind measurements on a mesoscale grid with station separations of 312 m to 10 km. *Boundary-Layer Meteorol.* 60:309–324.

Hanna, S. R., G. A. Briggs and R. P. Hosker, 1982. *Handbook on Atmospheric Diffusion.* U. S. Department of Energy Report DOE/TIC–11223, 110 pp.

Hanna, S. R., R. Britter, E. Argenta and J. Chang, 2012. The Jack Rabbit chlorine release experiments: Implications of dense gas removal from a depression and downwind concentrations. *J. Hazardous. Mat.* 213:406–412.

Harman, I. N. and J. J. Finnigan, 2007. A simple unified theory for flow in the canopy and roughness sub-layer. *Boundary-Layer Meteorol.* 123:339–363.

Harriott, P. and R. M. Hamilton, 1965. Solid–liquid mass transfer in turbulent pipe-flow, *Chem. Eng. Sci.* 20:1073–1078.

Harvey, G. W., 1966. Microlayer collection from the sea surface: a new method and initial results. *Limnology and Oceanog.* 11:608–613.

Haugen, D. A., J. C. Kaimal and E. F. Bradley, 1971. An experimental study of Reynolds stress and heat flux in the atmospheric surface layer. *Quart. J. Roy. Meteorol. Soc.* 97:168–190.

Heffter, J. L., J. F. Schubert and G. A. Meade, 1984. Atlantic coast unique regional atmospheric tracer experiment, October. *NOAA Tech. Memo.* ERL ARL–130. 67 pp.

Held, A., A. Nowak, W. Birmili, A. Wiedensohler, R. Forkel and O. Klemm, 2004. Observations of particle formation and growth in a mountainous forest region in central Europe. *J. Geophys. Res.* 109:doi:10.1029/2004/D005346.

Heller, J. M., 2011. Oil well fires of Operation Desert Storm—defining troop exposures and determining health risks. *Mil. Med.* 176:36–51.

Hewson, R. W., 1952. International Symposium on Atmospheric Turbulence in the Boundary Layer. *Geophysical Research Papers No. 19,* Air Force Cambridge Research Center FCRC Technical Report 53–9, 530 pp.

Heymann, E. and A. Yoffe, 1942. The stability of monomolecular films of hydrocarbon oils, containing spreaders, on water surfaces. *Trans. Faraday Soc.* 38:408–417.

Hicks, B. B., 1970. The measurement of atmospheric fluxes near the surface: a generalized approach. *J. Appl. Meteorol.* 9:386–388.

Hicks, B. B., 1972a: Propeller anemometers as sensors of atmospheric turbulence. *Boundary-Layer Meteorol.* 3:214–228.

Hicks, B. B., 1972b. Some evaluations of drag and bulk transfer coefficients over water bodies of different sizes. *Boundary-Layer Meteorol.* 3:201–213.

Hicks, B. B., 1973. Eddy fluxes over a vineyard. *Ag. Meteorol.* 12:203–215.

Hicks, B. B., 1975a. On the limiting surface temperature of cooling ponds. *Radiological and Environmental Research Division Annual Report,* Argonne National Laboratory ANL–75–3 Part 4, 42–53.

Hicks, B. B., 1975b. On the limiting surface temperature of exposed water bodies. *J. Geophys. Res.* 80:5077–5081.

Hicks, B. B., 1976. Wind profile relationships from the "Wangara" experiment. *Quart. J. Roy. Meteorol. Soc.* 102, 535–551.

Hicks B. B., 1978. Some limitations of dimensional analysis and power laws, *Boundary-Layer Meteorol.* 14:1451–1458.

Hicks, B. B., 1981. An examination of turbulence statistics in the surface boundary layer, *Boundary-Layer Meteorol.* 21:389–402.

Hicks, B. B. 1985. Behavior of turbulence statistics in the convective boundary layer. *J. Clim. Appl. Meteorol.* 24:607–614.

Hicks, B. B., 1988. ARCH: Some introductory notes to an issue of Boundary-Layer Meteorology dedicated to Arthur James Dyer, *Boundary-Layer Meteorol.* 43:1–8.

Hicks, B. B., 1992. An overview of the US studies of the Kuwait oil fire plume dispersion. pp 25–29 in "Second WMO Meeting of Experts to Assess the Response to the Kuwait oil fires," *World Meteorol. Org.* Geneva, Switzerland.

Hicks, B. B., 2005. Urban dispersion for the 21st century. *WIT Transactions on the Built Environment* 82:555–563.

Hicks, B. B., 2022. On kinematic isolation in stable stratification: the CASES-99 tower observations. *Boundary-Layer Meteorol.* 183:67–77.

Hicks, B. B. and D. D. Baldocchi, 2020. Measurement of fluxes over land: capabilities, origins, and remaining challenges. *Boundary-Layer Meteorol.* 177:365–394.

Hicks, B. B. and A. J. Dyer, 1970. Measurements of eddy–fluxes over the sea from an off–shore oil rig. *Quart. J. Roy. Meteorol, Soc.* 96:523–528.

Hicks, B. B. and N. S. Eash, 2021. On the effective surface temperature of a natural landscape: Infrared or not infrared. *Boundary-Layer Meteorol.* 180:353–362.

Hicks, B. B. and H. S. Goodman, 1971. The eddy–correlation technique of evaporation measurement using a sensitized quartz–crystal hygrometer. *J. Appl. Meteorol.* 10:221–223.

Hicks, B. B. and M. L. Wesely, 1977. Momentum fluxes over adjacent soybean and maize fields. pp 44–48 in *Argonne National Laboratory Report ANL-77-65* Part IV, 178 pp.

Hicks, B. B., D. D. Baldocchi, T. P. Meyers, R. P. Hosker and D. R. Matt, 1987. A preliminary multiple resistance routine for deriving dry deposition velocities from measured quantities. *Water Air and Soil Pollut.* 36:311–330.

Hicks, B. B., W. J. Callahan, W. R. Pendergrass III, R. J. Dobosy and E. Novakovskaia, 2012. Urban turbulence in space and time. *J. Appl. Meteorol. and Climatol.* 52:205–218.

Hicks, B. B., R. L. Drinkrow and G. Grauze, 1974. Drag and bulk transfer coefficients associated with a shallow water surface. *Boundary-Layer Meteorol.* 6:287–297.

Hicks, B. B., N. S. Eash, D. L. O'Dell and J. N.Oetting, 2020. Augmented Bowen ratio analysis –I: Site adequacy, fetch and heat storage (ABRA). *Agric. Forest Meteorol.* 290:108035.

Hicks, B. B., G. D. Hess and M. L. Wesely, 1979. Analysis of flux-profile relationships – an alternative view. *Quart J. Roy. Meteorol. Soc.* 105:1074–1077.

Hicks, B. B., G. D. Hess, M. L. Wesely, T. Yamada, P. Frrenzen, R. L. Hart, D. L. Sisterson, P. E. Hess, F. C. Kulhanek, R. C. Lipschutz and G. A. Zerbe, 1981. The Sangamon field experiments: observations of the diurnal evolution of the planetary boundary layer over land. *Argonne National Laboratory Report ANL/RER-81-1,* 184 pp.

Hicks, B. B., R. P. Hosker, T. P. Meyers and J. D. Womack, 1991. Dry deposition inferential measurement techniques – I. Design and tests of a prototype meteorological and chemical system for determining dry deposition. *Atmos. Environ.* 25A:2345–2359.

Hicks, B. B. C. H. Hunter and A. H. Weber, 2016. On dispersion above a forest — Measurements and methods. *J. Air. Waste. Management Assoc.* 66:768–785.

Hicks, B. B. P. Hyson and C. J. Moore, 1975. A study of eddy fluxes over a forest. *J. Appl. Meteorol.* 14:58–66.

Hicks, B. B., N. Lichiheb, D. L. O'Dell, J. N. Oetting, N. S. Eash, M. Heuer and L. Myles, 2021. A statistical approach to surface renewal: The virtual chamber concept. *Agrosyst. Geosci. Environ.* doi.org/10.1002/agg2.20141.

Hicks, B. B., D. R. Matt, R. T. McMillen, J. D. Womack, M. L. Wesely, R. L. Hart, D. R. Cook, S. E. Lindberg, R. G. de Pena and D. W. Thomson. 1989. A field investigation of sulfate fluxes to a deciduous forest. *J. Geophys. Res.* 94:13003–13011.

Hicks, B. B., D. B. O'Dell, N. S. Eash and T. Sauer, 2015a. Nocturnal intermittency in surface CO_2 concentrations in sub–Saharan Africa. *Agric. Forest Meteorol.* 200:129–134.

Hicks, B. B. W. R. Pendergrass, B. D. Baker, R. D. Saylor, D. L. O'Dell, N. S. Eash and J. T. McQueen, 2018. On the relevance of $\ln(z_0/z_{0T}) = kB^{-1}$. *Boundary-Layer Meteorol.* 167:285–301.

Hicks, B. B., W. R. Pendergrass III, C. A. Vogel and R. S. Artz, 2014a. On the drag and heat of Washington, DC, and New York City. *J. Appl. Meteorol.* 53:1454–1470.

Hicks, B. B., W. R. Pendergrass, C. A. Vogel, R. N. Keener and S. M. Leyton, 2014b. On the micrometeorology of the Southern Great Plains 1: Legacy relationships revisited. *Boundary-Layer Meteorol.* 151:389–405.

Hicks, B. B., W. R. Pendergrass, C. A. Vogel, R. N. Keener and S. M. Leyton, 2015b. On the micrometeorology of the Southern Great Plains 2: Turbulence statistics. *Boundary-Layer Meteorol.* 154:351–366.

Hicks B. B., R. D. Saylor and B. D. Baker, 2016. Dry deposition of particles to canopies — A look back and the road forward. *J. Geophys. Res. Atmos.* 121 doi:10.1002/2015JD024742.

Hicks, B. B., M. L. Wesely, R. L. Coulter, E. L. Hart, J. L. Durham, R. E. Speer and D. H. Stedman, 1982. An experimental study of sulfur deposition to grassland. pp 933–942 in "*Precipitation Scavenging, Dry Deposition and Resuspension*" (H. R. Pruppacher, R. G. Semonin and W. G. N. Slinn, eds.), Elsevier, New York.

Hicks, B. B., M. L. Wesely, J. L. Durham and M. A. Brown, 1982. Some direct measurements of atmospheric sulfur fluxes over a pine plantation. *Atmos. Environ.* 16:2899–2903.

Hicks, B. B., M. L. Wesely and C. M. Sheih, 1974. Eddy–correlation measurements over a cooling pond with limited fetch. *Argonne National Laboratory Annual Report ANL–75–3,* Part IV, 4–18.

Hill, G. R. and M. D. Thomas, 1933. Influence of leaf destruction by sulfur dioxide and by clipping on the yield of alfalfa. *Plant. Physiol.* 8:233–245.

Högström, U., 1964. An experimental study on atmospheric diffusion, *Tellus* 16:205–251.

Högström, U., R. Taesler, S. Karlsson, L. Enger and A. O. S. Högström, 1978. The Uppsala urban meteorology project. *Boundary-Layer Meteorol.* 15:69–80.

Hopkins, W. G., 1999. *Introduction to plant physiology*. Wiley, New York, 512 pp.

Horst, T. W., 1975. *The resuspension of material dispersed by atmospheric diffusion from a point source.* Battelle Pacific Northwest Laboratories Report BNWL-1964 UC-70, 55 pp.

Hothorn, T., K. Hornik and A. Zeileis, 2006. Unbiased recursive partitioning: A conditional reference framework. *J. Comput. Graphical Statistics* 15:651–674.

Howell, J. F. and J. Sun, 1999. Surface-layer fluxes in stable conditions. *Boundary-Layer Meteorol.* 90:495–520.

Hu, E. and S. Vogt, 1986. *Evaluation of tetroon flights and turbulent diffusion under weak wind conditions during the field experiment SIESTA.* KFK (Germany) Report 4126. 35 pp.

Hubbard, D. W. and E. N. Lightfoot, 1966. Correlation of heat and mass transfer data for high Schmidt and Reynolds numbers. *I/Ec Fundamentals* 5:370–379.

Hubbard, K. G., X. Lin and E. A. Walter–Shea, 2001. The effectiveness of ASOS. MMTS, Gill, and CRS air temperature radiation shields. *J. Atmos. Ocean. Technol.* 18:851–864.

Huebert, B. J. and C. H. Robert, 1985. The dry deposition of nitric acid to grass. *J. Geophys. Res. Atmos.* 90:2085–2090.

Husain, T., 1994. Kuwaiti oil fires—Source estimates and plume characterization. *Atmos. Environ.* 28:2149–2158.

Husain, T., 1995. *Kuwaiti oil fires: Regional environmental perspectives.* Pergamon, London, 292 pp.

Hutchison, B. A., D. R. Matt, R. T. McMillen, L. J. Gross, S. J. Tajchman and J. M. Norman, 1986. The architecture of an east Tennessee deciduous forest canopy. *J. Ecol.* 74:635–646.

Hwang. P. A. and M. A. Sletten, 2008. Energy dissipation of wind-generated waves and whitecap coverage. *J. Geophys. Res.* 113 C02012, doi:10.1029/2007JC004277.

Hyson, P. and B. B. Hicks, 1975. A single-beam infrared hygrometer for evaporation measurement. *J. Appl. Meteorol.* 14:301–307.

Inoue, E., 1950. On the turbulent diffusion in the atmosphere. *J. Meteorol. Soc.* Japan 28:441–456.

Inoue, E., 1955a. Studies of the phenomena of waving plants ("HONAMI") caused by the wind, Part 1. Mechanism and characteristics of waving plants phenomena. *J. Agr. Meteorol.* Japan 11:18–22.

Inoue, E., 1955b. Studies of the phenomena of waving plants ("HONAMI") caused by the wind, Part 2. Spectra of waving plants and plants' vibration. *J. Agr. Meteorol.* Japan 11:87–90.

Inoue, E., 1963. On the turbulent structure of airflow within plant canopies. *J. Meteorol. Soc.* Japan 41:317–326.

Irmak, S. and D. Mutiibwa, 2009. On the dynamics of stomatal resistance: relationships between stomatal behavior and micrometeorological variables and performance of Jarvis–type parameterization. *Trans. ASABE* 52:1923–1939.

Irvine, M. R., B. A. Gardiner and M. K. Hill, 1997. The evolution of turbulence across a forest edge. *Boundary-Layer Meteorol.* 84:47–496.

Irving, P., 1990. *Acid Deposition: State of Science and Technology.* In four volumes. U.S. National Acid Precipitation Assessment Program, Washington, DC.

Irwin, J. S., 1983. Estimating plume dispersion-a comparison of several sigma schemes. *J. Appl. Meteorol. and Climatol.* 22:92–114.

Irwin, J. S. and S. R. Hanna, 2005. Characterising uncertainty in plume dispersion models. *Int. J. of Environ. and Pollut.* 25:16–24.

Islitzer, N. E., 1961. Short-range atmospheric dispersion measurements from an elevated source. *J. Atmos. Sci.* 18:443–450.

Islitzer, N. E., 1962. Role of meteorology following the nuclear accident in southeast Idaho. [Accident of January 3, 1961]. pp. 75–93 in *Proc. Third Conference on Nuclear Chemistry,* Gatlinburg, TN, 9–11 Oct 1962, US Atomic Energy Commission TID 7641.

Islitzer, N. E. and R. K. Dumbauld, 1963. Atmospheric diffusion-deposition studies over flat terrain. *J. Air Water Pollut.* 7:999–1022.

Iziomon, M. G., H. Mayer and A. Matzarakis, 2003. Downward atmospheric longwave irradiance under clear and cloudy skies: measurement and parameterization. *J. Atmos. Solar-Terrest. Phys.* 65:1107–1116.

Izumi, Y. and J. S. Caughey, 1976. Minnesota 1973 atmospheric boundary layer experiment data report. *Air Force Cambridge Research Laboratory Environ. Res. Papers* 547, AFCRL-TR-0038, 28 pp.

Jarvis, P. G., 1976. The interpretation of the variations in leaf water potential and stomatal conductance found in canopies in the field. *Phil. Trans. R. Soc. Lon. B.* 273:595–610.

Jarvis, P. G. and K. G. McNaughton, 1986. Stomatal control of evapotranspiration: scaling up from leaf to region. *Advances Ecol. Res.* 15:1–49.

Johnson, O., 1959. An examination of the vertical wind profile in the lowest layers of the atmosphere. *J. Meteorol.* 16:144–148.

Kaimal, J. C. and J. A. Businger, 1963. A continuous-wave sonic anemometer–thermometer, *J. App. Meteorol.* 2:156–167.

Kaimal, J. C. and J. E. Gaynor, 1991. Another look at sonic thermometry. *Boundary-Layer Meteorol.* 56:401–410.

Kaimal, J. C., J. C. Wyngaard and D. A. Haugen, 1968. Deriving power spectra from a three–component sonic anemometer. *J. Appl. Meteorol.* 7:827–837.

Kaimal, J. C., J. C. Wyngaard, D. A. Haugen, O. R. Coté, Y. Izumi, S. J. Caughey and C. J. Readings, 1976. Turbulence structure in the convective boundary layer. *J. Atmos. Sci.* 33:2152–2169.

Kanda M., M. Kanega, T. Kawai and R. Moriwaki, 2007. Roughness lengths for momentum and heat derived from outdoor urban scale models. *J. Appl. Meteorol.* 46:1067–1079.

Kanemasu, E. T., M. L. Wesely, B. B. Hicks and J. L. Heilman, 1979. Techniques for calculating energy and mass fluxes. pp. 156–182 in "*Modifications of the Aerial Environment of Crops*," (B. J. Barfield, ed.) Am. Soc. Agric. Eng. Monograph #2, ASAE, St. Joseph, Michigan.

Kantha, L. H. and C. A. Clayson, 1980. *Numerical Models of Oceans and Oceanic Processes.* Academic Press, New York, 972 pp.

Kanwisher, J., 1963. On the exchange of gases between the atmosphere and the sea. *Deep Sea Res.* 10:195–207.

Kara, A. B., E. J. Metzger and M. A. Bourassa, 2007. Ocean current and wave effects on wind stress drag coefficient over the global ocean, *Geophys. Res. Letters* 34:L01604.

Katul, G. D. Cava, D. Poggi, J. Albertson and L. Mahrt, 2004. Stationarity, homogeneity, and ergodicity in canopy turbulence, pp. 161–180 in "*Handbook of Micrometeorology*," Lee et al. (eds.), Kluwer Academic, Dordrecht.

Katul, G., G. T. Grönholm, S. Launiainen and T. Vesala, 2010. Predicting the dry deposition of aerosol–sized particles using layer–resolved canopy and pipe-flow analogy models: Role of turbophoresis. *J. Geophys. Res.* 115, D12202.

Katul, G., G. Kuhn, J. Schieldge and C–I. Hsieh, 1997. The ejection-sweep character of scalar fluxes in the unstable surface layer. *Boundary-Layer Meteorol.* 83:1–26.

Kazansky, A. B. and A. S. Monin, 1956. Turbulence in the inversion layer near the surface. *Akad. Nauk. SSSR Ser. Geophys.* 1:79–86.

Kline, S. J., W. C. Reynolds, F. A. Schraub and P. W. Runstadler, 1967. The structure of turbulent boundary layers. *J. Fluid. Mech.* 30:741–773.

Klipp, C. and L. Mahrt, 2004. Flux–gradient relationship, self–correlation and intermittency in the stable boundary layer. *Quart. J. Roy. Meteorol. Soc.* 130:2087–2104.

Kljun, K., P. Calanca, M. W. Rotach and H. P. Schmid, 2004. A simple parameterisation for flux footprint predictions, *Boundary-Layer Meteorol.* 112:503–523.
Kochendorfer J., T. P. Meyers, J. Frank, W. J. Massman and M. W. Heuer, 2012. How well can we measure the vertical wind speed? Implications for fluxes of energy and mass, *Boundary-Layer Meteorol.* 145:383–398.
Kochendorfer J., T. P. Meyers, J. Frank, W. J. Massman and M. W. Heuer, 2013. Reply to comment by Mauder on "How well can we measure the vertical wind speed? Implications for fluxes of energy and mass," *Boundary-Layer Meteorol.* 147:337–345.
Kondo, J., Y. Fujinawa and G. Naito, 1973. High-Frequency Components of Ocean Waves and Their Relation to the Aerodynamic Roughness, *J. Phys. Oceanogr.* 3:197–202.
Koopman, R. P., F. Gavelli, B. B. Hicks, R. P. Hosker, D. P. Bacon, G. D. Kaiser and J. E. Cockayne, 2012. *Analysis of the Jack Rabbit field trials – modeling.* Final Report on Army Contract W011R-10-D-0017, Aberdeen Proving Ground, Maryland, 151 pp.
Kormann, R. and F. X. Meixner, 2001. An analytical footprint model for non–neutral stratification. *Boundary-Layer Meteorol.* 99:207–224.
Kunz, J. and B. Jähne, 2018. Investigating small-scale air=dea exchange processes via thermography. *Front. Mech. Eng.* 4:1–11.
Kustas, W. P., B. J. Choudhury, K. E. Kunkel and L. W. Gay, 1989. Estimate of the aerodynamic roughness parameters over an incomplete canopy cover of cotton. *Ag. Forest Meteorol.* 46:91–105.
Lal, D. and M. Baskaran, 2012. Applications of cosmogenic isotopes as atmospheric tracers. pp 575–589 in "*Handbook of Environmental Isotope Geochemistry*" (M. Baskaran, ed.), Springer, 653 pp.
Lalic, B. and D. T. Mihailovic, 2013. An empirical relation describing leaf–area density inside the forest for environmental modeling. *J. Appl. Meteorol.* 43:641–645.
Lamb, D. and V. Bowersox, 2000. The national atmospheric deposition program: an overview. *Atmos. Environ.* 34:1661–1663.
Large, W. G. and S. Pond, 1981. Open ocean momentum flux measurements in moderate to strong winds. *J. Phys. Ocean.* 11:324–336.
Launiainen, S., G. G. Katul, T. Grönholm and T. Vesala, 2013. Partitioning ozone fluxes between canopy and a forest floor by measurements and a multi–layer model. *Ag. Forest Meteorol.* 173:85–99.
Kavrac, N. and S. Dzeroski, 2001. *Relational data mining.* Springer, Berlin, 398 pp.
Lapworth, A., 2005. The diurnal variation of the marine surface wind in an offshore flow. *Quart. J. Roy. Meteorol. Soc.* 131:2367–2387.
Law, C. S., E. Breviere, G. de Leeuw, V. Garcon, C. Guieu, D. J. Kieber, S. Kontradowitz, A. Paulmier, P. K. Quinn, E. S. Saltzman, J. Stefels and R. von Glasow, 2013. Evolving research directions in Surface Ocean-Lower Atmosphere (SOLAS) science. *Environ. Chem.* 10:1–16.
Leclere, M. Y. and G. W. Thurtell, 1990. Footprint prediction of scalar fluxes using a Markovian analysis. *Boundary-Layer Meteorol.* 52:247– 258.
Lee, T. R., S. F. De Wekker, S. Pal, A, E. Andrews and J. Kofler, 2015. Meteorological controls on the diurnal variability of carbon monoxide mixing ratio at a mountaintop monitoring site in the Appalachian Mountaines. *Tellus B.* 67: https://doi.org/10.3402/tellusb.v67.25659.
Lee, V. and W. J. Massman, 2011. A perspective on thirty years of the Webb, Pearman and Leuning density corrections. *Boundary-Layer Meteorol.* 139:37–59.
Legrand, S. V. Legat and E. Deleersnijder, 2000. Delaunay mesh generation for an unstructured–grid ocean general circulation model. *Ocean Modelling.* 2:17–28.
Leighton, P. A., W. A. Perkins, S. W. Grinnell and F. X. Webster, 1963. The fluorescent particle atmospheric tracer. *J. Appl. Meteorol and Climatol.* 4:334–348.
Le Moigne, P., 2012. *SURFEX scientific documentation.* 237 pp. available as SURFEX Volume 7.2, #2.
Lempfert, R. G. K., 1931. The scientific work of the Meteorological Office, Cardington. *Quart. J. Roy. Meteorol. Soc.* 57:119–131.
Lenschow, D. H., 1986. Aircraft measurements in the boundary layer. pp 39–55 in "*Probing the Atmospheric Boundary Layer*" (D. H. Lenschow, ed.). Amer. Meteorol. Soc.
Lenschow, D. H., J. C. Wyngaard and W. T. Pennell, 1980. Mean-field and second-moment budgets in a baroclinic, convective boundary layer. *J. Atmos. Sci.* 37:1313–1326.

Lettau, H., 1950. A re–examination of the "Leipzig Wind Profile" considering some relations between wind and turbulence in the frictional layer. *Tellus* 2:125–129.

Lettau, H., 1951. Theory of surface temperature and heat transfer oscillations near a level ground surface. *Trans. Amer. Geophys. Union* 32:189–200.

Lettau, H., 1952. Summary of Discussion of the "*International Symposium on Atmospheric Turbulence in the Boundary Layer*" (E. W. Hewson, ed.), 501–520.

Lettau, H. H. and B. Davidson, 1957. *Exploring the Atmosphere's First Mile.* Pergamon Press, New York, 578 pp.

Leuning, R. E. van Gorsel, W. J. Massman and P. R. Isaac, 2012. Reflections on the surface energy imbalance problem. *Ag. For. Meteorol.* 156:65–74.

Lewis, J. M., D. W. Martin, R. M. Rabin and H. Moosmuller, 2010. Soumi – pragmatic visionary. *Bull. Amer. Meteorol. Soc.* 91:559–577.

Li, D., 2021. The O'KEYPS equation and 60 years beyond. *Boundary-Layer Meteorol.* 179:19–42.

Li, Z. and V. P. Aneja, 1992. Regional analysis of cloud chemistry at high elevations in the eastern United States. *Atmos. Environ.* 26A:2001–2017.

Liang, J., L. Zhang, Y. Wang, X. Cao, Q,. Zhang, H. Wang and B. Zhang, 2014. Turbulence regimes and the validity of similarity theory in the stable boundary layer over complex terrain of the Loess Plateau, China. *J. Geophys. Res.* 119:6009–6021.

Lichiheb, N., M. Heuer, B. B. Hicks, R. Saylor, R. Vargas, A. Vazquez–Lule, K. St. Laurent and L. Myles, 2021. Atmospheric ammonia measurements over a coastal salt marsh ecosystem along the mid–Atlantic U.S. *J. Geophys Res: Biogeosciences* 126:doi.org/10.1029/2019jG005522.

Linacre, E. T., 1994. Estimating US Class A pan evaporation from few climate data. *Water International* 19:5–14.

Liss, P. S. and R. A. Duce, 1997. '*The Sea Surface and Global Change.*' Cambridge University Press, 536 pp.

Liu, B. Y. H. and J. K. Agarwal, 1974. Experimental observatioin of aerosol deposition in turbulent flow. *J. Aerosol Sci.* 5:149–155.

Lorenz, E. N., 1963. Deterministic nonperiodic flow. *J. Atmos. Sci.* 20:130–141.

Lorenz, E. N., 1993. *Chaotic Behavior in Systems.* Washington University Press, 345 pp.

Lovelock, J. E. and G. J. Ferber, 1982. Exotic tracers for atmospheric studies. *Atmos. Environ.* 16:1467–1471.

Lovelock. J. E. and L. Margulis, 1974. Atmospheric homeostasis by and for the biosphere: the Gaia hypothesis. *Tellus* 26:2–10.

Lovett, G. M. and S. E. Lindberg, 1984. Dry deposition and canopy exchange in a mixed oak forest as determined by analysis of throughfall. *J. Appl. Ecol.* 21:1013–1027.

Luvall, J. C. and H. R. Holbo, 1989. Measurements of short-term thermal responses of coniferous forest canopies using thermal scanner data. *Remote Sens. Environ.* 27:1–10.

Mahrt, L., 1981. The early evening boundary layer transition. *Quart. J. Roy. Meteorol. Soc.* 107:329–343.

Mahrt L., 2010. Variability and maintenance of turbulence in the very stable boundary layer, *Boundary-Layer Meteorol.* 135:1–18.

Mahrt, L. and D. Vickers, 2002. Contrasting vertical structures of nocturnal boundary layers. *Boundary-Layer Meteorol.* 105:351–363.

Mahrt, L., L. Pfister and C. K. Thomas, 2020. Small-scale variability in the nocturnal boundary layer. *Boundary-Layer Meteorol.* 174:81–98.

Malek, E. and G. E. Bingham, 1993. Comparison of the Bowen ratio-energy balance and the water balance methods for the measurement of evapotranspiration, *J. Hydrol. 1*46:209–220.

Malhi Y., 1996. The behaviour of the roughness length for temperature over heterogeneous surfaces. *Quart. J. Roy. Meteorol. Soc.* 122:1095–1125.

Martinot–Lagarde, A., A. Fauquet, and F. N. Frenkiel, 1952. The IMFL anemoclinometer — and instrument for the investigation of atmospheric turbulence. pp 207 – 218 in "*International Symposium on Atmospheric Turbulence in he Boundary Layer*" (E. W. Hewson, ed.) Air Force Cambridge Research Center, Cambridge, MA. 530 pp.

Marzban C., S. Sandbathe and J. D. Doyle, 2014. Model tuning with canonical correlation analysis. *Mon. Wea. Rev.* 142:2018–2027.

Mason, P. J., 1988. The formation of an areally-averaged roughness length. *Quart. J. Roy. Meteorol. Soc. 114:399–420.*
Masson, V., 2006. Urban surface modeling and the meso-scale impact of cities. *Theor. Appl. Climatol.* 84:35–45.
Matsushima, D., 2005. Relations between aerodynamic parameters of heat transfer and thermal-infrared thermometry in the bulk surface formulation. *J. Meteorol, Soc. Japan* 83:373–389.
Matt, D. R. and T. P. Meyers, 1993. On the use of the inferential technique to estimate dry deposition of SO_2. *Atmos. Environ.* 27:493–501.
Matt, D. R., R. T. McMillen, J. D. Womack and B. B. Hicks, 1987. A comparison of estimated and measured SO2 deposition velocities. *Water Air & Soil Pollut.* 36:331–347.
Mauder, M., 2013. A comment on "How well can we measure the vertical wind speed? Implications for fluxes of energy and mass" by Kochendorfer et al. *Boundary-Layer Meteorol.* 147:329–335.
Mauder, M. and T. Foken, 2011. *Documentation and Instruction Manual of he Eddy–Covariance Software Package TK3.* Report #46, Micrometeorology Department, Uni. Bayreuth, 60 pp.
Mawson, D., 1915. The Australasian Antarctic expedition, 1911–14. *Scottish Geograp. Mag.* 31:337–360.
Mazurek, J. and L. Smolarek, 2013. Oil spill models: a state of the art of the grid map as a function of wind, current and oil parameters. *Inter. J. Marine Navigation and Safety of Sea Transportation.* 1:19–23.
McCamley, N. J., 2006. *Secret History of Chemical Warfare.* Pen and Sword Books, Barnsley, Yorkshire UK, 188 pp.
McIlroy, I. C., 1979. Micrometeorological measurements. pp 35–53 in "*The Koorin Expedition: atmospheric boundary layer data over tropical savannah land*" (R. H. Clarke and R. R. Brook, eds.),
McMillen, R. T., 1986. A Basic program for eddy correlation in non–simple terrain. *NOAA Technical Memorandum ERL ARL–147.* 59 pp.
McNaughton, K. G., 1989. Micrometeorology of shelter belts and forest edges. *Phil. Trans. Roy. Society London.* B 324:351–368.
McNaughton, K. G., 2009. The rise and fall of Monin–Obukhov theory. *AsiaFlux Newsletter* 30:1–5.
Megonigal, J. P., P. E. Brewer and K. L. Knee, 2020. Radon as a natural tracer of gas transport through trees. *The New Phytologist* 225:1470–1475.
Meidner, H., 1987. Three hundred years of research into stomates. pp 7–28 of "*Stomatal Function*" (Zeiger, E. Farquhar, G. D. and Cowan, I. R. eds.). Stanford Univ. Press, Stanford, CA, 580 pp.
Mestayer, P. G., P. Durand, P. Augustin, S. Bastin, J. -M. Bonnefond, B. Bénech, B. Campistron, A. Coppalle, H. Delbarre, B. Dousset, P. Drobinski, A. Druilhet, E. Fréjafon, C. S. B. Grimmond, D. Groleau, M. Irvine, C. Kergomard, S. Kermadi, J. -P. Lagouarde, A. Lemonsu, F. Lohou, N. Long, V. Masson, C. Moppert, J. Noilhan, B. Offerle, T. R. Oke, G. Pigeon, V. Puygrenier, S. Roberts, J. -M. Rosant, F. Sanïd, J. Salmond, M. Talbaut and J. Voogt, 2005. The urban boundary-layer field campaign in Marseille (ubl/clu-escompte): set-up and first results. *Boundary-Layer Meteorol.* 114:315–365.
Mészáros, R., Á. Leelőssy, T. Kovács and I. Lagzi, 2016. Predictability of the dispersion of Fukushima-derived radionuclides and their homogenization in the atmosphere. *Scientific Reports* 6:1-10.
Meyers, T. P., P. Finkelstein, J. Clarke, T. G. Ellested and P. F. Sims, 1988. A multi–layer model for inferring dry deposition using standard meteorological measurements. *J. Geophys. Res.: Atmospheres* 103:22465–22661.
Meyers, T. P., B. B. Hicks, R. P. Hosker Jr. J. D. Womack and L. C. Satterfield, 1991. Dry deposition inferential techniques – II. Seasonal and annual deposition rates of sulfur and nitrate. *Atmos. Environ.* 25A:2361–2370.
Meyers, T. P. and S. E. Hollinger, 2004. An assessment of storage terms in the surface energy balance of maize and soybean. *Ag. and Forest Meteorol.* 125:105–115.
Meyers, T. P. and K. T. Paw U, 1986. Testing of a higher-order closure model for modeling airflow within and above plant canopies. *Boundary-Layer Meteorol.* 37:297–311.
Mildner, P., 1932. Uber Reibung in einer spezeillen Luftmasse, *Beitr. Phys. Fr. Atmosph.* 19:151.
Miles, J. W., 1986. Richardson's criterion for the stability of stratified shear flow. *Phys. Fluids* 29:3470, doi.org/10.1063/1.865812.
Miller, S. D., C. Marandino and E. S. Saltzman, 2010. Ship-based measurement of air-sea CO_2 exchange by eddy covariance. *J. Geophys. Res.: Atmospheres* 115:doi:10.1029/2009JD012193.

Mitchell, K., M. Ek, V. Wong, D. Lohmann, V. Koren, J. Schaake and Q. Duan, 2005. *The community Noah land-surface model (LSM) user's guide,* version 2.7. 1. NOAA/NCEP Doc 26.

Mizushina, T., F. Ogino, Y. Oka and H. Fukuda, 1971. Turbulent heat and mass transfer between wall and fluid streams at large Prandtl and Schmidt numbers. *Inter. J. Heat Mass Transfer* 14:1705–1716.

Mohnen, V. A., 1989. *Mountain Cloud Chemistry Project: wet, dry, and cloud water deposition.* State University of New York, Albany, Report PB–89–148597/XAB.

Möller, U. and G. Schumann, 1970. Mechanisms of transport from the atmjosphere to the Earth's surface. *J. Geophys. Res.* 75:3013–3019.

Monahan, E. C. and M. Lu, 1990. Acoustically relevant bubble assemblages and their dependence on meteorological parameters. *J. Ocean. Eng.* 15:340–349.

Monahan, E. C. and I. Ó. Muircheartaigh, 1981. Optimal power-law description of oceanic whitecap coverage dependence on wind speed. *J. Phys. Ocean.* 10:2094–2099.

Monin A. S. and A. M. Obukhov, 1954. Basic relationships of turbulent mixing in the surface layer of the atmosphere, *Acad. Nauk. SSSR Trud. Geofys. Inst.* 24:163–187.

Monteith, J. L., 1965. Evaporation and Environment. In *19th Symposia of the Society for Experimental Biology,* University Press, Cambridge, 19:205–234.

Monteith, J. L., 1973. *Principles of Environmental Physics.* Edward Arnold Press, London, 285 pp.

Monteith, J. L. and M. H. Unsworth, 1990. *Principles of Environmental Physics,* 2nd ed. Edward Arnold Press, London, 291 pp.

Montgomery, R. B., 1948. Vertical eddy flux of heat in the atmosphere. *J. Meteorol.* 5:265–274.

Moon, I–J., I. Ginis and T. Hara, 2004. Effect of surface waves on Charnock coefficient under tropical cyclones. *Geophys. Res. Letters,* 31:L.20302.

Moore, D. R., J. Toomre, E. Knobloch and N. O. Weiss, 1983. Period doubling and chaos in partial differential equations for thermosolutal convection. *Nature,* 303:663–667.

Moses, H. and J. E. Peason, 1965. Radon–222 concentration–height–time relationships. *Argonne National Laboratory Report ANL–7060,* pp. 120–128.

Mudd, J. B. and T. T. Kozlowski, eds. 1975. *Responses of Plants to Air Pollution.* Academic Press, London, 383 pp.

Munn, R. E., 1966. *Descriptive Micrometeorology.* Academic Press, New York, 245 pp.

Munro, D. S. and T. R. Oke, 1975. Aerodynamic Boundary-Layer adjustment over a crop in neutral stability. *Boundary-Layer Meterol.* 9:53–61.

Murphy, A. H. and E. S. Epstein, 1989. Skill scores and correlation coefficients in model verification. *Monthly Weather Rev.* 117:572–582.

Muschinski, A., R. Frehlich, M. Jensen, R. Hugo, A. Hoff, F. Eaton and B. Balsley, 2001. Fine scale measurements of turbulence in the lower troposphere: an intercomparison between a kite– and balloon–borne, and a helicopter–borne measurement system. *Boundary-Layer Meteorol.* 98:219–250.

Nappo, C. J., 2002. *An Introduction to Atmospheric Gravity Waves.* Academic Press, New York, 276 pp.

Nappo, C. J., J. Cuxart, W. Blumen, X. Lee and X–Z. Hu, 2002. Intermittent turbulence associated with a density current passage in the stable boundary layer. *Boundary-Layer Meteorol.* 105:199–219.

Ngan, F. and A, F. Stein, 2017. A long-term WRF meteorological archive for dispersion simulations: Application to controlled tracer experiments. *J. Appl. Meteorol. and Climatol.* 56:2203-2220.

Nho-Kim, E.-Y., M. Michou and V.-H. Peuch, 2004. Parameterization of size-dependent particle dry deposition velocities for global modeling. *Atmos. Environ.* 38:1933–1933.

Nicholson, K. W., 1988. A review of particle resuspension. *Atmos. Environ.* 22:2639–2651.

Nobel, P. S., 1999. *Physiochemical and Environmental Plant Physiology*, 2nd ed. Academic Press, San Diego, California.

Nodop, K., R. Connolly and F. Girardi, 1998. The field campaigns of the European Tracer Experiment (ETEX): overview and results. *Atmos. Environ.* 32:4095–4108.

Noilhan, J. and S. Planton, 1989. A simple parameterization of land surface processes for meteorological models. *Monthly Weather Rev.* 117:536–549.

Norman, J. M., 1979. Modeling the complete crop canopy. pp 249–277 in '*Modifications of the Aerial Environment of Crops,*' B. J. Barfield, ed. Am. Soc. Agric. Eng. Monograph #2, ASAE, St. Joseph, Michigan.

NRC, 1997. *Toxicological assessment of the Army's zinc cadmium sulfide dispersion tests.* National Academy Press, Washington, DC. 358 pp.

Obukhov, A. M., 1946. Turbulence in an atmosphere with anon–uniform temperature. In Russian in *Trudy Inst. Geophys []. USSR 1;* Translation by R. G. Taylor, 2/2/1951; also translation in Boundary-Layer Meteorol. 2:7–29 (1971).

O'Dell, D., N. S. Eash, B. B. Hicks, J. N. Oetting, T. J. Sauer, D. M. Lambert, J. Logan, W. C. Wright and J. A. Zahn, 2018. Reducing CO_2 flux by decreasing tillage in Ohio: Overcoming conjecture with data. *J. Agric. Sci.* 10:3 doi:10.5539/jas.v10n3p1.

O'Dell, D., T. J. Sauer, B. B. Hicks, C. Thierfelder, D. M. Lambert, J, Logan and N. S. Eash, 2015. A short–term assessment of carbon dioxide fluxes under contrasting agricultural and soil management practices in Zimbabwe. *J. Agric. Sci.* 7:32 doi/10.5539/jas.v7n3p32.

Ohmura, A., 1982. Objective criteria for rejecting data for Bowen Ratio flux calculations. *J. Appl. Meteorol.* 21:595–598.

Oke, T. R., *1974: Review of urban climatology, 1968-1973, WMO Tech. Note No. 134,* WMO No. 383. Geneva: World Meteorological Organization.

Oke, T. R., H. A. Cleugh, S. Grimmond, H. P. Schmid and M. Roth, 1989. Evaluation of spatially-averaged fluxes of heat, mass and momentum in the urban boundary layer. *Weather and Climate* 9:14-21.

Oliphant, A. J., C. S. B. Grimmond, H. N. Zutter, H. P. Schmid, H.-B. Su, S. L. Scott, B. Offerle, J. C. Randolph and J. Ehman, 2004. Heat storage and energy balance fluxes for a temperate deciduous forest. *Ag. Forest Meteorol.* 126:185–201.

Orgill, M. M. and R. I. Schreck, 1985. An overview of the ASCOT multi–laboratory field experiments in relation to drainage winds and ambient flow. *Bull. Amer. Meteteorol. Soc.* 66:1263–1277.

Otte, T. L., G. Pouliot, J. E. Pleim, J. O. Young, K. L. Schere, D. C. Wong, P. C. S. Lee, M. Tsidulko, J. T. McQueen, P. Davidson, R. Mathur, H–Y. Chuang, G. DiMego and N. L. Seaman, 2005. Linking the Eta model with the Community Multiscale Air Quality (CMAQ) modeling system the build a national air quality forecasting system. *Weather and Forecasting* 20:367–384.

Owen, P. R. and W. R. Thomson, 1963. Heat transfer across rough surfaces. *J. Fluid Mech.* 15:321–334.

Panofsky, H. A., 1961. An alternative derivation of the diabatic wind profile. *Quart. J. Roy. Meteorol. Soc.* 87:109–110.

Parkhurst, R. S., T. C. Winter, D. O. Rosenberry and A. M. Sturrock. 1998. Evaporation from a small prairie wetland in the Cottonwood Lake area, North Dakota–an energy–budget study. *Wetlands* 18:272–287.

Parson, D. and A. F. Bunker, 1952. A recording computer for the direct measurement of the turbulent heat exchange in the atmosphere. pp. 369–379 in "*International Symposium on Atmospheric Turbulence in the Boundary Layer*" (E. W. Hewson, ed.), *Geophysical Research Papers #19,* Air Force Cambridge Research Center, Cambridge Mass. (December, 1952), 530 pp.

Pasquill, F., 1956. Meteorological research at Porton. *Nature* 177:1148–1150.

Pasquill, F., 1961. The estimation of the dispersion of windborne material. *Meteorol. Mag.* 90:33–49.

Pascal–Delannoy, F., B. Sorli and A. Boyer, 2000. Quartz crystal microbalance (QCM) used as a humidity sensor. *Sensors and Actuators* 84:285–291.

Paulson, C. A., 1970. The mathematical representation of wind speed and temperature in the unstable atmospheric surface layer. *J. Appl. Meteorol.* 9:857–861.

Pearson, J. E. and G. E. Jones, 1965. Emanation of radon 222 from soils and its use as a tracer. *J. Geophys. Res.* 70:5279–5290.

Pendergrass, W. R., T. P. Meyers and R. White, 2017. ATDD Convective Initiation Program T–Profiler Observations, Belle Mina, Alabama. *NOAA Technical Report OAR ARL–271*, 60 pp.

Penman, H. L., 1942. Theory of porometer used in the study of stomatal movements in leaves. *Proc. Roy. Soc.* London 130:416–434.

Penman, H. L., 1948. Natural evaporation from open water, bare soil and grass. *Proc. Roy. Soc.* London A193:120–146.

Penman, H. L., 1963. Vegetation and hydrology. *Tech. Comm. No. 53,* Commonwealth Bureau of Soils, Harpenden, England. 125 pp.

Perez, P. J., F. Castellvi, M. Ibañez and J. I. Rosell, 1999. Assessment of reliability of Bowen ratio method for partitioning fluxes. *Agric. Forest Meteorol.* 97:141–150

Petroff, A., A. Mailliat, M. Amielh and F. Anselmet, 2008. Aerosol dry deposition on vegetative canopies. Part I: Review of present knowledge. *Atmos. Environ.* 42:3625–3653.

Petroff, A. A. and L. Zhang, 2010. Development and validation of a size-resolved particle dry deposition scheme for application in aerosol transport models. *Geosci. Model. Dev.* 3:753–769.

Philip, J. R., 1996. One-dimensional checkerboards and blending heights. *Boundary-Layer Meteorol.* 77:135–151.

Placet, M., 1990. *Emissions involved in acidic deposition processes: Methodology and results.* AWMA Conference report CONF–900676–6, OSTI # 6971407.

Plate, E. J., 1971. *Aerodynamic Characteristics of Atmospheric Boundary Layers.* US Atomic Energy Commission, Springfield, Virginia, TID–25465, 190 pp.

Plate, E. J., 1999. Methods of investigating urban wind fields – physical models. *Atmos. Env.* 33:3981–3989.

Pleim, J. and L. Ran, 2011. Surface flux modeling for air quality applications. *Amosphere* 2:271–301.

Porch, W. M., 1982. *Ground-based optical-remote-sensing results from 1981 ASCOT field experiment.* Lawrence Livermore National Laboratory Report UCID-19377, ASCOT-82-2, 26 pp.

Porch, W. M., W. E. Clemens and R. L. Coulter, 1991. Nighttime valley waves. *J. Appl. Meteorol.* 30:141–156.

Poulos, G. S., W. Blumen, D. C. Fritts, J. K. Lundquist , J. Sun, S. P. Burns, C. Nappo, R. Banta, R. Newsom, J. Cuxart, E. Terradellas , B. Balsley and M. Jensen, 2002. CASES–99: A comprehensive investigation of the stable nocturnal boundary layer. *Bull. Amer. Meteorol. Soc.* 83:551–581.

Pournazeri, S., M. Princevac and A. Venkatram, 2012. Scaling of building affected plume rise and dispersion in water channels and wind tunnels – revisit of an old problem. *J. Wind. Eng. Ind. Aerodyn.* doi:10.1016/j.jweia.2012.01.006.

Powell, M. D., P. J. Vickery and T. A. Reinhold, 2003. Reduced drag coeffoicient for high wind speeds in tropical cyclones. *Nature* 422:279–283.

Prandtl, L., 1932. Meteorologische anwendung der strőmungslehre [Meteorological application of fluid mechanics]. *Beitr. Phys. Atmosphäre,* 19:188–202.

Priestley, C. H. B., 1955. Free and forced convection in the atmosphere near the ground, *Quart. J. Roy. Meteorol. Soc.* 81:139–143.

Priestley C. H. B. 1959. *Turbulent Transfer in the Lower Atmosphere.* University of Chicago Press, 130 pp.

Priestley, C. H. B. and R. J. Taylor, 1972. On the assessment of surface heat flux and evaporation using large–scale parameters. *Mon. Weather Rev.* 100:81–92.

Priestley, C. H. B. and W. C. Swinbank, 1947. Transport of heat and other properties by atmospheric turbulence. *Nature* 159:61.

Pudykiewicz, J., 1989. Simulation of the Chernobyl dispersion with a 3-D hemispheric tracer model. *Tellus B* 41:391–412.

Randerson, D., 1976. Overview of regional–scale numerical models. *Bull. Amer. Meteorol. Soc.* 57:797–804.

Rannik, Ű., M. Aubinet, O. Kurbanmuradov, K. K. Sabelfeld, T. Markkanen and T. Vesala, 2000. Footprint analysis for measurements over a heterogeneous forest. *Boundary-Layer Meteorol.* 97:137–166.

Raupach, M. R., 1978. Infrared fluctuation hygrometry in the atmospheric surface layer. *Quart. J. Roy. Meteorol. Soc.* 104:309–322.

Raupach, M. R. and J. J. Finnigan, 1988. 'Single-layer models of evaporation from plant canopies are incorrect but useful, whereas multilayer models are correct but useless': Discuss. *Aust. J. Plant Physiol.* 15:705–716.

Raupach, M. R., J. B. Stewart and A. S. Thom, 1979. Comments on the paper 'Analysis of flux-profile relationships above tall vegetation - an alternative view' by B. B. Hicks, G. D. Hess and M. L. Wesely (*Q.J.* 105, 1074–1077), *Quart. J. Roy. Meteorol. Soc.* doi/10.1002/qj.49710544624.18.

Ray, J. D., D. H. Stedman and G. J. Wendel, 1986. Fast chemiluminescent method for measurement of ambient ozone. *Anal. Chem.* 58:598–600.

Record, F. A. and H. E. Cramer, 1958. Preliminary analysis of Project Prairie Grass diffusion measurements. *J. Air Pollut. Contr. Assoc.* 8:240–248.

Regener, V. H., 1957. The vertical flux of atmospheric ozone. *J. Geophys. Res.* 62:221–228.

Reynolds, O., 1883. The law of resistance in parallel channels. *Trans. R. Soc. London,* 35:935–982.

Rich, S., P. E. Waggoner and H. Tomlinson, 1970. Ozone uptake by bean leaves. *Science* 169:79–80.

Richter, I. A., 1980. N*otebooks — Leonardo da Vinci.* Oxford University Press, New York.

Rider, N. E., J. R. Philip and E. F. Bradley. 1963. The horizontal transport of heat and moisture — a micrometeorological study. *Quart. J. Roy. Meteorol. Soc*. 89:507–530.

Ritchie, J. T. and E. Burnett, 1968. A precision weighing lysimeter for row crop water use studies. *Agro. J.* 60:545–549.

Robinson, J. B., 1998. Bias in Daily Pan Evaporation Data. *Ag. Sci.* 1(4):22–24.

Rogak, S. N., R. C. Flagan and H. V. Nguyen, 1993. The mobility and structure of aerosol agglomerates. *Aerosol Sci. and Technol.* 18:25–47.

Rosenberg, N. J., 1973. *Microclimate: the biological environment.* John Wiley and Sons, New York. 495 pp.

Rotach, M. A. and D. Zardib, 2007. On the boundary-layer structure over highly complex terrain: Key findings from MAP. *Quart. J. Roy. Meteorol. Soc.* 133: 937–948.

Rouil, L., C. Honoré, R. Vautard, M. Beekman, B. Bessanet, L. Malherbe, F. Meleux, A. Dufuor, C. Euchegaray, J–M. Flaud, L. Menut, D. Martin, A. Peuch, V–H. Peuch and N. Poisson, 2009. PREV'AIR: an operational forecasting and mapping system for air quality in Europe. *Bull. Amer. Meteorol. Soc.* 90:71–83.

Rudnik, P., 1964. FLIP: An oceanographic buoy. *Science,* 146:1268.

Ryall, D. B. and R. H. Maryon, 1997. Validation of the UK Met Office's NAME model against the ETEX dataset. pp 151–154 in "*Proceedings ETEX Symposium on Long Range Atmospheric Transport*" (K. Nodop, ed.), European Commission EUR 17 346.

Saiki, E. M., C–H. Moeng and P. P. Sullivan, 2000. Large–eddy simulation of the stably stratified planetary boundary layer. *Boundary-Layer Meteorol.* 95:1–30.

Saint–Guilhem, R., 1962. The principles of dimensional analysis. *Mem. Sciences Mathematiques,* 152 . (Translation by F. A. Gifford, 1967, available as US Environmental Sciences Research Administration, Atmospheric Turbulence and Diffusion Laboratory Report IERTM–ATDL 3, 86 pp.).

Salmond, J. A., 2005. Wavelet analysis of intermittent turbulence in a very stable nocturnal boundary layer: Implications for the vertical mixing of ozone. *Boundary-Layer Meteorol.* 114:463–488.

Sauer, T. J. and R. Horton, 2005. Soil heat flux. In: J. L. Hatfield and J. M. Baker (Eds), Micrometeorology in Agricultural Systems. *American Society of Agronomy Monograph* 47:131–154.

Saxena, V. K. and N–H. Lin, 1990. Cloud chemistry measurements and estimates of acidic deposition on and above cloudbase coniferous forest. *Atmos. Environ.* 24A:329–352.

Saylor, R. D., G. M. Wolfe, T. P. Meyers and B. B. Hicks, 2014. A corrected formulation of the Multilayer Model (MLM) for inferring geseous dry deposition to vegetated surfaces. *Atmos. Environ.* 92:141–145.

Saylor R. D., B. D. Baker, P. Lee, D. Tong, L. Pan and B. B. Hicks, 2019. The particle dry deposition component of total deposition from air quality models: right, wrong or uncertain? *Tellus B:* 71:1445379, https://doi.org/10.1080/16000889.2018.1550324.

Schmid, H. P., 1994. Source areas for scalars and scalar fluxes. *Boundary-Layer Meteorol.* 67:293–318.

Schmitt, C. G. and A. J. Heymsfield, 2010. The dimensional characteristics of ice crystal aggregates from fractal geometry. *J. Atmos. Sci.* 67:1605–1616.

Schuepp, P. H., M. Y. Leclerc, Y. I. Macpherson and R. L. Desjardins, 1990. Footprint prediction of scalar fluxes from analytical solutions of the diffusion equation. *Boundary-Layer Meteorol.* 50:355–373.

Schulze, E. –D., O. L. Lange, L. Kappen, U. Buschbom and M. Evenari, 1973. Stomatal responses to changes in temperature at increasing water stress. *Planta* 110:29–42.

Scollo, S., P, del Carlo and M. Coltelli, 2007. Tephra fallout of 2001 Etna flank eruption: Analysis of the deposit and plume dispersion. *J. Volcanol. Geotherm. Res.* 160:147–164.

Scorer, R. S., 1958. "*Natural Aerodynamics,*" Pergamon, London, 313 pp.

Scrase, F. J., 1930. *Some characteristics of eddy motion in the atmosphere.* Meteorological Office Geophysical Memoirs, Air Ministry, London, No. 52.

Scrase, F. J. and P. A. Sheppard, 1944. The errors of cup anemometers in fluctuating winds. *J. Sci. Instrum.* 21:160–165.

Schroeder, M., 1991. *Fractals, Chaos, Power Laws.* Freeman, New York, 429 pp.

Sehmel, G. A., 1970. Particle deposition from turbulent airflow. *J. Geophys. Res*. 75:1766–1781.

Sehmel, G. A., 1979. *Deposition and resuspension processes.* Report PNL–SA–6746, Battelle Pacific Northwest Laboratory, Richland, Washington.

Sehmel, G. A., 1980a. Particle and gas dry deposition: A review. *Atmos. Environ.* 14:983–1012.
Sehmel, G. A., 1980b Particle resuspension: A review. *Env. International* 4:107–127.
Sehmel, G. A. and W. H. Hodgson, 1978. *A model for predicting dry deposition of particles and gases to environmental surfaces.* Report BNWL–SA–5124, Battelle Pacific Northwest Laboratory, Richland, Washington, 13 pp.
Sehmel, G. A. and F. D. Lloyd, 1974. *Particle resuspension rates.* Report PNL–SA–6721, Battelle Pacific Northwest Laboratory, Richland, Washington.
Sehmel, G. A., W. H. Hodgson and S. L. Sutter. 1973. *Dry deposition of particles.* Report BNWL–1850, Battelle Pacific Northwest Laboratory, Richland, Washington.
Seeger, M., R. Smeets, J. Yan, H. Ito, M. Claessens, E. Dullni, L. Falkingham, C. M. Franck, F. Gentils, W. Hartmann, Y. Kieffel, S. Jia, G. Jones, J. Mantilla, S. Pawar, M. Rabie, P. Robin–Jouan, H. Schellekens, J. Spencer, T. Uchii, X. Li and S. Yanabu, 2017. Recent trands in development of high voltage circuit breakers with SF6 alternative gases. *Plasma Phys. Technol.* 20:1–5.
Sellers, W. D., 1962. Simplified derivation of the diabatic wind profile. *J. Atmos. Sci.* 19:180–181.
Sellers, P., F. Hall, H. Margolis, B. Kelly, D. Baldocchi, G. den Hartog, J. Cihlar, M. G. Ryan, B. Goodison, P. Crill, K. J. Ranson, D. Lettenmaier and D. E. Wickland, 1995. The Boreal Ecosystem–Atmosphere Study (BOREAS): An overview and early results from the 1994 field year. *Bull. Amer. Meteorol. Soc.* 76:1549–1577.
SethuRaman, S., 1977: The observed generation and breaking of atmospheric gravity waves over the ocean, *Boundary Layer Meteorol.* 12:331–349.
Shannon, J. D., 1981. A model of regional long–term average sulfur atmospheric pollution, surface removal and net horizontal flux. *Atmos. Env.* 15:689–701.
Shaw, R. H., 1985. On diffusive and dispersive fluxes in forest canopies. pp 407–49 in '*The Forest Atmosphere Interaction*' (B. A. Hutchison and B. B. Hicks, eds.), D. Reidel, Dordrecht, 684 pp.
Shaw, R. H., K. T. Paw U. and U. Weigang Gao, 1998. Detection of temperature ramps and flow structures at a deciduous forest site. *Agric. Forest Meteorol.* 47:123–138.
Sherwood, S. I., 1992. The greening of American monuments: the role of atmospheric chemistry in the corrosion of outdoor bronzes. pp 33–72 in "*The Conservation of Bronze Sculpture in the Outdoor Environment*" (T. Drayman–Weisser, ed.), Nat. Assoc. Corrosion Engineers, Houston, Texas. 390 pp.
Sherwood, S. I., 1995. Kitson "Hikers" capture a half–century of corrosion history. *Cult. Res. Management* 18:23–25.
Shrivastava, P., 1992. *Bhopal: Anatomy of a Crisis.* Sage Publications, Cambridge, MA,
Shuepp, P. H., M. Y. Leclerc, M. J. Macpherson and R. L. Desjardins, 1990. Footprint prediction of scalar fluxes from analytical solutions of the diffusion equation. *Boundary-Layer Meteorol.* 50:355–373.
Shuttleworth, W. J. and J. S. Wallace, 1985. Evaporation from sparse crops – an energy combination theory. *Quart. J. Roy. Meteorol. Soc.* 111:839–855.
Sickles, J. E. and D. S. Shadwick, 2002. Precision of atmospheric dry deposition data from the Clear Air Status and Trends Network. *Atmos. Environ.* 36:5671–5686.
Sieburth, J. McN., 1983. Microbiological and organic-chemical processes in the surface and mixed layers. pp 121–172 in "*Air-Sea Exchange of Gases and Particles*" (P. S. Liss and W. G. N. Slinn eds.), Riedel, New York.
Siegel, A. F., 1982. Geometric data analysis: an interactive graphics program for shape comparison. pp 103–122 of '*Modern Data Analysis*' (R. L. Launer and A. F. Siegel, eds.), Academic Press, New York.
Singh, M. P. and S. Ghosh, 1987. Bhopal gas tragedy: model simulation of the dispersion scenario. *J. Hazardous Materials* 17:1–22.
Slatyer, R. O. and I. C. McIlroy, 1961. *Practical Microclimatology.* UNESCO, 355 pp.
Slinn, W. G. N., 1982. Predictions for particle deposition to vegetative canopies. *Atmos. Environ.* 16:1785–1704.
Sinclair, T. R., L. H. Allen and E. R. Lemon, 1975. An analysis of errors in the calculation of energy flux densities above vegetation by a Bowen-ratio profile method. *Boundary-Layer Meteorol.* 8:129–139.
Smith, S. D., 1988. Coefficients for sea surface wind stress, heat flux and wind profiles as a function of wind speed and temperature. *J. Geophys. Res.* 93:467–472.

Smith, S. D., R. J. Anderson, W. A. Oost, C. Kraan, N. Maat, J. DeCosmo, K. B. Katsaros, K. L Davidson, K. Bumke, L. Hassae and H. M. Chadwick, 1992. Sea surface wind stress and drag coefficients: The Hexo results. *Boundary-Layer Meteorol.* 60:109–142.

Smith, J., R. Pinkel and R. A. Weller, 1987. Velocity structure in the mixed layer during MILDEX. *J. Phys. Oceanog.* 17:425–439.

Sofiev, M., P. Siljamo, I. Valkama, M. Ilvonen and J. Kukkonen, 2006. A dispersion modelling system SILAM and its evaluation against ETEX data. *Atmos. Environ.* 40:674–685.

Spicer, T. O., S. B. Fox and B. B. Hicks, 2021. Preliminary assessment of chlorine reactivity with environmental materials accounting for boundary layer and maximum deposition effects. *Atmos. Environ.* doi.org/10.1016/j.atmosenv.2021.118274.

Stein, A. F. and J. C. Wyngaard, 2001. Fluid modeling and the evaluation of inherent uncertainty. *J. Appl. Meteorol.* 40:1769–1774.

Stewart, J. B. and W. R. Rose, 1977. Substantiation of the Priestley and Taylor parameter $\alpha = 1.26$ for potential evaporation in high latitudes. *J. Appl. Meteorol.* 16:649–650.

Stewart, J. B. and A. S. Thom, 1973. Energy budgets in a pine forest. *Quart. J. Roy. Meteorol. Soc.* 99:154–170.

Stewart J. B., W. P. Kustas, K. S. Humes, W. D. Nichols, M. S. Moran and H. A. R. de Bruin, 1994. Sensible heat flux–radiometric surface temperature relationship for eight semi–arid areas. *J. Appl. Meteorol.* 33:1110–1117.

Stohl, A., M. Hittenberger and G. Wotawa, 1998. Validation of the lagrangian particle dispersion model FLEXPART against large-scale tracer experiment data. *Atmos. Environ.* 32:4245–4264,

Stramska, M. and T. Petelski, 2003. Observations of oceanic whitecaps in the north polar waters of the Atlantic. *J. Geophys. Res.* 108(C3):3086, doi:10.1029/2002JC001321.

Straume, A. G., E. N. Koffi and K. Nodop, 1998. Dispersion Modeling Using Ensemble Forecasts Compared to ETEX Measurements. *J. Appl. Meteorol. Climatol.* 37:1444–1456.

Stull, R. B., 1988. *An Introduction to Boundary Layer Meteorology.* Kluwer Academic Publishers, Dordrecht, 666 pp.

Stunder, B. J. B. and R. R. Draxler, 1989. Across North America Tracer Experiment (ANATEX) Volume II: Aircraft-Based Sampling. *NOAA Tech. Memo ERL ARL*–1177. 113 pp.

Sun, J., S. P. Burns, D. H. Lenschow, R. Banta, R. Newsom, R. Coulter, S. Frasier, T. Ince, C. Nappo, J. Cuxart, W. Blumen, X. Lee and X–Z. Hu, 2002. Intermittent turbulence associated with a density current passagein the stable boundary layer. *Boundary-Layer Meteorol.* 105:199–219.

Sun, J., S. P. Burns, D. H. Lenschow, R. Banta, R. Newsom, R. Coulter, S. Frasier, T. Ince, C. Sun, J. L. Mahrt, R. M. Banta and Y. L. Pichugina, 2012. Turbulence regimes and turbulence intermittency in the stable boundary layer during CASES–99. *J. Atmos. Sci.* 69, 338–351.

Suomi, V. E., 1957. Energy budget studies and development of the sonic anemometer for spectrum analysis. *AFCRC Tech. Report 56–274,* Dept. of Meteorol. University of Wisconsin. 91 pp.

Sutton, M. A., C. E. R. Pitcairn and D. Fowler, 1993. The exchange of ammonia between the atmosphere and plant communities. *Advances Ecol. Res.* 24:301–393.

Sutton, O. G., 1947. The problem of diffusion in the lower atmosphere. *Quart. J. Roy. Meteorol. Soc.* 73:317–318.

Sutton, O. G., 1952. *Micrometeorology.* McGraw–Hill, New York, 333 pp.

Sutton, O. G., 1953. In discussion (led by H. Lettau) of '*International Symposium on Atmospheric Turbulence in the Boundary Layer* (E. W. Hewson, ed.), p509.

Swinbank, W. C., 1951. The measurement of the vertical transfer of heat, water vapor, and momentum by eddies in the lower atmosphere. *J. Meteorol.* 8:135–145.

Swinbank, W. C., 1952a. The measurement of the vertical transfer of heat, water vapor, and momentum by eddies in the lower atmosphere, with some results. pp. 355–367 in "*International Symposium on Atmospheric Turbulence in the Boundary Layer*" (E. W. Hewson, ed.), Geophysical Research Papers #19, Air Force Cambridge Research Center, Cambridge Mass. (December, 1952), 530 pp.

Swinbank, W. C., 1952b. The criterion of atmospheric turbulence. *Quart. J. Roy. Meteorol. Soc.* 78:420–425.

Swinbank, W. C., 1960. Wind profile in thermally stratified flow. *Nature,* 186:463–464.

Swinbank, W. C., 1963. Long wave radiation from clear skies. *Quart. J. Roy. Meteorol. Soc.* 89:339–348.

Swinbank, W. C., 1964. The exponential wind profile. *Quart. J. Roy. Meteorol. Soc.* 90:119–135.

Swinbank, W. C and A. J. Dyer, 1967. An experimental study in micrometeorology, *Quart. J. Roy. Meteorol. Soc.* 93:494–500.

Sykes, R. I., S. F. Parker, D. S. Herin and W. S. Lewellen, 1993. Numerical simulation of ANATEX tracer data using a turbulence closure model for long-range dispersion. *J. Appl. Meteorol. Climatol.* 32:929–947.

Tani, N., 1960. The wind on the cultivated field, *J. Agr. Soc. Japan* 16:89–93.

Tanner, C. B., 1960. Energy balance approach to evapotranspiration from crops. *Soil Sci. Soc. Amer. Proc.* 24:1–9.

Tanner, R. L., T. D'Ottavio, R. Garber and L. Newman, 1980. Determination of ambient aerosol sulfur using a continuous flame photometric detection system. *Atmos. Environ.* 14:121–127.

Taylor, G. I., 1927. Turbulence. *Quart. J. Roy. Meteorol. Soc.* 53:201–211.

Taylor, P. A., 1988. Turbulent wakes in the atmospheric boundary layer. pp. 270–292. In W. L. Steffen and O. T. Denmead (eds.) *Flow and Transport in the Natural Environment: Advances and Applications,* Springer–Verlag, Berlin, Germany.

Taylor, R. J., 1958. A linear, unidirectional anemometer of rapid response. *J. Sci. Instr.* 33:47–52.

Taylor, R. J. and E. K. Webb, 1955. A mechanical computer for micrometeorological research. *CSIRO Meteorological Physics Technical Paper #6,* 16 pp.

Ter Horst, J. H., D. Bedeaux and S. Kjeistrup, 2011. The role of temperature in nucleation processes. *J. Chem. Phys.* 134:054703; doi.org/10.1063/1.3544689.

Tetons, O., 1930. Uber einege meteorologische Begreffe. *Z. Beophys.* 6:297–309.

Thom, A. S., 1971. Momentum absorption by vegetation. *Quart. J. Roy. Meteorol. Soc.* 97:414–428.

Thurtell, G. W., C. B. Tanner and M. L. Wesely, 1970. Three–dimensional pressure–sphere anemometer system. *J. Appl. Meteorol.* 9:379–385.

Touma, J. S., 1977. Dependence of the wind profile power law on stability at various locations. *J. Air Poll. Contr. Ass.* 27:863–866.

Tsvang, L. R., M. M. Fedorov, B. A. Kader, S. L. Zubkovskii, T. Foken, S. H. Richter and Y. A. Zeleny, 1991. Turbulent exchange over a surface with chessboard-type inhomogeneities. *Boundary-Layer Meteorol.* 55:141–160.

Urban, J., M. Ingwers, M. A. McGuire and R. O. Teskey, 2017. Stomatal conductance increases with rising temperature, *Plant Signaling & Behavior,* 12:8 10.1080/15592324.2017.1356534.

van Aalst, R. M., 1986. Dry deposition of aerosol particles. pp. 933 – 949 in "*Aerosols*" (S. D. Lee et al. eds.), Lewis Publishers, Michigan, 1221 pp.

Vandemark, D., P. D. Mourad, S. A. Bailey, T. L. Crawford, C. A. Vogel, J. Sun and B. Chapron, 2001. Measured changes in ocean surface roughness due to atmospheric boundary layer rolls. *J. Geophys. Res.: Oceans,* 106:4639–4654.

van de Wiel, B. J. H., R. J. Ronda, A. F. Moene, H. A. R. De Bruin and A. A. M. Holtslag, 2003. Intermittent turbulence and oscillations in the stable boundary layer over land. Part III: A classification for observations during CASES–99. *J. Atmos. Sci.* 60, 2509–2522.

van Dop, H., G. Graziani and W. Klug, 1999. ETEX: a European tracer experiment; observations, dispersion modelling and emergency response. *NATO Science Series* 57:137–150.

van Vleet, E. S. and P. M. Williams, 1980. Sampling sea surface films: A laboratory evaluation of techniques and collecting materials. *Limnol. Oceanog.* 25:764–770.

Vautz, W., S. Pahl, H. Pilger, M. Schilling and D. Klockow, 2003. Deposition of trace substances via cloud droplets in the Atlantic Rain Forest of the Serra Do Mar, São Paulo State, SE Brazil. *Atmos. Environ.* 37:3277–3287.

Venkatram, A. and J. Pleim, 1999. The electrical analogy does not apply to modeling dry deposition of particles. Atmos. *Environ.* 33:3075–3076.

Verhoef, A., H. A. R. de Bruin and B. J. J. M. van den Hurk, 1997. Some practical notes on the parameter kB^{-1} for sparse vegetation. *J. Appl. Meteorol.* 36:560–572.

Verhoef, W., L. Jia, Q. Xiao and Z. Su, 2007. Unified optical-thermal four-stream radiative transfer theory for homogeneous vegetation canopies. *IEEE Trans. Geosci. & Remote Sensing* 45:1808–1822.

Vialet-Chabrand, S. R. M., J. S. A. Matthews, L. McAusland, M. R. Blatt, H. Griffiths and T. Lawson, 2017. Temporal dynamics of stomatal behavior: modeling and implications for photosynthesis and water use. *Plant Physiology* 174:603–613.

Vogt, S. and P. Thomas, 1988. *Investigation by tetroon of the atmospheric transport during the city-climate experiment Karlsruhe.* KFK (Germany) Report 4337, 70 pp.

Wabba, G., D. Johnson, F. Gao and J. Gong, 1995. Adaptive tuning of numerical weather prediction models: Randomized GCV in three and four dimensional data assimilation. *Mon. Wea. Rev.* 123:3358–3369.

Warner, J., 1972. The structure and intensity of turbulence in air over the sea. *Quart. J. Roy. Meteorol. Soc.* 98:175–186.

Warner, J., 1973. Spectra of the temperature and humidity fluctuations in the marine boundary layer. *Quart. J. Roy. Meteorol. Soc.* 99:82–88.

Watson, A. J., E. J. Bock, B. Jahne, W. E. Asher, N. M. Frew, L. Hasse, G. M. Korenowski, L. Merlivat, L. F. Phillips, P. Schluessel, D. K. Woolf and P. S. Liss, 1997. Report 1 – Physical processes in the microlayer and the air–sea exchange of trace gases. pp. 1–34 of '*The Sea Surface and Global Change*' (P. S. Liss and R. A. Duce eds.), Cambridge University Press, 536 pp.

Webb, E. K., 1965. Aerial Microclimate. *Meteorol. Monographs* 6:27–58.

Webb, E. K., 1970. Profile relationships: The log–linear range and extension to strong stability. *Quart. J. Roy. Meteorol. Soc*. 96:67–90.

Webb, E. K., 1982. Profile relationships in the superadiabatic surface layer. *Quart. J. Roy. Meteorol. Soc.* 108:661–688.

Webb, E. K., G. I. Pearman and R. Leuning, 1980. Correction of flux measurements for density effects due to heat and water vapour transfer. *Quart. J. Roy. Meteorol. Soc.* 106:85–100.

Weber, A. H., R. J. Kurzeja and C. H. Hunter, 2012. *Roughness Lengths for the Savannah River Site.* SRNLSTI-2012-00016. Aiken, SC: Savannah River National Laboratory.

Weng, H. and K.–M. Lau, 1994. Wavelets, period doubling, and time–frequency localization with application to organization of convection over the tropical West Pacific. J. Atmos. Sci. 51:2523–2541.

Went, F. W., 1960. Blue hazes in the atmosphere. *Nature* 187:641–643.

Wentzel, M., H. Gorzawski, K. -H. Naumann, H. Saathoff and S. Weinbruch, 2003. Transmission electron microscopical and aerosol dynamical characterization of soot aerosols. *Aerosol Sci.* 34:1347–1370.

Wesely, M. L., 1970. *Eddy correlation measurements in the atmospheric surface layer over agricultural crops.* Ph.D. Dissertation, University of Wisconsin.

Wesely, M. L., 2007. Parameterization of surfaces resistances to gaseous dry deposition in regional–scale numerical models. *Atmos. Environ.* 41:52–63.

Wesely, M. L. and R. L. Hart, 1985. Variability of short–term eddy correlation estimates of mass exchange. pp 591– 612 in '*The Forest Atmosphere Interaction*' (B. A. Hutchison and B. B. Hicks, eds.) D. Reidel, Dordrecht, 684 pp.

Wesely, M. L. and B. B. Hicks, 1977. Some factors that affect the deposition rates of sulfur dioxide and similar gases on vegetation. *J. Air Pollut. Contr. Assoc.* 27:1110–1116.

Wesely, M. L., D. R. Cook, R. L. Hart and R. E. Speer. 1985. Measurements and parameterization of particle sulfur deposition over grass. *J. Geophys. Res.* 90:213–2143.

Wesely, M. L., J. A. Eastman, D. R. Cook and B. B. Hicks, 1978. Daytime variations of ozone eddy fluxes to maize. *Boundary-Layer Meteorol.* 15:361 – 373.

Wesely, M. L., J. A. Eastman, D. H. Stedman and E. D. Yalvac, 1982. An eddy–correlation measurement of NO_2 flux to vegetation and comparison to O_3 flux. *Atmos. Environ.* 16:815–820.

Wesely, M. L., B. B. Hicks, W. P. Dannevik, S. Frisella and R. B. Husar, 1977. An eddy–correlation measurement of particulate deposition from the atmosphere. *Atmos. Environ.* 11:561–563.

White, R., C. A. Vogel and W. R. Pendergrass, 2013. *Flux Observations Ocotillo Texas (FOOT) Experiment.* NOAA Atmospheric Turbulence and Diffusion Division Technical Memorandum, in press.

Wiederhold, P. R., 1997. *Water Vapor Measurement.* Marcel Dekker Inc. New York, 365 pp.

Wilczac, J. M., S. P. Oncley and S. A. Stage, 2001. Sonic anemometer tilt correction algorithms. *Boundary-Layer Meteorol.* 99:127–150.

Wilczak, J. M. and J. E. Tillman, 1980.The three–dimensional structure if convection in the atmospheric surface layer. *J. Atmos. Sci.* 37:2424–2443.

Willis, G. E. and J. W. Deardorff, 1979: Laboratory observations of turbulent penetrative–convection planforms, *J. Geophy. Res.* 84:295–302

Wilson, K. B., D. D. Baldocchi, M. Aubinet, P. Berbigier, C. Bernhofer, H. Dolman, E. Falge, C. Field, A. Goldstein, A. Grelle, T. Halldor, D. Hollinger, G. Katul, B. E. Law, A. Lindroth, T. P. Meyers, J. Moncrieff, R. Monson, W. Oechel, J. Tenhunen, R. Valentini, S. Verma, T. Vesala and S. Wofsky, 2002. Energy partitioning between latent and sensible heat flux during the warm season at FLUXNET sites. *Water Resources Res.* 38:30–1 to 30–11.

WMO–GAW, 2017. *WMO Global Atmosphere Watch (GAW) Implementation Plan: 2016– 2023.* GAW Report # 228, WMO, Geneva, 76 pp.

Wylie, R. G. and T. Lalas, 1969. *Measurement of temperature and humidity.* World Meteorol. Org, (Geneva) Technical Note #194.

Xiong, C. and S. K. Friedlander, 2001. Morphological properties of atmospheric aerosol particles. *Proc. Natl. Acad. Sci.* 98:1185111856.

Yamamoto, G. 1959. Theory of turbulent transfer in non–neutral conditions. *J. Meteorol. Soc. Japan* 37:60–67.

Yang, B., Y. Qian, G. Lin, L. R. Leung, P. J. Rasch, G. J. Zhang, S. A. McFarlane, C. Zhao, H. Wang, M. Wang and X. Liu, 2013. Uncertainty quantification and parameter tuning in the CAM5 Zhang-McFarlane convection scheme and impact of improved convection on the global circulation and climate. *J. Geophys. Res, Atmos*. 118:395–415.

Zeiger, E., G. D. Farquhar and I. R. Cowan, 1987. *Stomatal Function.* Stanford Univ. Press, Stanford, CA. 580 pp.

Zhang L., S. Gong, J. Padro and L. Barrie, 2001. A size–segregated particle dry deposition scheme for an atmospheric aerosol module. *Atmos. Environ.* 35:549–560.

Zhao, Z–K., C–X. Liu, G–F. Dai, Q–T. Song and W–H. Lv, 2015. Typhoon air–sea drag coefficients in coastal regions. *J. Geophys. Res. Oceans,* 120:716–727.

Zilitinkevich, S. S., 2009. The effect of stratification on the roughness length and displacement height. *Boundary-Layer Meteorol.* 129:179–190.

Zilitinkevich, S. S., I. Mammarella, A. A. Baklanov and S. M. Joffre, 2008. The effect of stratification on the aerodynamic roughness length and displacement height. *Boundary-Layer Meteorol.* 129:179–190.

About the Author

Bruce B. Hicks
Metcorps, Norris, TN USA
Adjunct Professor, Agriculture, University of Tennessee,
Knoxville, TN USA
Email: hicks.metcorps@gmail.com

As a freshly minted graduate of the University of Tasmania, Hicks took a job (on probation) as a lowly Experimental Officer with the Meteorological Physics Chapter of the Australian Commonwealth Scientific and Industrial Research Organization. He served as the assistant to the man who would become one of his mentors and personal friends — Arthur James ("Arch") Dyer. Dyer and Hicks shared an office, the third in a long corridor of offices housing scientists who were unknown to him at that initial time. The first office was that of the Chief Scientist of the Chapter — C. H. B. ("Bill") Priestley. Number two was administrative. Whenever Priestley had need to talk with somebody he wandered down the corridor until he found a welcoming conversationalist. Most often, he stopped with Arch and Hicks witnessed many revealing discussions. Further down the corridor were Eric Webb, Len Deacon, Reg Taylor, Ian McIlroy and David Angus. Bill Swinbank was in a different area adjunct to the main corridor — not by accident, it seems. Upstairs were the synoptic luminaries: Reg Clarke, Andrzej Berson, Keith Ball, Peter Funk and Ragnath Kulkarni.

With formal training in Physics and Mathematics at the University of Tasmania and with an emphasis on statistics, Hicks soon developed a healthy skepticism about much of the micrometeorology of the time. The instrumentation on which the research relied was rudimentary and was subject to noise that evidenced itself as instrumental error. The Monin-Obukhov Similarity Theory (MOST) constituted the gospel of the day. Hicks's interest was piqued by conviction that much of MOST was self-fulfilling, a perspective that has not faded.

A nearby research area, at Edithvale, a few km from Aspendale, was the focus of preliminary testing of new devices, such as the dry- and wet-bulb thermometry for which he was responsible. Subsequent field testing necessitated countless hours of continuous monitoring of meters while visually observing the sensors. (In Australia, any sensor is vulnerable to attack by insects, especially flies.) Observing atmospheric turbulence as it happened served to exacerbate the insecurities that grew out of statistical issues related to dimensional analysis and the interpretation of the Buckingham Π theorem.

In later years, while associated with Argonne National Laboratory, the eddy covariance path was followed with increased enthusiasm, first to start understanding how to extend surface micrometeorology into the vertical (the Sangamon series) and second to measure the rates of deposition of air pollutants (dry deposition). In 1981, he was appointed to lead the famous Atmospheric Turbulence and Diffusion Laboratory in Oak Ridge, Tennessee, and in 1990 he moved to Washington, DC, to become Director of the Air Resources Laboratory of the National Oceanic and Atmospheric Administration. Since his departure from the federal research community in 2006 he has prospered in a new environment of research freedom which has culminated in the writing of the document now presented. This presentation is of observations, results and conclusions arising from 60 years of active field work unencumbered by dogma imposed by people lacking modern instrumentation. Most of the observations used here resulted from the field research programs conducted by the Atmospheric Turbulence and Diffusion Division (in Oak Ridge, Tennessee) of the Air Resources Laboratory of the US National Oceanic and Atmospheric Administration, with Will Pendergrass being the main contributor.

Index

D

K

L

M

N

O

P

T

Z